a programmed
course in
BASIC ELECTRONICS

a programmed course in BASIC ELECTRONICS
second edition

The Prince Project
New York Institute of Technology
Alexander Schure, Project Director

McGraw-Hill Book Company/Gregg Division

New York St. Louis Dallas San Francisco Auckland Düsseldorf
Johannesburg Kuala Lumpur London Mexico Montreal New Delhi
Panama Paris São Paulo Singapore Sydney Tokyo Toronto

Library of Congress Cataloging in Publication Data
New York Institute of Technology.
    Programmed course in basic electronics.

    Includes index.
    1. Electronics—Programmed instruction.   I. Schure,
Alexander, (date)   II. Title.
TK7816.N47   1976   621.381′07′7   75-35844
ISBN   0-07-046391-3

A PROGRAMMED COURSE IN BASIC ELECTRONICS

Copyright © 1976, 1964 by McGraw-Hill, Inc. All rights reserved. Chapters 1, 2, 3, 4, 5, 6, 7, 10, 11, 12, and 15 taken from A PROGRAMMED COURSE IN BASIC TRANSISTORS, copyright © 1976, 1964 by McGraw-Hill, Inc. All rights reserved. Printed in the United States of America. No part of this publication may be reproduced, stored in a retrieval system, or transmitted, in any form or by any means, electronic, mechanical, photocopying, recording, or otherwise, without the prior written permission of the publisher.

567 KPKP 898

The editors for this book were Gordon Rockmaker
and Alice V. Manning,
the designer was Charles A. Carson,
and the production supervisors were Phyllis Lemkowitz
and Regina R. Malone.
It was set in Palatino by Kingsport Press, Inc.
It was printed and bound by Kingsport Press, Inc.

CONTENTS

Preface	vii
Introduction	ix
To the User	xi

Chapter			
	1	Semiconductor fundamentals	1
	2	Transistor fundamentals I	38
	3	Transistor fundamentals II	63
	4	Transistor parameters, equivalent circuits, and gain calculations	92
	5	Transistor bias stability	116
	6	Using transistor characteristic curves and charts	145
	7	Reading transistor specifications	185
	8	First principles of electron tubes	198
	9	Operating characteristics of electron tubes	221
	10	Audio amplifiers I	247
	11	Audio amplifiers II	292
	12	Wideband amplifiers	327
	13	Radio-frequency amplifiers	355
	14	Oscillator principles	376
	15	*LC* oscillator circuits	395
	16	Crystal oscillators	412
	17	Modulation fundamentals	431
	18	Detection and detectors	454
	19	Superheterodyne principles	472
	20	Superheterodyne stage analysis	487
	21	Basic power suppliers	517
	22	Analysis of a superheterodyne receiver	534

Answers to Criterion Check Tests	548
Index	553

PREFACE

This textbook is designed for an introductory course in electronics. The only academic prerequisite is a background in fundamental electricity as covered in the companion volume *A Programmed Course in Basic Electricity*. Both texts are structured in linear programmed form. They may be used independently or in sequence. But there are two other major prerequisites for the successful completion of this course: desire and discipline. An inner drive to learn is often a more important ingredient than a high I.Q. Armed with determination and perseverance, the person with a strong desire to learn and the discipline to keep at the job simply cannot fail.

The books had their beginnings in 1958. At that time, faculty members of the electrical technology department of New York Institute of Technology undertook the development of an integrated series of programmed learning materials relating to electronics. In conferences held with training directors and curriculum specialists, an agreement was reached on the topics contained in those segments of electronic training considered essential in the majority of industrial and formal educational programs. The conferees further agreed that variations in background would require structuring the programs on several levels, with one major program being fundamental in approach. The overall project was termed the PRINCE Project (*P*rogrammed *R*einforced *I*nstruction *N*ecessary to *C*ontinuing *E*ducation). This volume is part of the series dealing with the basics of electricity, electronics, semiconductors, transistors, and pulse circuits.

By early 1961, preliminary versions of the programs had been prepared and tested by the programming teams. The next step required extensive field testing under various instructional conditions. In May of 1961, the Institute requested the assistance of the Educational Coordinating Committee of the Electronics Industries Association, offering the developed programs to industry on a cooperative data-exchange basis. Participating companies were asked to make their test data available for use in refining, revising, and developing a validated teaching instrument. Through the efforts of the committee, and through the interest of other industrial groups, a number of companies became aware of the project and agreed to aid in the validation. Among these companies were:

Aerovox Corporation, *New Bedford, Massachusetts*
The Boeing Company, *Renton, Washington*
Corning Glass Works, *Corning, New York*
E.I. DuPont de Nemours & Company, *Wilmington, Delaware*
Eastman Kodak Company, *Rochester, New York*

General Dynamics/Pomona, *Pomona, California*
General Electronic Company, *Philadelphia, Pennsylvania*
Hycon Manufacturing Company, *Monrovia, California*
International Telephone and Telegraph, *Belleville, New Jersey*
Lockheed Aircraft Corporation, Missiles and Space Division,
    Sunnyvale, California
McDonnell Aircraft Corporation, *Saint Louis, Missouri*

Concurrent with the industrial validation, a validation program in a number of technical institutes and community colleges was undertaken with the cooperation of McGraw-Hill, Inc. Thus, all material was thoroughly tested in the preliminary version titled *Basic Electronics*.

The data gathered from these exchanges were returned to the programming teams as the study progressed. Revisions were then made and the restructured programs retested and put in final form.

The second edition of *A Programmed Course in Basic Electronics* underwent the same vigorous classroom testing process.

Grateful acknowledgment is hereby extended to the training directors of the participating companies, the programming teams and faculty of New York Institute of Technology; and the many others who, through their efforts, assisted in the preparation of this edition.

ALEXANDER SCHURE

introduction

A student entering the field of electronics is almost immediately assailed by a battery of new terms and phrases which have not been previously defined. To avoid the confusion that tends to accompany these first steps in the new endeavor and to help the student progress smoothly through the first chapter, we have included this introductory section.

Rather than group the new terms and phrases in glossary form, we define and explain them here in the order in which they occur in the normal sequence of frame-to-frame progress. We start with the transistor, the nucleus of modern electronic devices.

Transistor: A solid-state device which functions in a number of different ways depending upon the circuit in which it appears. It may be used to build up voltages or currents (amplification), generate alternating currents at almost any desired frequency (oscillation), change alternating voltage or current to direct voltage or current (rectification), or perform a number of other electronic tasks. All of these will be discussed in detail in the appropriate section of this book. Transistors have largely replaced electronic tubes in today's technology because they are much smaller, require less power, produce less heat, and generally have a longer life.

Current carrier: A positive or negatively charged body which can move through certain types of matter, transferring an electric charge from one point to another. The electron is the familiar current carrier in metals. You will be introduced to another type of current carrier, the "hole," which is not an actual "body" in the usual sense of the word. But this must wait until later.

Semiconductor: An element, compound, or mixture of materials through which current carriers may move, often in very special ways. A semiconductor, as its name implies, has more resistivity than most metals but much less resistivity than insulators like glass or plastics.

Crystal set: A primitive radio receiver dating back to the earliest days of the art. It contained no tubes or transistors.

Demodulation: The process by which a radio receiver separates an audio signal from a carrier signal. Voice and music are transmitted from one point to another without wires by means of a radio wave. This wave consists of two parts: (1) a very high-frequency component that serves as a vehicle or "carrier"; and (2) a lower-frequency component that contains the voice or music information, and that rides on the carrier. The process of adding the *audio* component (voice or music) to the *radio* component (carrier) is called modulation. This occurs at the transmitter. When the composite signal reaches its destination, the radio receiver separates the audio from the radio component in the process of demodulation. Demodulation is essentially a rectification process in which alternating current is converted to pulsating direct current.

Diode: A device containing two electrodes or internal active connections. Diodes are normally of about the same physical size as small transistors.

Thermionic heater: The heating element in an electron tube which produces the energy required to eject electrons from a nearby element called the cathode. In some electron tubes, the thermionic heater in itself also serves as the cathode, or source of electrons.

Junction: An area or surface where two dissimilar kinds of semiconductor meet in intimate contact.

N-type, P-type: Used to distinguish between semiconductor materials which function in 2 different ways. The *N* represents *negative* and the *P* represents *positive*. The reasons for this nomenclature will become clear as you move through the chapter.

to the user

PROGRAMMED INSTRUCTION

No matter how deeply you probe the nature of electronics, you will always be dealing with certain basic principles. The understanding necessary to become an electronics technician comes from a grasp of the fundamental concepts of electricity.

As your skill in electronics grows, you will realize how important these basic ideas are. To understand the complex, you should know about the simple. Much of this basic information is in the program which follows.

The information in this course is organized in a relatively new manner. As you work with the materials in the lessons that follow, you will find that the information you are asked to learn has been arranged to provide for your participation in the instruction. This method of teaching is called *programmed learning*.

In presenting the information, the subject matter is divided into small units called frames. Most of the time, each frame requires that the statement in the frame be completed. With a little thought you should be able to provide the correct answer, or response. The correct answer is enclosed in parentheses and appears at the beginning of the frame which follows. You should look at the correct answer only after you have written what you feel to be the proper one.

In order to introduce you to this method, the following section is programmed. Cover the answers with a cardboard or paper sheet. Lower the sheet only *after* you have written your answer. Use reasonable judgment to decide whether your response is the same as the printed answer. Now try the sample frames which follow.

1
Self-instructional materials will be given to you in the same form as the frames which follow. Your first feelings may be that you are taking a test. You are not! Understand this clearly—programmed learning is a teaching method and is much more than just a _____. (Complete the statement.)

2
(*test*) The purpose of this book is *to teach you* in just the same way as if you were receiving individual instruction. So, your self-instructional text, called a program, acts as your private teacher. We can now say: A self-instructional book is called a _____.

3

(*program*) A program presents the information to be learned in small bits, a few sentences at a time. You will either complete a statement, find information on a diagram, or make a choice. For example, you might be asked to complete this statement:

   What you learn from each frame of this program will not be graded because the program is a (test, teaching device).

4

(*teaching device*) Let's move on now. We've said the program presents information in small amounts at a time. These small segments are called frames. Each numbered statement in the program is a frame.

   This is the fourth _____ in this series.

5

(*frame*) Thus, we see each frame leads us further into the subject by giving a little additional information or by searching for information learned in that frame or a previous one. After receiving this information, you check the next frame to determine, immediately, if you are _____.

5

(*correct*)

INFORMATION

Some frames or chapter introductions give you information and do not ask you to make a response. Information and summaries given in this way are important. Read them carefully. Then, just go on to the next frame. Do so now.

6

Let's summarize: Programmed learning involves breaking up the subject matter into small units, called _____.

7

(*frames*) Each frame will require a(n) _____ from you.

8

(*response* or *answer*) In the frame below, you will find the correct _____.

9

(*response* or *answer*) From time to time you will be given supplementary material in the form of problems and exercises. You will be required to do these _____ and _____.

10
(*problems, exercises*) Proceed at the rate best suited to you. One of the great advantages of programmed material is that each student can proceed at his own _____.

11
(*rate* or *pace*) Because you are working at your own rate, there will be no _____ limit for your completion of "daily lessons."

12
(*time*) Remember the difference between the words *teach* and *learn*. The program will teach, but only you, the _____, can learn.

13
(*student*) You must, therefore, make the effort to _____ from the program.

14
(*learn*) No matter how a course is taught, the responsibility for learning is with you, the _____.

15
(*student*) The program is a teaching device and not a _____. Since this is so, there is no reason to look ahead for an answer. If you do so, you will merely be cheating yourself of the opportunity of le__ing.

16
(*test, learning*) Extra *learning* steps are built into the program. Completion summaries and self tests will help you _____.

17
(*learn*) You are to complete each of these self tests. Be sure to answer every question. You may then compare your answer with the _____ answer shown.

18
(*correct*) Remember, when you reach the c_____tion summaries, you are to fill them in and check your answers. You are also required to answer all the questions in the _____ tests.

18
(*completion, self*)

You are now ready to proceed with the program.

TO THE USER **xiii**

1 semiconductor fundamentals

OBJECTIVES

Upon completion of a section you should be able to achieve the objectives listed for it. The objectives are listed in the order in which the subject material is presented in each section.

(1) Describe the operation of a simple crystal radio receiver. **(2)** Compare semiconductor devices with vacuum tubes. **(3)** Sketch a point-contact diode, junction diode, and transistor. **(4)** Contrast the point-contact devices with the junction devices. **(5)** List and give examples of the 3 states of matter. **(6)** Define and give examples of elements and compounds. **(7)** Diagram the structure of the water molecule. **(8)** Describe and sketch the structure of a typical atom. **(9)** State the conditions for an electron flow to occur in a substance. **(10)** Discuss electrical conductivity of matter, including the 3 categories of conductivity, examples in each category, and dependence on temperature. **(11)** Discuss the crystalline structure and electrical conductivity of pure semiconductors. **(12)** Describe impurities in semiconductors, including definitions of donor impurities, acceptor impurities, and *N-type* and *P-type* semiconductor materials. **(13)** Discuss the properties of a hole. **(14)** Sketch the current flow in a P-type semiconductor, showing the charge carriers in the semiconductor and in the external circuit. **(15)** Diagram and discuss an experiment which proves holes are the current carriers in P-type semiconductors. **(16)** Describe the buildup of the PN junction barrier. **(17)** Define and describe diffusion current, hole-electron recombination, depletion region, barrier height. **(18)** Indicate how the junction barrier stops the diffusion current. **(19)** Diagram and compare the magnitude of current flow of the reverse-biased PN junction and the forward-biased PN junction. **(20)** Sketch and describe the volt-ampere characteristic of a junction diode. **(21)** Define *minority* carriers. **(22)** List the majority and minority carriers for P-type and N-type semiconductors. **(23)** Draw the volt-ampere characteristic for a PN junction diode. **(24)** Indicate the principal current carriers in the forward-biased and reverse-biased regions.

INTRODUCTION

1-1
In a transistor, the flow of current carriers occurs in a material known as a *semiconductor*. The transistor's ability to control _____ carriers makes it potentially the most useful single element in modern signal communication and entertainment equipment.

1-2
(*current*) In increasing numbers, _____ are being applied in radio, television, radar, facsimile, telephone, teletypewriter, and computer assemblies.

1-3
(*transistors*) The first semiconductor "crystals" to be used in radio appeared in early receivers where they were used to recover the audio-frequency component from a received carrier. Thus, the first use for semiconductor crystals was that of demodulation or _____ of an amplitude-modulation (AM) signal.

1-4
(*detection* or *rectification, etc.*) In these early "crystal sets," the semiconductor material was usually lead sulfide or galena. A galena-metal interface has the property of high resistance for one direction of current flow and low resistance for the other. Thus, a galena crystal can be used to perform the function of _____ since it changes alternating current (ac) to pulsating direct current (dc).

1-5
(*rectification* or *detection, etc.*) A simple crystal set circuit is shown (Fig. 1-5). The required resonant assembly is formed by the combination of L and _____.

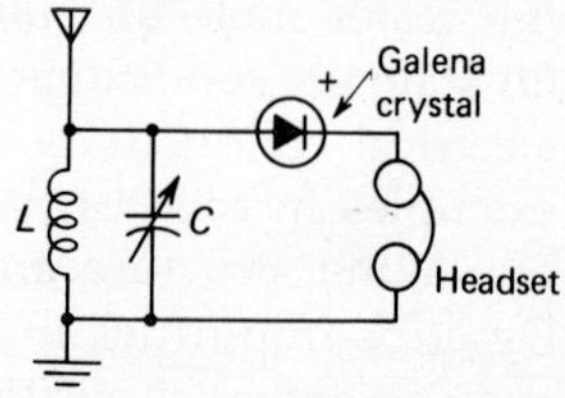

Fig. 1-5

1-6
(C) The station to be heard is selected by adjusting the variable capacitor C. Detection or rectification is then accomplished by use of the _____ crystal, and reproduction of the audio is handled by the headset.

1-7
(*galena*) A semiconductor crystal rectifier is universally symbolized as shown (Fig. 1-7). The triangular element bears no identifying marks, but the "flat-plate" element here is identified by a _____ sign; such a sign is not always included.

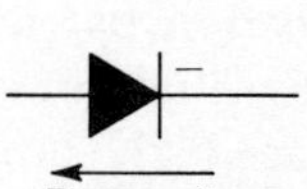

Electron flow
(low resistance direction)

Fig. 1-7

1-8
("−") Note the arrow under the symbol. Electrons flow easily in the direction indicated by this arrow. Thus, electrons flow easily from the flat-plate element toward the _____ element.

1-9
(*triangular*) The reason for the "−" designation used on the flat-plate element can be seen from Fig. 1-9. In this circuit, a(n) _____ generator is shown containing a crystal rectifier and a load.

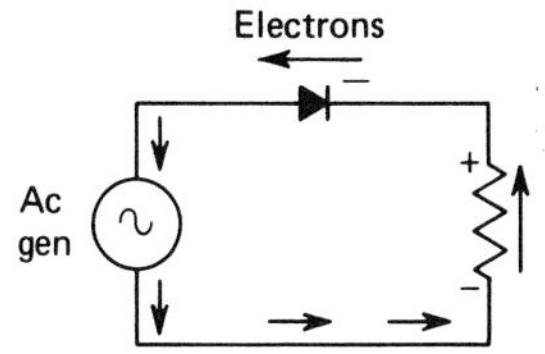

Fig. 1-9

1-10
(*ac*) The direction of _____ flow is shown by the arrows; we might call this the direction of electron conductivity.

1-11
(*electron*) In Fig. 1-9, it is seen that the flat-plate element is in series with the resistor. Hence, the triangular element is positive and the flat plate is labeled _____.

1-12
("−") During World War II, efficient semiconductor rectifiers were widely used (Fig. 1-12). By this time, galena had been abandoned as a semiconductor and was replaced by a better material. As shown in Fig. 1-12, this new semiconductor material is called _____.

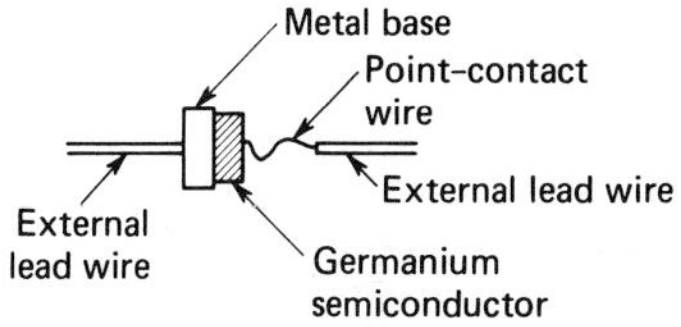

Fig. 1-12

1-13
(*germanium*) The rectifier in Fig. 1-12 is called a *point-contact diode*. On one side of a metal base an external lead is welded; on the other side of the metal base the wafer of _____ semiconductor is welded.

1-14
(*germanium*) A short piece of spring wire is oriented so that its point presses against the germanium semiconductor. This wire is called the _____-_____ wire.

1-15
(*point-contact*) The point-contact diode has important advantages over the diode vacuum tube. First, it may be made very _____ in size and can thus be adapted to miniaturized equipment.

SEMICONDUCTOR FUNDAMENTALS 3

1-16
(*small*) Second, the point-contact diode has no thermionic heater. Hence, it requires no heater voltage or current, and does not dissipate or waste _____.

1-17
(*power* or *energy, etc.*) Third, the contact area between the point-contact wire and the germanium wafer is very small. This makes the capacitance between these two elements very _____ as well.

1-18
(*small*) Thus, the point-contact diode has a very small shunt capacitance. A small _____ capacitance is desirable to avoid bypassing high-frequency signals when the diode is in the nonconducting part of the input cycle.

1-19
(*shunt*) A point-contact *transistor* is shown in Fig. 1-19. It is constructed like the diode except that it has _____ point-contact wires instead of 1 as in the diode.

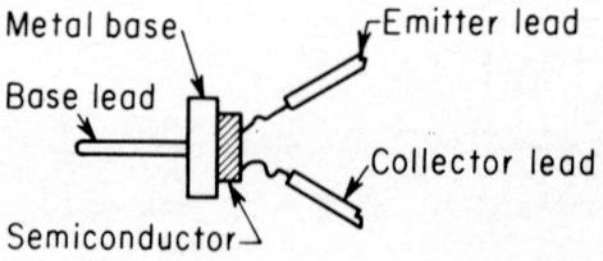

Fig. 1-19

1-20
(*2*) It is now a 3-terminal device instead of a 2-terminal device like the diode. The lead coming from the welded joint at the metal base is called the _____ lead.

1-21
(*base*) One of the point-contact leads is called the *emitter lead*. The other point-contact lead is called the _____ lead.

1-22
(*collector*) About 1 year after the announcement of the point-contact germanium diode, the _____ diode shown in Fig. 1-22 was described in the literature (1948–1949).

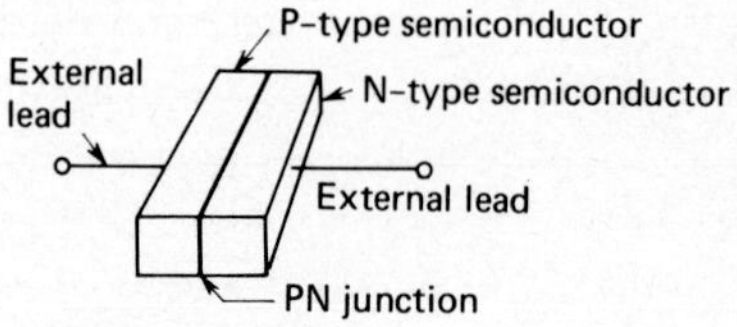

Fig. 1-22   The junction diode

1-23
(*junction*) The junction diode consists of a junction between 2 dissimilar wafers of semiconductor material. For reasons to be discussed later, one of these types is called *P-type* and the other is called _____-type.

4 A PROGRAMMED COURSE IN BASIC ELECTRONICS

1-24
(*N*) The large junction contact surface does not heat as easily as the point-contact. It would be expected, then, that the junction diode could handle _____ power than the point-contact diode.

1-25
(*more* or *higher*, etc.) At the same time, the large junction area tends to make the shunt capacitance of the junction diode _____ than the shunt capacitance of the point-contact type.

1-26
(*higher* or *larger*, etc.) A larger shunt capacitance tends to _____ the by-passing effect of the junction transistor at high frequencies.

1-27
(*increase*) The junction transistor was announced concurrently with the junction diode. As illustrated in Fig. 1-27, there are _____ different types of junction transistors.

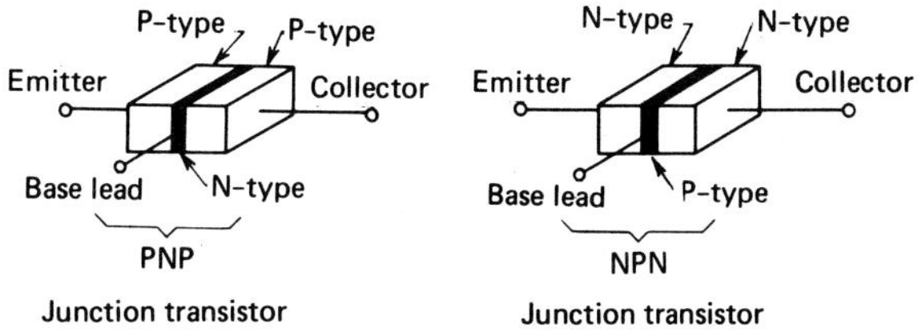

Fig. 1-27   Junction transistor   Junction transistor

1-28
(*2*) In one of these types, a thin wafer of N-type semiconductor is sandwiched between 2 slabs of _____-type semiconductor.

1-29
(*P*) This type of junction transistor is called a(n) _____ type.

1-30
(*PNP*) In the second type of transistor, a thin wafer of P-type semiconductor is sandwiched between 2 slabs of _____-type semiconductor.

1-31
(*N*) This type of junction transistor is called a(n) _____ type.

1-32
(*NPN*) In both types of junction transistors, the thin central section is called the *base*, while one of the outer slabs is the _____ and the other is the *emitter*.

1-33
(*collector*) Wherever a P-type joins an N-type, we have a PN junction. In the PNP transistor, there are two PN junctions; in the NPN transistor, there are also _____ PN junctions.

1-34
(2) The junction transistor permits more accurate prediction of performance than the point-contact transistor. This must be considered as a definite _____ that the junction transistor possesses over the point-contact type.

1-35
(*advantage*) The large contact areas permit more heat dissipation in the junction transistor. Hence, a junction transistor can be expected to handle _____ power than the point-contact type. This and other advantages have led to the complete dominance of the junction devices.

1-36
(*more*) Transistors can accept a small-signal current and build it up into a larger output-signal current. This tells us that a transistor may serve as a current _____.

1-37
(*amplifier*) Transistors can accept a direct-current (dc) power-source input and convert this into an alternating power output. They do this by breaking into oscillation in the proper circuit. Thus, a transistor may serve as an _____.

1-38
(*oscillator*) A transistor can superimpose audio power on RF carrier power, in the process of modulation. Thus, a transistor may be used as a _____.

1-39
(*modulator*) In the reverse process, a transistor may be used to separate the audio component from the carrier component of a received signal. In this case, the transistor is acting as a _____.

1-40
(*detector* or *demodulator, etc.*) A transistor may also be used to change the *shape of a wave*. It may, for example, convert a sine wave into a square wave. In such applications, the _____ becomes part of a wave-shaping circuit.

1-41
(*transistor*) A transistor does not require heater power; hence its *power efficiency* is higher than that of a vacuum _____.

1-42
(*tube*) A transistor does not require warmup time. This is a definite _____ over the ordinary vacuum tube.

1-43
(*advantage*) The average life of a transistor has been estimated as 8 years under *continuous use*. The life of the average vacuum tube is much _____ than this.

1-44
(*shorter*) A transistor is much _____ in size than a vacuum tube and, therefore, lends itself better to miniature construction.

1-45
(*smaller*) Because the power efficiency of a transistor is higher than that of an equivalent vacuum tube, the heat produced by an operating transistor is substantially _____ than the heat produced by a tube.

1-46
(*less* or *smaller, etc.*) The voltage and current requirements of transistors are much smaller than those of tubes. Thus, the associated parts (transformers, resistors, capacitors, etc.) may also be much _____ in size.

1-46
(*smaller*)

MATTER AND ITS STRUCTURE

1-47
Matter may be defined as that which has mass and occupies space. A brick has mass and occupies space; hence a brick is an example of matter. Air has mass and occupies space; hence air must be classified as _____.

1-48
(*matter*) Matter occurs in 3 states: solid, liquid, and gas. A brick is solid matter; water is liquid matter; air is matter in its _____ state.

1-49
(*gaseous*) Matter is found in nature in the form of *elements* and *compounds*. A substance is considered to be an element if it cannot be decomposed by chemical methods into simpler substances. For example, oxygen cannot be decomposed into simpler substances, nor can it be built up from simpler substances; hence _____ is considered to be an element.

1-50
(*oxygen*) Hydrogen is an element because no amount of ordinary chemical treatment can break hydrogen down into anything of _____ nature.

1-51
(*simpler*) Copper is a(n) _____ because it is impossible to produce copper by combining 2 or more other substances.

1-52
(*element*) Bromine is an element usually existing in the liquid state. When bromine is treated in any chemical process whatever, it does not change to another element; nor can bromine be "created" by combining 2 or more other _____.

1-53
(*substances* or *elements, etc.*) At standard temperature and pressure (STP) hydrogen and oxygen are gaseous elements; bromine is a liquid element; copper and iron are _____ elements at STP.

1-54
(*solid*) A compound is a substance containing more than 1 element in *chemical combination*. Water is formed by the chemical union of oxygen and hydrogen; hence water is a(n) _____.

1-55
(*compound*) Carbon dioxide is formed by the chemical union of carbon and oxygen. Both carbon and oxygen are elements, but carbon dioxide is a(n) _____.

1-56
(*compound*) Sodium carbonate is a compound. It is formed by the chemical union of sodium, carbon, and oxygen. It is an example of a compound formed from the union of 3 different _____.

1-57
(*elements*) A *molecule* is the smallest particle of matter that retains all the properties of the original substance. If a drop of water is divided until the smallest possible particle that is still water is obtained, this particle is a _____.

1-58
(*molecule*) The molecule is very small. When a stone is crushed to the finest possible dust, each particle of dust contains thousands of separate _____.

1-59
(*molecules*) Molecules are composed of atoms. An atom is the smallest part of an element that can take part in a chemical reaction. All atoms of a given element have approximately the same mass as all other _____ of the same element.

1-60
(*atoms*) There is a different atom for each element. Thus, an atom of oxygen differs widely from a(n) _____ of hydrogen.

1-61
(*atom*) Altogether, there are over 105 known elements. This means that there must be over _____ different kinds of atoms.

1-62
(*105*) All molecules are composed of 1 or more atoms. For example, as shown in Fig. 1-62, the helium molecule contains only _____ helium atom.

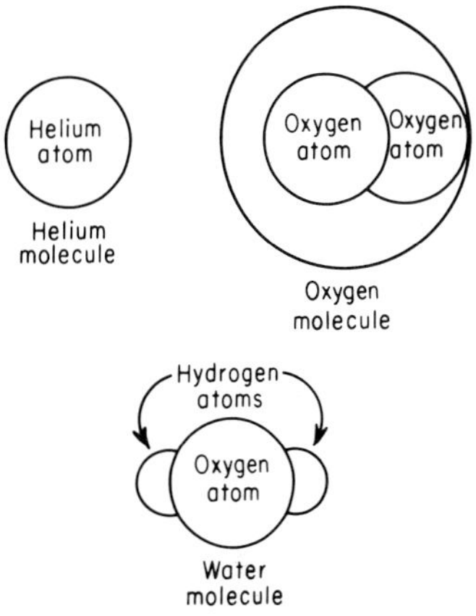

Fig. 1-62

1-63
(*1*) On the other hand, a molecule of oxygen consists of _____ oxygen atoms linked together as shown in Fig. 1-62.

1-64
(*2*) Water is a compound. As seen in Fig. 1-62, the water _____ consists of 2 hydrogen atoms linked to a single oxygen atom.

1-65
(*molecule*) An atom can be further subdivided into still smaller particles. These particles are called subatomic particles. Electrons, protons, and neutrons are three of the important _____ particles.

1-65
(*subatomic*)

ATOMIC STRUCTURE

1-66
The hydrogen atom (Fig. 1-66) is the simplest atom of all. It contains 1 elementary negative charge called an electron, and 1 elementary positive charge called a _____.

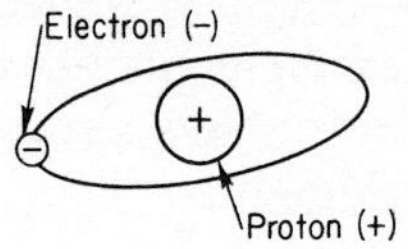

Fig. 1-66

1-67
(*proton*) The electron may be thought of as moving around the _____ in an approximately elliptical orbit.

1-68
(*proton*) The atom shows zero *net* charge. This signifies that the magnitude of the negative charge on the electron must equal the magnitude of the _____ charge on the proton since these cancel each other to yield a zero net charge.

1-69
(*positive*) All electrons are identical in mass and charge no matter where they originate. Thus, an electron from a hydrogen atom has exactly the same mass and _____ as an electron from any other of the 105 or more atoms.

1-70
(*charge*) The charge associated with the electron is the smallest negative electrical charge yet discovered. Similarly, the charge associated with the proton is the _____ positive electrical charge yet discovered.

1-71
(*smallest*) All atoms except that of hydrogen contain another primary subatomic particle called the *neutron*. The neutron is electrically neutral; that is, it has neither a net positive nor a net _____ charge.

1-72
(*negative*) The central body of an atom is called the *nucleus*. The single proton of the hydrogen atom may be called the _____ of the hydrogen atom.

1-73
(*nucleus*) Figure 1-73 shows the nucleus of a *helium* atom. This nucleus contains 2 protons and _____ neutrons.

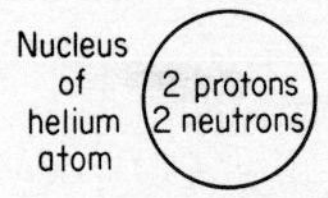

Fig. 1-73

1-74
(2) In Fig. 1-74, the complete helium atom is shown. Its nucleus is symbolized as "2+, 2N," meaning 2 protons and 2 neutrons. Circling around the nucleus are 2 _____.

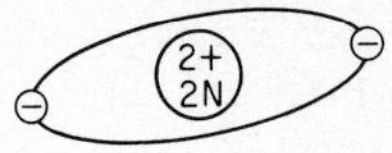

Fig. 1-74

10 A PROGRAMMED COURSE IN BASIC ELECTRONICS

1-75
(*electrons*) Note again that the net charge on the atom must be zero because the atom contains as many elemental negative charges (electrons) as it contains elemental positive charges (_____).

1-76
(*protons*) Since the neutron has no charge, it neither adds to nor _____ from the net charge of the atom as a whole.

1-77
(*subtracts*) It is found that *all* atoms in their normal state are electrically neutral or balanced. This means that *all* normal atoms contain exactly as many circling electrons as there are _____ in the nucleus.

1-78
(*protons*) For example, note the structure of the aluminum atom in Fig. 1-78. (Electron orbits are shown as circles for clarity.) The electrons occupy 3 rings or orbits. There are 2 in the first ring, 8 in the second ring, and _____ in the third, or *outer*, ring.

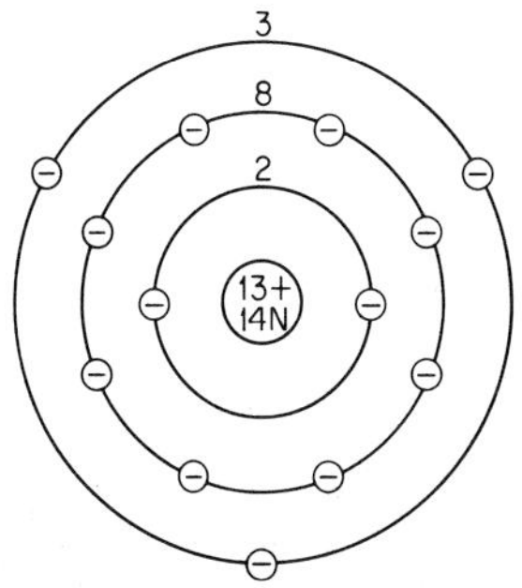

Fig. 1-78   Aluminum

1-79
(3) The nucleus of the aluminum atom contains _____ protons and 14 neutrons.

1-80
(13) Counting electrons in the 3 rings, we find there are _____ electrons altogether circling the nucleus.

1-81
(13) Thus, the numbers of electrons and protons in the whole atom are equal, and the atom must be electrically _____.

1-82
(*neutral*) The neutrons contribute nothing to the net charge on the atom because the charge on a neutron is _____.

SEMICONDUCTOR FUNDAMENTALS 11

1-83
(*zero*) The 3 outer orbit electrons are loosely held to the nucleus. Any 1 or more of these may leave a given atom and wander at random from atom to atom. Electrons that move at random through the material are called *free* electrons. Free electrons always come from _____ orbits.

1-84
(*outer*) In examining the atomic structure of any atom, it is important to note the number of electrons in the outer orbit. For this reason, let us reexamine the aluminum atom and observe that there are _____ electrons in the outer ring.

1-85
(3) Phosphorus is another familiar element. Its atomic structure is given in Fig. 1-85. This atom, as discussed previously, is electrically _____ because it contains the same number of electrons as protons, 15 of each.

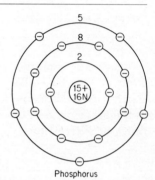

Fig. 1-85 Phosphorus

1-86
(*neutral*) It should be mentioned, too, that the nucleus of phosphorus contains _____ neutrons.

1-87
(*16*) Finally, observe that the *outer* orbit contains _____ loosely held electrons.

1-88
(5) The germanium atom shown in Fig. 1-88 is very important in transistor fabrication. The number of protons in the germanium nucleus is _____.

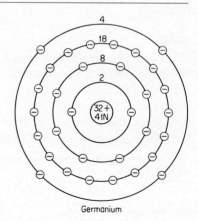

Fig. 1-88 Germanium

12 A PROGRAMMED COURSE IN BASIC ELECTRONICS

1-89
(32) The number of neutrons in the germanium atom is _____.

1-90
(41) The number of electrons revolving around the germanium nucleus is _____.

1-91
(32) These electrons occupy a total of _____ distinct orbits.

1-92
(4) The total number of outer electrons that might become free electrons under the proper conditions is _____.

1-93
(4) Of the 3 types of particles discussed, the electron is the smallest in mass, while the other 2 particles are *approximately* the same in mass. Thus, the mass of a proton is approximately the same as the mass of a _____.

1-94
(*neutron*) The proton is approximately 1,850 times as massive as the electron. Thus, the neutron is approximately _____ times as massive as the electron.

1-95
(*1,850*) All electrons, no matter where they are found, are like all other electrons. Thus, the single electron of the hydrogen atom (Fig. 1-66) is identical to any one of the electrons in the germanium atom. Also, free _____ are identical to electrons in inner orbits.

1-96
(*electrons*) The electrical charge on the proton is exactly equal and opposite to the electrical charge on the electron. That is, the _____ and the electron are exactly equal amounts of opposite kinds of electricity.

1-97
(*proton*) The charge on the electron is believed to be the smallest charge that can exist. One coulomb of charge contains about 6.28×10^{18} elemental electron charges. Since 1 coulomb (C) per second (s) may be defined as an ampere (A), then when 1 A flows, a total of _____ electrons must pass each point in the conductor in 1 s.

1-98
(6.28×10^{18}) In a normal atom, the number of protons in the nucleus equals the number of electrons revolving about the nucleus. This state of balance between charges accounts for the fact that a normal atom is electrically _____.

1-99
(*neutral*) It is possible to remove electrons from the atoms of some substances. When electrons are removed (from the outer orbits), the atom then has a deficiency of electrons and, therefore, carries a net _____ charge.

1-100
(*positive*) It is also possible to add electrons to a substance. When this occurs, the body has more electrons than protons and, therefore, carries a net _____ charge.

1-101
(*negative*) If the haphazard movement of free electrons is controlled so that electrons move generally in the same direction, an electron flow or *electron drift* occurs. This electron flow or drift gives rise to an electric _____.

1-101
(*current*)

CONDUCTORS, INSULATORS, AND SEMICONDUCTORS

1-102
All materials may be placed in 1 of 3 major categories: conductors, insulators, or _____.

1-103
(*semiconductors*) The category in which we place a given substance depends upon its ability to allow electron flow or electron drift. A substance that permits a large electron flow through it must have a large number of free _____.

1-104
(*electrons*) A substance that permits a large electron flow is classified as a *conductor*. Thus, most good conductors have _____ numbers of free electrons.

1-105
(*large*) Metals that have large numbers of free electrons like silver, copper, and aluminum are, therefore, excellent _____.

14 A PROGRAMMED COURSE IN BASIC ELECTRONICS

1-106
(*conductors*) An insulator is a material that has few loosely held electrons in outer orbits. Thus, an insulator has _____ free electrons.

1-107
(*few*) No material known is a perfect insulator. That is, no material known has infinite resistance. When the resistance is very high, however, as in glass, dry wood, rubber, mica, polystyrene, etc., the substance is classed as a(n) _____.

1-108
(*insulator*) To establish the limits of resistance for conductors, we consider a cube of the material measuring 1 centimeter (cm) on each side (Fig. 1-108). If such a cube of a given material has a resistance of less than 3×10^{-6} ohm (Ω), it is considered to be a very good _____.

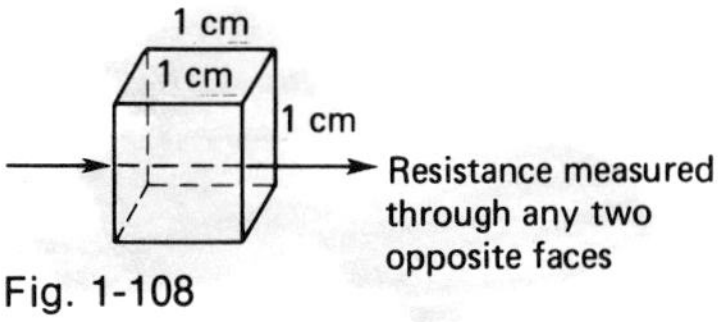

Fig. 1-108

1-109
(*conductor*) Glass, dry wood, rubber, etc., all have resistances that measure several million ohms per centimeter cube. For this reason, such substances are classified as very good _____.

1-110
(*insulators*) Between the extremes of good *conductors* and good *insulators* are a number of materials which are neither good conductors nor good insulators. These are the materials we classify as _____.

1-111
(*semiconductors*) Under standard conditions, pure germanium has a resistance of about 60 Ω/cm³. Pure silicon has a resistance of 60,000 Ω/cm³. Both these figures are between the extremes for conductors and insulators; hence germanium and silicon are both classified as _____.

1-112
(*semiconductors*) In making transistors, certain carefully selected impurities are added to germanium and silicon. This causes their resistance to become about 2 Ω/cm³. Thus, transistor semiconductors have a _____ resistance than the pure materials.

1-113
(*lower*) The resistance of pure semiconductors *decreases* very rapidly with rising temperature. Hence, the resistance of pure germanium at 0 degrees Celsius (°C, formerly centigrade) is considerably _____ than the resistance of the same material at room temperature.

1-114
(*higher*) If transistor silicon (silicon with proper type and amount of impurity) is heated from room temperature to the temperature of boiling water, its _____ may be expected to decrease sharply.

1-114
(*resistance*)

INTRODUCTION TO CRYSTALS

1-115
Most solids under microscopic examination show a regular geometric structure known as a *crystal pattern*. Even rocks and ordinary metals show distinctive patterns; hence they may be said to have a _____ type of structure.

1-116
(*crystal* or *crystalline*) Snowflakes, although formed in an almost infinite number of overall geometric patterns, contain only 60° angles. This means that all individual ice _____ must be essentially alike.

1-117
(*crystals*) Some materials, like ordinary table salt, form cubes; other materials form long needles, rhomboids, hexagons, or other geometric forms. This shows that each material has a characteristic _____ pattern.

1-118
(*crystal* or *crystalline*) X-ray analysis has shown that the *atoms* in a single crystal are arranged in a specific pattern; that any given atom is not related to only 1 or 2 other atoms, but rather is related to a number of adjacent, equidistant _____.

1-119
(*atoms*) To appreciate these relationships, we need some additional conventions in drawing. In Fig. 1-119, an abbreviated picture of the germanium atom is shown. It has 32 protons, 41 neutrons, and _____ electrons.

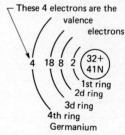

Fig. 1-119

1-120
(*32*) Of this total number of electrons, there are 28 in inner orbits and _____ in the loosely bound outer orbit.

16 A PROGRAMMED COURSE IN BASIC ELECTRONICS

1-121
(4) The 4 electrons in the outer orbit are called the _____ electrons.

1-122
(*valence, pronounced vay-lentz*) To simplify the drawings of germanium atoms, we shall use the convention in Fig. 1-122. The central body containing the 32 protons, 41 neutrons, and 28 inner electrons will be called the germanium _____.

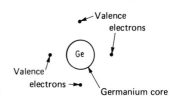

Fig. 1-122

1-123
(*core*) It will be symbolized by the chemical shorthand expression for germanium. This symbol is _____.

1-124
(*Ge*) The 4 small circles represent the _____ electrons.

1-125
(*valence*) A single crystal of pure germanium is represented in the two-dimensional drawing shown in Fig. 1-125. Altogether, we have shown _____ cores.

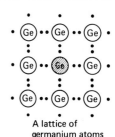

Fig. 1-125

A lattice of germanium atoms

1-126
(9) Each core of germanium is surrounded by _____ valence electrons in Fig. 1-125.

1-127
(4) Note that the electrons of each adjacent pair of cores are shown close to each other. It has been shown that the rotation of 1 valence electron of 1 atom is coordinated with the rotation of 1 valence electron of an adjacent _____.

1-128
(*atom*) This coordinated rotation sets up a condition in which these two electrons form what is known as an "electron-pair _____" (Fig. 1-128).

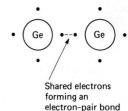

Fig. 1-128

Shared electrons forming an electron-pair bond

SEMICONDUCTOR FUNDAMENTALS **17**

1-129
(*bond*) The electron-pair bond causes an attraction between cores that holds the atoms together. The bound atoms thus form a structure called a _____ (Fig. 1-125).

1-130
(*lattice*) Confined as we are by a 2-dimensional page, we must show a flattened picture of the true situation. Actually, the crystal lattice is 3-dimensional so that some cores are in front of and some cores are _____ others.

1-131
(*behind*) One core in Fig. 1-125 has been shaded to show that it is in the center of the diagram. It is seen that this core is equidistant from _____ (how many?) other germanium cores.

1-132
(*4*) An ordinary germanium crystal contains millions of cores and electron-pair bonds. Thus, all cores are equidistant from _____ (how many?) adjacent germanium cores.

1-133
(*4*) But each core is bound to the adjacent cores by 4 electron-pair _____.

1-134
(*bonds*) These bonds are quite strong. An electron that forms part of an electron-pair bond is held firmly in place by the bonding action. Thus, an electron that is part of a paired bond cannot _____ around freely within the atomic structure of the lattice.

1-135
(*move*) In a conductor such as copper, the free electrons that move through the material are *charge carriers*. These electrons can _____ freely because they do not form a crystal lattice.

1-136
(*move*) Since the electrons in paired bonds do not have freedom of motion, they cannot act as _____ carriers.

1-137
(*charge*) Charge carriers are necessary in a material if it is to serve as an electrical conductor. Since pure germanium does not have an abundance of charge carriers, it is, therefore, a _____ conductor.

1-138
(*poor*) The same is true of other crystalline semiconductors such as carbon and silicon. Unless some steps are taken to break or weaken the electron-pair bonds, the substance cannot behave as a good _____.

1-138
(*conductor*)

IMPURITIES

1-139
Certain impurities can join the crystal lattice structure of germanium and other semiconductors. Figure 1-139 shows a core called a _____ ion core surrounded by 5 valence electrons.

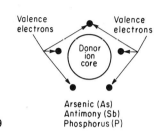

Fig. 1-139

1-140
(*donor*) All impurities that have 5 valence electrons are called *donor* impurities. Elements commonly used as donor impurities are arsenic, _____, and antimony.

1-141
(*phosphorus*) All electrons are alike in charge and mass. If an arsenic atom (or some other donor atom) replaces a germanium atom in a lattice, the _____ of the donor atom will take part in the coordinated rotation just as though they belonged to the germanium core.

1-142
(*electrons*) However, only 4 of the total of _____ (how many?) valence electrons will find other electrons from adjacent atoms with which they can form electron-pair bonds.

1-143
(*5*) As shown in Fig. 1-143, 4 of the 5 arsenic (As) valence electrons have formed electron-pair _____ with nearby germanium atoms.

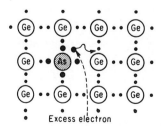

Fig. 1-143

SEMICONDUCTOR FUNDAMENTALS

1-144
(*bonds*) This leaves 1 _____ electron of the original 5 valence electrons free to wander through the lattice.

1-145
(*excess*) It will be recalled, however, that a normal atom such as arsenic is electrically neutral because the number of circling electrons is equal to the number of _____ in the nucleus.

1-146
(*protons*) If the loosely held excess electron (Fig. 1-143) wanders away from the arsenic core, this means that the core is losing 1 elemental _____ charge.

1-147
(*negative*) Thus, the arsenic atom must be left with 1 excess _____ charge.

1-148
(*positive*) Since a charged atom is called an *ion*, the donor atom, having *donated* 1 electron to the crystal lattice, is now a donor _____.

1-149
(*ion*) Although the donor ion now bears 1 excess elemental positive charge, the crystal lattice still contains the donated electron in its free state. Since the charge on the donor ion (1 elemental charge) is equal and opposite to the charge on the electron, the crystal itself is still electrically _____.

1-150
(*neutral*) Germanium that contains donor impurities is called *N-type germanium*. The letter *N* refers to the _____ charge carried by the free electron.

1-151
(*negative*) Figure 1-151 shows a germanium crystal in which one of the germanium atoms has been replaced by an atom of _____.

Trivalent (acceptor) elements are indium (In), gallium (Ga), and aluminum (Al)

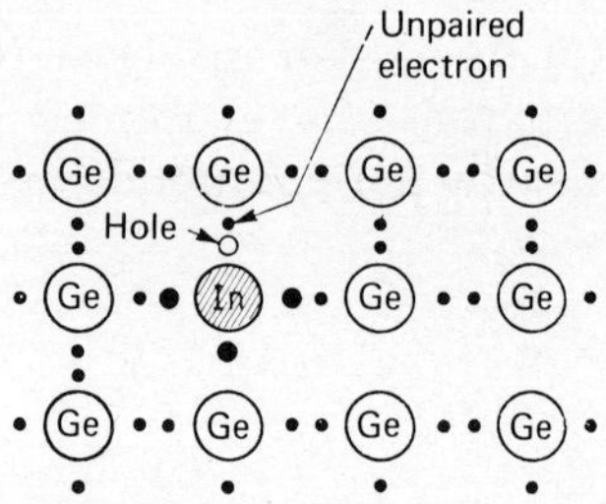

Fig. 1-151

1-152
(*indium*) The three black dots around the indium core indicate that this element has _____ valence electrons.

1-153
(*3*) Gallium (Ga) and aluminum (Al), like indium, also have 3 _____ electrons surrounding the core.

1-154
(*valence*) When such elements are used as impurities in transistor germanium, they can *accept* rather than *donate* electrons to the lattice. For this reason, these impurities are known as _____ impurities.

1-155
(*acceptor*) Each of the 3 valence electrons of the indium atom forms electron-pair bonds with the electrons of neighboring _____ atoms.

1-156
(*germanium*) Since an acceptor impurity has only 3 valence electrons, one of the germanium atoms near the acceptor has 1 electron that cannot form an electron-pair bond. This electron has been labeled an _____ electron in Fig. 1-151.

1-157
(*unpaired*) Note the open circle just above the indium core in Fig. 1-151. This position would *normally* be filled by an electron with which the unpaired electron could form a bond. Since this electron does not exist in this lattice, the empty spot is called a _____.

1-158
(*hole*) Figure 1-158 shows a simplified view of the same lattice shown in Fig. 1-151. Because of thermal agitation or applied electric fields, it is possible for an electron from some neighboring germanium atom to break loose from its core and migrate toward the _____ as shown by the arrow in Fig. 1-158.

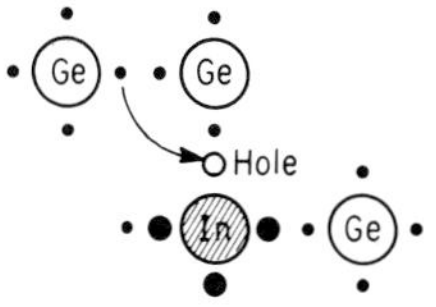

Fig. 1-158

1-159
(*hole*) Figure 1-159 illustrates the completed migration of an electron from the upper left Ge core to the spot formerly occupied by the hole. The old hole has been filled, but a new _____ now appears to the right of the upper lefthand Ge core.

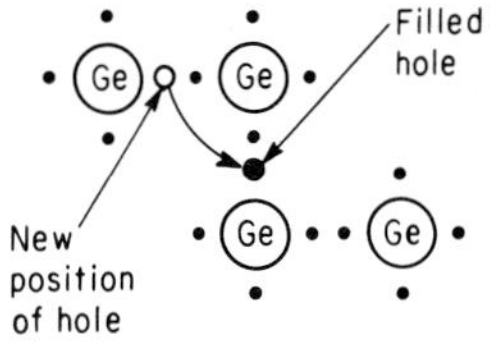

Fig. 1-159

1-160
(*hole*) Now the acceptor atom (In) has one more electron than it has when it is a neutral atom. Thus, the indium atom now has 1 excess _____ charge.

1-161
(*negative*) Since it is a *charged* atom, we no longer call it an atom. As we have seen, a charged atom is called an ion. Hence, where an indium atom existed before, an indium _____ now exists.

1-162
(*ion*) Furthermore, the Ge atom that now has a hole has *lost* 1 electron to the indium. Hence, the Ge atom that lost the electron is left with a net _____ charge.

1-163
(*positive*) The loss of a valence electron to an acceptor impurity causes the formation of a hole in the original position of the lost electron. Thus, the presence of a hole has exactly the same effect on the net charge of an atom as the loss of an _____.

1-164
(*electron*) The presence of a hole gives the Ge atom a net charge of *one positive* (+1). Thus, rather than consider that the atom has lost an electron, we might find it more convenient to consider that the atom has gained a _____.

1-165
(*hole*) As valence electrons move from atom to atom in the manner just described, we may, therefore, picture that holes are _____ from atom to atom in the opposite direction.

1-166
(*moving*) The whole Ge crystal, however, has 1 indium ion that is *negative* (Fig. 1-159) by 1 elemental charge and 1 hole that is positive by 1 elemental charge. Hence, the whole crystal may be considered as electrically _____.

1-167
(*neutral*) Transistor germanium containing acceptor impurities is referred to as *P-type germanium*. The letter *P* refers to the _____ charge of the hole.

1-167
(*positive*)

PROPERTIES OF HOLES

1-168
A hole has been identified as a location which has been vacated by an electron from the valence ring of an atom. After the hole has formed, the atom becomes an ion with a net positive charge of 1. It is convenient to think of a hole as an actual particle having a _____ charge.

1-169
(*positive*) As valence electrons move from point to point in a crystal lattice, the holes they produce may be considered as moving in the opposite direction. Thus, the movement of holes is equivalent to the movement of positively _____ particles.

1-170
(*charged*) Conductors such as copper and aluminum do not have electron-pair bonds. A hole can form only when an electron leaves an electron-pair bond. Thus, copper and aluminum cannot form _____ within their substances.

1-171
(*holes*) Thus, holes can exist only in certain materials such as germanium and silicon. Germanium and silicon are semiconductors. Hence, holes are usually formed only in _____.

1-172
(*semiconductors*) An electron located between 2 oppositely charged bodies (*a* and *b* in Fig. 1-172) tends to move toward the positively charged body. Thus, in Fig. 1-172, the electron is shown moving toward the body identified as _____.

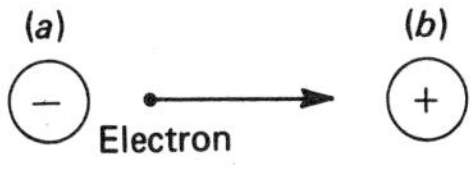

Fig. 1-172

1-173
(*b*) This follows the law of electric charges which says that like charges repel and _____ charges attract.

1-174
(*unlike*) A hole is a positively charged *particle*. Hence, a hole will move toward the more _____ of the two bodies, as indicated in Fig. 1-174.

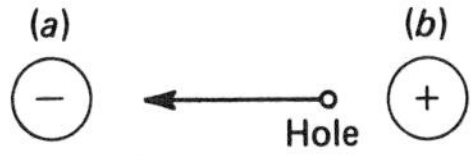

Fig. 1-174

1-175
(*negative*) An electron moving across a magnetic field is deflected by the field. Given a certain deflection direction for an electron moving across a given field, a hole moving through the same magnetic field will be deflected in the _____ direction.

SEMICONDUCTOR FUNDAMENTALS

1-176
(*opposite*) In electronics, an electron is considered permanently indestructible. When a hole is filled by an electron from an adjacent electron-pair bond, the _____ is considered as having moved from one position to another.

1-177
(*hole*) However, when a hole is filled by a *free* or *excess* electron, a new hole position is *not* created. Thus, when a hole is filled by an _____ electron, the hole *ceases to exist*.

1-178
(*excess*) In normal conductors, the current is due to the movement of free _____ from the minus toward the plus terminals of the generator or battery.

1-179
(*electrons*) Figure 1-179 shows an electron current. An electron current flowing in a wire produces circular lines of magnetic force so that the wire is surrounded by a _____ field.

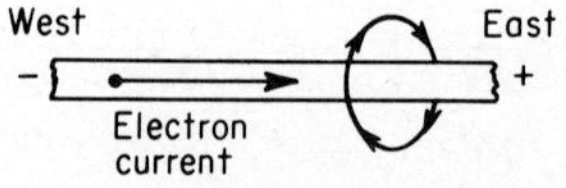

Fig. 1-179

1-180
(*magnetic*) The direction of this field is clockwise looking at the end of the wire from the east toward the west terminals (lefthand rule for wires). Instead of an electron current, Fig. 1-180 shows a current consisting of _____ moving through a semiconductor.

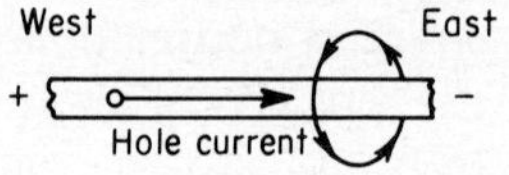

Fig. 1-180

1-181
(*holes*) Looking from the east toward the west once again, this time in Fig. 1-180, we see that the magnetic field is *counterclockwise* for the hole current. Thus, the direction of the field produced by a hole current is _____ that produced by an electron current.

1-182
(*opposite*) These experiments prove that an electric current in a semiconductor may be caused by the movement of either holes or _____.

1-183
(*electrons*) They show, furthermore, that electrons and holes flow in _____ directions when the semiconductor is placed in an electric field having a given direction (Figs. 1-172 and 1-174).

1-184
(*opposite*) Also, the _____ fields produced by electrons and holes flowing in a given direction through a semiconductor are opposite in sense.

1-184
(*magnetic*)

HOLES AS CURRENT CARRIERS

1-185
In these frames we offer experimental proof that holes may act as current carriers although we normally think of electrons as current carriers. In Fig. 1-185 we see a slab of _____-type germanium.

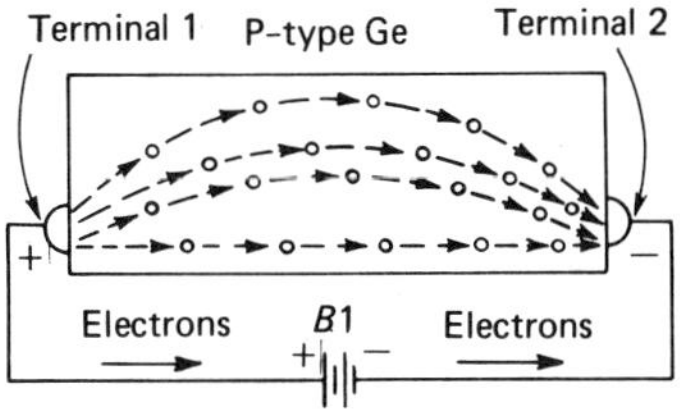

Fig. 1-185

1-186
(*P*) As we have found out, in P-type germanium there may be a large number of _____ because acceptor impurities have broken electron-pair bonds to complete their outer rings of electrons.

1-187
(*holes*) Since holes carry a positive charge, they will be attracted toward terminal _____, as shown by the arrows in Fig. 1-185 (terminal 1 is positive because of *B*1).

1-188
(*2*) There are plenty of free electrons at terminal 2 because of its connection to the negative end of the battery *B*1. Thus, when a hole reaches terminal 2, a(n) _____ from this terminal enters the germanium and fills the hole.

1-189
(*electron*) At the same time, terminal 1, being very positive, exerts a strong electrostatic force on the electrons in the electron-pair bonds of nearby Ge atoms. An electron-pair bond breaks, and one of the electrons thus freed immediately goes into terminal _____.

1-190
(*1*) The breaking of this bond and the loss of an electron to the terminals creates a new _____.

SEMICONDUCTOR FUNDAMENTALS 25

1-191
(*hole*) The newly created hole promptly begins to migrate toward terminal _____.

1-192
(2) This action provides a continuous flow of _____ in the external circuit.

1-193
(*electrons*) And, at the same time, the same action provides a continuous flow of _____ in the germanium.

1-194
(*holes*) Note: In Fig. 1-185 the _____ flow in an arc over most of the paths.

1-195
(*holes*) Physicists have measured the time required for the holes to flow over the various paths. From the diagram, it is quite clear that it would take a _____ time for holes to flow over the curved paths than over the straight.

1-196
(*longer*) Let us note this fact, then study Fig. 1-196. The same arrangement is used as before except that _____ plates have been placed above and below the germanium slab.

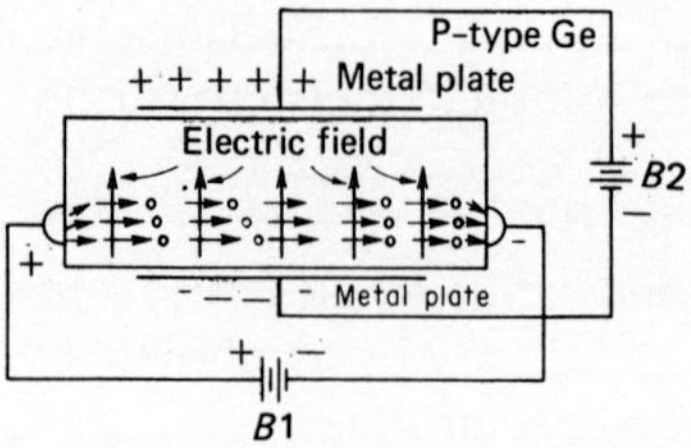

Fig. 1-196

1-197
(*metal*) These plates are connected to a second battery B2 which causes the top plate to assume a _____ potential with respect to the bottom plate.

1-198
(*positive*) The time required for the current carriers to go from one terminal to the other is again measured and compared with the earlier timing. It is found that the time is *shorter*, hence the paths must be nearly _____ lines rather than arcs.

1-199
(*straight*) This indicates that the electric field produced by B2 must have driven the current carriers nearer to the _____ plate in order to straighten out the arcs.

1-200
(*bottom* or *negative*) But if the current carriers had been electrons, they would have been driven closer to the _____ metal plate.

1-201
(*top* or *positive*) If this had been the case, the transit time of the carriers from one terminal to the other would have been _____ rather than decreased.

1-202
(*increased*) This was not the case. Hence, the current carriers could not have been negative electrons. This leaves only one other possibility. The current carriers must be positive _____.

1-202
(*holes*)

PN JUNCTION BARRIER

1-203
Figure 1-203 shows a slab of P-type germanium. For simplicity, the electron-pair bonds have been omitted, and only the holes and the _____ ions have been indicated.

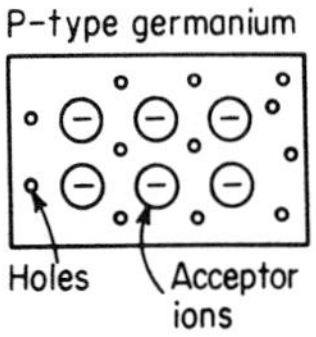

Fig. 1-203

1-204
(*acceptor*) Figure 1-204 illustrates a slab of _____-_____ germanium.

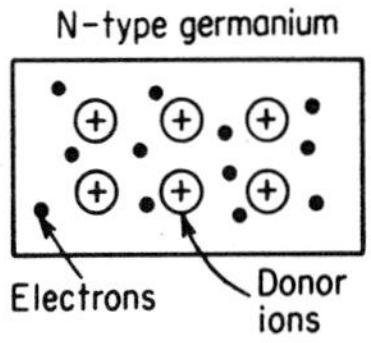

Fig. 1-204

1-205
(*N-type*) This has been similarly simplified, showing only the electrons and donor ions and omitting the _____-_____ bonds.

1-206
(*electron-pair*) For clarity, we have shown a large number of donor and acceptor ions in the 2 types of germanium. Actually, in transistor germanium, there is an extremely small percentage of impurity ions. The concentration is about 1 atom of impurity material to 10,000,000 atoms of _____.

1-207

(*germanium*) Heat energy is always present. This thermal energy causes rapid, vibratory motion of the constituents of matter. If we could look into the Ge slabs we would see the germanium cores and impurity ions vibrating rapidly because of _____ energy.

1-208

(*thermal* or *heat*) The cores and ions, however, do not leave their lattice positions. If they did, they might be responsible for the production of a current. Since they are fixed in their lattice positions, they cannot produce a _____.

1-209

(*current*) The holes and electrons do move from atom to atom because of thermal energy. But this motion is random and haphazard so that the _____ and electrons are evenly distributed throughout their respective Ge slabs.

1-210

(*holes*) This haphazard movement occurs in the absence of any externally applied electric or magnetic field. This motion is called *diffusion*. Any electron or hole motion that may occur because of an external field, therefore, will be referred to as current rather than _____.

1-211

(*diffusion*) Figure 1-211 shows the 2 slabs from the previous figures joined end to end. Since holes and electrons carry opposite _____, some of the holes and some of the electrons near the junction begin to move toward one another.

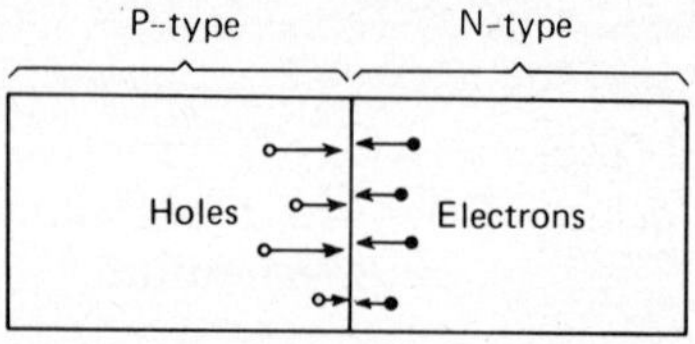

Fig. 1-211

1-212

(*charges*) As the charged particles meet, they *recombine*, thus eliminating each other. In each recombination, 1 hole and 1 excess _____ are eliminated.

1-213

(*electron*) One might think that holes would continue to flow from the P-type Ge to the _____-type Ge, causing continuous recombinations.

1-214

(*N*) And one might also think that electrons would continue to flow from the N-type toward the P-type past the junction to produce additional _____.

28 A PROGRAMMED COURSE IN BASIC ELECTRONICS

1-215
(*recombinations*) But this does not happen. See Fig. 1-215. After relatively few recombinations, the process stops because a number of uncompensated ions are left facing each other across the junction. Thus, a hole approaching the N-type slab across the junction is repelled by _____ ions on the other side.

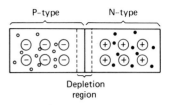

Fig. 1-215

1-216
(*positive*) Similarly, an electron going from the N-type toward the P-type Ge across the junction faces the repelling force of the _____ ions on the other side.

1-217
(*negative*) These repelling forces produce what is known as a *barrier*. Because of this barrier, neither electrons nor holes can pass to cause new recombinations. This leaves a region devoid of current carriers. This region (Fig. 1-215) is called the _____ region.

1-218
(*depletion*) There is an electric field in the depletion region. This _____ is called the *junction barrier*.

1-219
(*electric field*) The presence of the junction barrier is the result of the ranging of negative acceptor ions on one side of the junction and positive _____ ions on the other side of the junction.

1-220
(*donor*) The donor ions prevent holes from moving across the barrier in the absence of an external electric field. The acceptor ions prevent _____ from moving across the barrier in the opposite direction in the absence of an external electric field.

1-221
(*electrons*) The physical distance from one side of the barrier to the other is called the *width* of the barrier. The width of the barrier depends upon the density of electrons (excess) and _____ in the crystals.

SEMICONDUCTOR FUNDAMENTALS **29**

1-222
(*holes*) Because the depletion region consists of positive and negative ions ranged on each side, there is a difference of potential across the depletion area. The potential difference from one side of the barrier to the other is called the *height* of the barrier and is determined by the intensity of the electric field produced by the oppositely charged ions. Barrier height, being a potential difference, is measured in volts (V). With no external battery connected, the height of the barrier is on the order of a few tenths of a _____.

1-222
(*volt*)

PN JUNCTION BIAS

1-223
Figure 1-223 shows the changes that take place when an external battery is connected with the indicated polarity to the ends of the crystal. As shown, the negative terminal of the external battery is connected to the P-type germanium and the positive terminal is connected to the _____-_____ germanium.

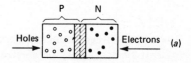
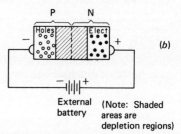

Fig. 1-223

(Note: Shaded areas are depletion regions)

1-224
(*N-type*) The external battery sets up an electric field along the length of the crystal. The holes, being positive charges, are attracted to the negative connection from the battery since _____ charges attract each other.

1-225
(*unlike*) For the same reason, the electrons in the N-type germanium are attracted to the _____ connection from the battery.

1-226
(*positive*) With the holes and electrons drawn *farther* from the junction, the depletion region (shown as a shaded area in the drawings) must become _____ than it was before, as indicated in Fig. 1-223b.

1-227
(*wider*) As we have seen, the width of the depletion region governs the *height* (potential) of the barrier. Thus, as the depletion region widens, the height (potential) of the _____ must also increase.

1-228
(*barrier*) The depletion region widens until the potential of the depletion region becomes equal to the potential of the external battery. When this occurs, the attractive forces on each side of the holes become equal, so that the net force tending to move the holes becomes _____.

1-229
(*zero*) The same is true of the electrons in the N-type section. The forces attracting electrons to the right and left become _____, so that the net force acting on the electrons is zero.

1-230
(*equal*) Since neither the holes nor electrons tend to move when the net force acting on them is zero, the current flowing through the crystal due to the external battery must be _____.

1-231
(*zero*) When an external battery is connected in the fashion shown in Fig. 1-223b, the PN junction is said to be *reverse-biased*. Thus, a condition of reverse bias results in a very small net _____ and current across the junction.

1-232
(*potential*) And, when the net potential is zero, the _____ flowing through the junction must also be zero.

1-233
(*current*) In Fig. 1-233b, we show the external battery reversed in polarity. This time the positive end of the battery is connected to the P-type and the _____ end to the N-type germanium.

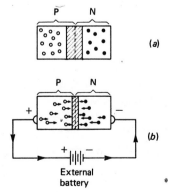

Fig. 1-233

1-234
(*negative*) Since the electric field is reversed as compared with the previous connection, we now see that both the holes and the _____ are acted upon by electric forces that tend to move them toward the barrier.

1-235
(*electrons*) Due to the energy acquired by holes and electrons in motion, they can penetrate more deeply into the _____ region at the junction; they then can continue through the depletion region and recombine on the other side.

1-236
(*depletion*) This penetration has the effect of causing the width of the depletion region to _____.

1-237
(*decrease*) Electrons and holes forced through the narrowed depletion region find each other and recombine. Each recombination causes the disappearance of an excess electron and a _____.

1-238
(*hole*) For each electron that is lost due to a recombination, another electron enters the N-type germanium from the _____ terminal of the external battery.

1-239
(*negative*) These electrons under the force of the externally applied field also drift toward the junction to undergo ultimate _____ with holes at this point. The recombination occurs in the depletion region only to a small extent, but a great deal occurs in the N and P regions.

1-240
(*recombination*) Each time a hole recombines with an electron at the junction, another _____ forms near the positive terminal of the battery when an electron breaks its bond and enters the positive terminal wire.

1-241
(*hole*) Since electrons are constantly entering the N-type germanium at the negative end and leaving the germanium at the _____ end, an electron current flows in the external circuit as shown in Fig. 1-233*b*.

1-242
(*positive*) As long as the external battery is connected in the manner shown in Fig. 1-233*b*, there is a current in the P-type germanium consisting of holes and a current in the N-type germanium consisting of _____.

32 A PROGRAMMED COURSE IN BASIC ELECTRONICS

1-243
(*electrons*) In the condition just discussed, the PN junction is said to be *forward-biased*. Thus, forward bias is a condition that results in a flow of _____ in the external circuit.

1-244
(*current* or *electrons*) If excess forward bias is used, the rate of flow of holes and electrons in the crystal causes a great increase in thermal agitation. Such thermal agitation may cause severe _____ to the crystal structure.

1-244
(*damage* or *harm, etc.*)

DIODE ACTION

1-245
Perfectly pure germanium containing absolutely no impurities would contain no donor or _____ ions.

1-246
(*acceptor*) When an impurity such as arsenic, which contains 5 valence electrons, is added to pure germanium, the arsenic is called a donor ion because it donates 1 free _____ to the lattice.

1-247
(*electron*) Donor ions form N-type germanium because the free electrons carry a Negative charge. In N-type germanium, the current carriers are _____ rather than holes.

1-248
(*electrons*) No purification process has ever been devised which would prevent *a few* acceptor ions from finding their way into N-type germanium during manufacture. Therefore, N-type germanium contains many free electrons, but also a few _____.

1-249
(*holes*) Similarly, it is impossible to prevent a few donor ions from finding their way into presumably perfect P-type germanium. Hence, although P-type germanium contains mostly holes as current carriers, it also contains a few free _____.

1-250

(*electrons*) Thus, in every piece of N-type germanium the *majority* carriers are electrons but, since there are a few holes present, these are called *minority* _____. The larger part of the minority carriers present are not due to impurities, but arise as a result of broken covalent bonds in the bulk material.

1-251

(*carriers*) Similarly, in every piece of P-type germanium the majority carriers are holes; since there are also a few electrons present, however, the _____ carriers are electrons.

1-252

(*minority*) Thus, every piece of transistor germanium contains both majority and minority carriers. The electric charge on the majority carriers is always opposite that of the _____ carriers.

1-253

(*minority*) This means that the direction of motion of the majority carriers through a crystal or crystal junction under the influence of a given electric field is always _____ the direction of flow of the minority carriers in the same field.

1-254

(*opposite*) For example, when a junction is *reverse-biased* for the majority carriers, it is *forward-biased* for the minority carriers. When the junction is forward-biased for the majority carriers, it is _____-_____ for the minority carriers. It should be remembered, however, that the minority carrier flow across the junction is due to diffusion and is always present.

1-255

(*reverse-biased*) Referring to Fig. 1-255, we can see the current flow in a crystal under conditions of forward and reverse _____.

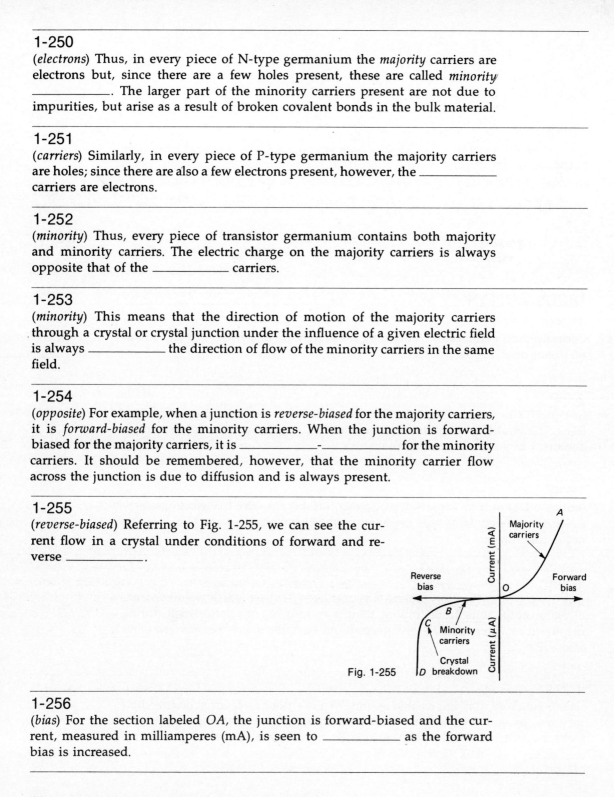

Fig. 1-255

1-256

(*bias*) For the section labeled *OA*, the junction is forward-biased and the current, measured in milliamperes (mA), is seen to _____ as the forward bias is increased.

1-257
(*increase*) As indicated, the forward-bias current is carried by the _____ carriers.

1-258
(*majority*) As we go into the reverse-bias region, we would expect the current to be zero if there were no minority carriers. But the current, as is clearly shown, is not _____.

1-259
(*zero*) This happens because the junction may be reverse-biased for the majority carriers, but it is _____-biased for the minority carriers that are inevitably present.

1-260
(*forward*) Hence, some current is obtained because of the presence of the minority carriers. This current is normally very small since, as shown on the vertical axis below point O, the current is measured in _____.

1-261
(*microamperes* or μA) If the reverse bias is carried beyond a definite critical point (point C), a condition called crystal _____ occurs.

1-262
(*breakdown*) Between point C and point D, after crystal breakdown occurs, the current may become very _____, especially if the protective resistance in the circuit is too small or nonexistent.

1-263
(*large*) If the current is held to a safe value, the crystal will not be damaged and will be restored to normal when the magnitude of the reverse bias is _____ to point B or O.

1-264
(*reduced*) If the current is allowed to become too large, or to flow for too long a time, the thermal agitation will become very great and the crystal is likely to be permanently _____.

1-264
(*damaged* or *spoiled*, etc.)

CRITERION CHECK TEST

____1-1 In a single-crystal radio receiver, the crystal is used to (*a*) amplify the radio signal, (*b*) tune the radio signal, (*c*) rectify the radio signal, (*d*) attenuate the radio signal.

____1-2 A point-contact diode consists of a (*a*) metal-metal junction, (*b*) metal-germanium junction, (*c*) germanium-silicon junction, (*d*) metal oxide–metal junction.

____1-3 Which of the following is not an element? (*a*) Water, (*b*) oxygen, (*c*) bromine, (*d*) copper.

____1-4 The 3 states of matter are (*a*) liquid, gas, element, (*b*) compound, element, gas, (*c*) gas, liquid, solid, (*d*) mixture, compound, solid.

____1-5 A molecule is made of one or more (*a*) compounds, (*b*) gases, (*c*) atoms, (*d*) electrons.

____1-6 There are approximately how many known elements? (*a*) 50, (*b*) 80, (*c*) 105, (*d*) 142.

____1-7 The water molecule has (*a*) 2 H atoms and 2 O atoms, (*b*) 1 H atom and 2 O atoms, (*c*) 2 H atoms and 1 O atom, (*d*) 1 H atom and 1 O atom.

____1-8 The charge of the electron is (*a*) equal to that of the proton, (*b*) twice that of the proton, (*c*) equal to that of the neutron, (*d*) $\frac{1}{1,840}$ that of the proton.

____1-9 A normal hydrogen atom has exactly (*a*) 1 electron and 1 neutron; (*b*) 1 proton, 1 neutron, and 2 electrons; (*c*) 1 proton and 2 electrons; (*d*) 1 proton and 1 electron.

____1-10 The nucleus of an atom contains (*a*) electrons and neutrons, (*b*) only protons, (*c*) protons and neutrons, (*d*) only neutrons.

____1-11 A neutral atom might have only (*a*) 3 neutrons and 3 electrons, (*b*) 2 protons and 3 electrons, (*c*) 3 electrons and 2 protons, (*d*) 4 protons and 4 electrons.

____1-12 The aluminum atom has how many rings of electrons? (*a*) 1, (*b*) 2, (*c*) 3, (*d*) 4.

____1-13 Free electrons always come from (*a*) the nucleus of an atom, (*b*) the inner rings of an atom, (*c*) the outer rings of an atom, (*d*) neutron breakup.

____1-14 The number of potential free electrons in germanium is (*a*) 1, (*b*) 2, (*c*) 3, (*d*) 4.

____1-15 An electric current is caused by the (*a*) regular drift of electric charges, (*b*) removal of electrons from a neutral atom, (*c*) addition of electrons to a neutral atom, (*d*) circling of electrons in orbits.

____1-16 A substance which does not conduct electricity is called (*a*) an insulator, (*b*) a semiconductor, (*c*) a liquid, (*d*) a metal.

____1-17 When a pure semiconductor is heated, its (*a*) resistance goes up, (*b*) resistance goes down, (*c*) resistance does not change, (*d*) atomic structure changes.

____1-18 Valence electrons (*a*) are found in the nucleus of an atom, (*b*) are the inner core electrons of an atom, (*c*) are always free electrons, (*d*) are the outer orbit electrons.

____1-19 The atomic core of germanium consists of the (*a*) nucleus and valence electrons, (*b*) protons and all electrons, (*c*) nucleus, (*d*) nucleus and all electrons exclusive of the valence electrons.

____1-20 An electron pair consists of (*a*) 2 valence electrons of the same atom, (*b*) 2 electrons from different orbits of an atom, (*c*) 2 electrons shared by 2 separate atoms, (*d*) 2 free electrons.

____1-21 The strength of a crystal lattice comes from (*a*) forces between nuclei, (*b*) electron-pair bonds, (*c*) forces between neutrons, (*d*) forces between protons.

____1-22 Pure semiconductors are poor conductors because (*a*) they have no valence electrons, (*b*) all valence electrons are in electron pairs, (*c*) they have too many free electrons, (*d*) there are fewer electrons than protons.

_____1-23 How many valence electrons do donor impurities possess? (*a*) 1, (*b*) 3, (*c*) 5, (*d*) 7.

_____1-24 A charged atom is called (*a*) a molecule, (*b*) a mixture, (*c*) a lattice, (*d*) an ion.

_____1-25 The extra valence electron of a donor atom becomes (*a*) a member of a pair bond, (*b*) an ion, (*c*) a free electron, (*d*) a bound electron.

_____1-26 A hole in a lattice is defined as (*a*) a free proton, (*b*) a free neutron, (*c*) the incomplete part of an electron-pair bond, (*d*) an acceptor ion.

_____1-27 The charge of a hole is (*a*) zero, (*b*) equal to that of the neutron, (*c*) equal to that of the electron, (*d*) equal to that of the proton.

_____1-28 As a general rule, holes are found only in (*a*) metals, (*b*) semiconductors, (*c*) insulators, (*d*) conductors.

_____1-29 A hole and electron in close proximity would tend to (*a*) repel each other, (*b*) attract each other, (*c*) have no effect on each other, (*d*) move at right angles to each other.

_____1-30 In semiconductor materials, current is due (*a*) only to holes, (*b*) only to free electrons, (*c*) only to valence electrons, (*d*) to holes or electrons.

_____1-31 In Fig. 1-179, the circular arrows indicate the direction of (*a*) hole current, (*b*) electron current, (*c*) deflected hole, (*d*) magnetic field.

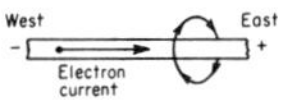

Fig. 1-179 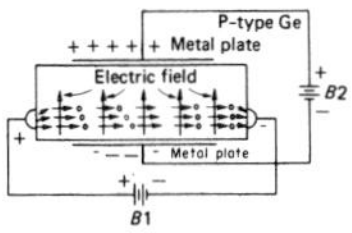   Fig. 1-196 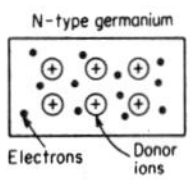   Fig. 1-204 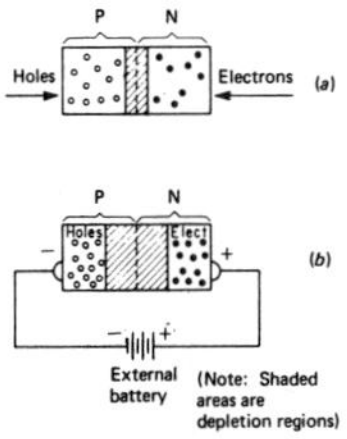   Fig. 1-223

_____1-32 In Fig. 1-196, the upper metal plate tends to (*a*) repel electrons, (*b*) repel holes, (*c*) increase the number of holes, (*d*) decrease the number of holes.

_____1-33 In Fig. 1-204, the circles with a plus sign represent (*a*) germanium atomic cores, (*b*) germanium nuclei, (*c*) impurity atoms, (*d*) positive electrons.

_____1-34 The random motion of holes and electrons due to thermal agitation is called (*a*) pressure, (*b*) impurities, (*c*) ionization, (*d*) diffusion.

_____1-35 Recombination refers to (*a*) creation of free electrons, (*b*) ionization of an impurity atom, (*c*) breaking of a pair bond, (*d*) annihilation of a hole and electron.

_____1-36 In the depletion region of a PN junction, there is a shortage of (*a*) acceptor ions, (*b*) donor ions, (*c*) holes and electrons, (*d*) Ge core atoms.

_____1-37 The donor ions at the junction barrier prevent (*a*) protons from moving across the barrier, (*b*) holes from moving across the barrier, (*c*) electrons from moving across the barrier, (*d*) acceptor ions from moving across the barrier.

_____1-38 In Fig. 1-223, the external battery has caused (*a*) holes to move to the left, (*b*) electrons to move to the left, (*c*) donor atoms to move to the left, (*d*) donor atoms to move to the right.

_____1-39 A reverse-biased PN junction has (*a*) a net hole current, (*b*) a net electron current, (*c*) almost no current, (*d*) a very narrow depletion region.

2 transistor fundamentals I

OBJECTIVES

(1) Discuss the bias conditions at the two junctions of a PNP (NPN) transistor. **(2)** Sketch a PNP transistor sandwich, labeling the ions, electrons, and holes. **(3)** Describe the depletion regions at the two PN junctions of a PNP transistor. **(4)** Discuss the flow of majority carriers at the base-emitter and base-collector junctions. **(5)** Tell the effect on collector current of making the base region very narrow. **(6)** Estimate the percentage of emitter current which reaches the collector. **(7)** Contrast the NPN transistor with the PNP transistor. **(8)** Distinguish between hole and electron currents. **(9)** Diagram a PNP transistor, defining, labeling, and categorizing as hole currents or electron currents I_E, I_C, I_B, I_{EC}, I_{EB}, and I_{CBO}. **(10)** Diagram an NPN transistor, labeling and categorizing as hole currents or electron currents I_E, I_C, I_B, I_{EC}, I_{EB}, and I_{CBO}. **(11)** Estimate the relative magnitude of I_{EB}, I_{CBO}, and I_C. **(12)** Draw and discuss a curve showing the dependence of collector current on emitter-base potentials. **(13)** Compute the forward current gain of a transistor in the common-base configuration. **(14)** Define the current amplification factor (α) and the common-base (CB), common-collector (CC), and common-emitter (CE) configurations. **(15)** Compute the voltage amplification of a transistor in the CB configuration. **(16)** Draw the symbol used to represent an NPN (PNP) transistor, stating the letter used to label transistors, the meaning of the arrow used in the symbol, and a mnemonic used to obtain proper transistor bias polarities. **(17)** Discuss the CB, CE, and CC transistor configurations, drawing a PNP transistor in each configuration, labeling bias batteries, external input and output resistances, and the grounded lead, and demonstrating that the bias batteries provide the correct polarities.

INTRODUCTION TO TRANSISTORS

2-1
Figure 2-1 shows 3 pieces of transistor germanium. Each of these contains the impurities required to form holes and electrons. The P-type germanium contains an impurity that forms _____ ions and _____ holes.

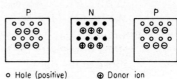

Fig. 2-1

2-2
(*negative, positive*) The N-type impurity is of a type that will form positive _____ and free electrons.

2-3
(*ions*) Recall that we indicate electrons by filled-in or solid circles, while holes are shown as hollow _____.

2-4
(*circles*) Also, the acceptor ions are shown as larger circles that enclose a "—" sign, while the donor ions are shown as larger circles that enclose a "_____" sign.

2-5
(+) Electrons are always associated with donor ions in the semiconductor; _____ are always associated with acceptor ions.

2-6
(*holes*) If the 3 pieces of transistor germanium are properly fused together (Fig. 2-6), two PN junctions are formed. At each of these junctions, just as in diode junctions, a _____ region forms.

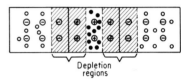

Fig. 2-6

2-7
(*depletion*) As we have seen, the depletion regions contain very few holes or _____.

2-8
(*electrons*) When 3 pieces of transistor germanium are fused as in Fig. 2-6, a transistor is formed. This type of transistor is called a PNP type because it is a sandwich of 2 pieces of _____-type germanium and only 1 piece of N-type germanium.

2-9
(*P*) Refer to Fig. 2-9. For the moment we shall ignore the righthand P-type section and the battery labeled *reverse bias*, considering only the PN junction at the left. The battery connected across this junction provides _____ bias.

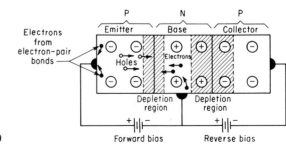

Fig. 2-9

TRANSISTOR FUNDAMENTALS | 39

2-10
(*forward*) Note that the N-type piece is called the *base* and that the lefthand P-type piece is called the _____.

2-11
(*emitter*) Because this junction is forward-biased, the _____ region is quite narrow.

2-12
(*depletion*) Holes diffuse into the depletion region from the emitter, and electrons diffuse into the depletion region from the _____.

2-13
(*base*) The junction is arranged so that there are many holes flowing into the base. Here they recombine with electrons. For each recombination, an electron enters the base from the negative side of the battery. Also, for each recombination, an electron breaks away from a bond and flows through the connecting wire to the _____ side of the battery.

2-14
(*positive*) We speak of a forward-biased PN junction as a *low-resistance* junction, because the depletion region is _____, thus causing the potential barrier to be small. This permits high currents for a given voltage.

2-15
(*narrow*) It is important to bear in mind that a forward-biased PN junction is considered to be a _____-resistance junction.

2-16
(*low*) Thus far, we have treated the emitter-base junction as though it were part of a simple diode. If this were the case, a relatively large external current would flow because of the recombinations of holes and _____ in the depletion region.

2-17
(*electrons*) If this were a simple diode there would, therefore, be a large number of electron-hole _____ taking place in the emitter P region and the base N region.

2-18
(*recombinations*) But, the addition of the third piece of germanium changes this picture completely as we shall see. The third piece of germanium is _____-type germanium, forming a second PN junction.

40 A PROGRAMMED COURSE IN BASIC ELECTRONICS

2-19
(P) Treating the righthand PN junction as a simple diode for the moment, we see that the battery connections are such as to provide _____ bias.

2-20
(*reverse*) This causes the depletion region to become considerably _____ than the other depletion region on the left.

2-21
(*wider*) Thus, the potential barrier is much greater at the righthand PN junction. Since no current due to majority carriers flows through this junction, the total external current (due only to minority carriers) must be very _____. (For the moment we are not noting the minority current caused by the emitter-injected carriers.)

2-22
(*small*) Since the current for a given voltage is small, we may consider this junction as being a _____-resistance junction in contrast to the lefthand junction, where a given voltage caused a relatively large current to flow.

2-23
(*high*) Thus, it is important to bear in mind that a reverse-biased PN junction is considered as a _____-resistance junction.

2-24
(*high*) Note also that the new section of P-type material on the right side (Fig. 2-9) will henceforth be called the _____ of the transistor.

2-25
(*collector*) Reviewing: The base-emitter junction is forward-biased; hence it is a low-resistance junction. The base-_____ junction is reverse-biased; hence it is a high-resistance junction.

2-26
(*collector*) We shall now see how the addition of the collector causes an important change in the action. We have thus far considered the 2 junctions separately and have been talking about them as though they formed 2 simple _____.

2-27

(*diodes*) Now we shall talk of the *entire* device as a *transistor*. In a practical transistor, the wafer of germanium used for the _____ is made much thinner than the emitter and collector slabs (Fig. 2-27).

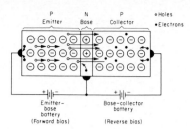

Fig. 2-27

2-28

(*base*) Referring to the diagram of Fig. 2-27, we may observe that by making the base very thin, the total number of positive _____ in the base is much less than the total number of negative ions in the collector.

2-29

(*ions*) Now look at the connections of the 2 batteries to each other and to the transistor. Since the "−" terminal of the emitter-base battery is connected to the "+" terminal of the base-collector battery, these 2 batteries are in _____ connection.

2-30

(*series*) This causes a large total potential to appear across the entire body of the transistor, from collector to emitter. A large potential will set up a _____ electric field.

2-31

(*large* or *strong*) This electric field has a direction such that positive charges will be acted upon to move from left to right. The mobile positive charges in this case are the _____ that come into existence in the P-type germanium at the left.

2-32

(*holes*) It should be remembered that the holes from the emitter also tend to diffuse into the base because of the narrowness of the _____ region when forward bias is used. (It should be remembered that the potentials are developed across the junctions, so there is little field in the base region and carriers move because of diffusion.)

2-33

(*depletion*) Now comes the departure from simple diode action. Because the base is very thin, and because a strong electric field is set up by the series batteries and the presence of a large number of _____ ions just to the right of the base, the following effect occurs.

42 A PROGRAMMED COURSE IN BASIC ELECTRONICS

2-34
(*negative*) The positive holes diffusing into the base do not recombine with electrons in the base in large numbers. Instead, as shown in Fig. 2-27, most of the holes diffuse right through the base to the _____ electrode of the transistor.

2-35
(*collector*) Thus, there are very few recombinations of holes and electrons in the base electrode. Most of the recombinations due to the holes that pass right through the thin base occur in the _____ electrode.

2-36
(*collector*) This means that the external current flowing in the emitter-base circuit is relatively small (even though this is a low-resistance junction), because there are very few _____ available for recombination with the electrons entering the base from the emitter-base battery. It should be noted that the base current is small although the emitter current may be large.

2-37
(*holes*) The holes pass on through the base into the collector where the recombinations occur. Since a large number of recombinations may take place (comparatively speaking), the emitter-to-collector current may be comparatively _____.

2-38
(*large*) In a typical transistor, only about 5 percent of the holes that diffuse into the base recombine with electrons in the base. This means that about _____ percent of the current from the emitter reaches the collector.

2-39
(*95*) Transistors may also be fabricated by making a "sandwich" with two N-type pieces on the outside and a _____-type piece between them.

2-40
(*P*) This type of transistor, called the NPN type, is different from the _____ type in only 2 ways.

2-41
(*PNP*) The first difference is: In the PNP type, the emitter-to-collector carrier is the hole; in the NPN type, the current carrier is the _____.

2-42
(*electron*) The other difference is: In the PNP type the battery polarities are those shown in Figs. 2-9 and 2-27. In the NPN type, the batteries are reversed in polarity. That is, the negative end of the emitter battery goes to the emitter and the _____ end of the base-collector battery goes to the collector.

TRANSISTOR FUNDAMENTALS | 43

2-43
(*positive*) The batteries will still be in series across the transistor, however, because the positive end of the emitter-base battery will be connected to the _____ end of the base-collector battery.

2-43
(*negative*)

CURRENTS IN A PNP TRANSISTOR

2-44
In this section, we learn some common symbols and abbreviations used in transistor work. In Fig. 2-44, a PNP transistor is illustrated. The common abbreviations for the 3 elements are: E for emitter, B for base, and C for _____.

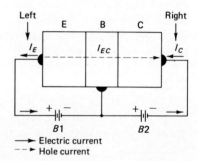

Fig. 2-44   → Electric current   --→ Hole current

2-45
(*collector*) As the legend in Fig. 2-44 indicates, a solid line will be used to designate the current direction when the carriers are electrons; a broken line will be used to indicate current direction when the carriers are _____.

2-46
(*holes*) As shown by the solid lines on the external wires in Fig. 2-44, the current carriers in *wires* are never holes; they are always _____.

2-47
(*electrons*) Another familiar convention we shall retain is this: When holes move to the right (as the current I_{EC} in Fig. 2-44) and reach a wire, we cannot show holes flowing out into the wire. Instead, we show electrons flowing in the _____ direction, that is, from the wire into the semiconductor.

2-48
(*opposite* or *other*, etc.) The same idea may be seen at the left in Fig. 2-44. Here, holes in the form of I_{EC} are flowing away from the wire in the semiconductor. Thus, we show _____ flowing away from the semiconductor in the wire, going toward the left.

2-49
(*electrons*) The current flowing from the negative end of the collector battery (B2 in Fig. 2-44) is called the collector current and is symbolized by _____.

2-50
(I_C) The current flowing from the emitter into the positive end of the base battery (B1 in Fig. 2-44) is called the emitter current and is symbolized by _____.

2-51
(I_E) Recall that very few recombinations of holes and electrons take place in the base, and that about 95 percent of the holes diffuse into the collector from the emitter during typical operation. This current, called the internal emitter-to-collector current, is symbolized in Fig. 2-44 by _____.

2-52
(I_{EC}) Figure 2-44 is only part of the total picture. Figure 2-52 continues the sequence. In this drawing, a new current symbolized by _____ appears.

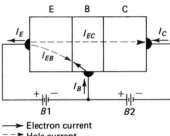

Fig. 2-52

→ Electron current
--▸ Hole current

2-53
(I_B) Note, in Fig. 2-52, that a hole current due to the small number of recombinations in the base flows from the left toward the right; the electrons that recombine with these holes in the base are shown by the solid arrow that meets the hole current arrow. The hole current is symbolized by _____.

2-54
(I_{EB}) Electrons must enter, going upward, into the base electrode to make up for those electrons that _____ with the holes in the base.

2-55
(*recombine*) This electron current going upward into the base is known as the base current and is symbolized by _____.

TRANSISTOR FUNDAMENTALS | 45

2-56
(I_B) Thus, Fig. 2-52 shows (a) the emitter-to-collector internal current as a flow of holes from left to right in the transistor and (b) the emitter-to-base current also as a flow of _____ from left to right.

2-57
(*holes*) However, the base current I_B is a flow of _____ upward into the base from the battery.

2-58
(*electrons*) Also, although the emitter-to-collector current I_{EC} comprises a flow of holes, the collector current I_C is shown as a flow of _____ into the semiconductor from right to left.

2-59
(*electrons*) To complete the picture, we go to Fig. 2-59. A new current symbolized by _____ has been added internally and is shown as a solid arrow indicating electron flow from the negative terminal of B2 going from right to left.

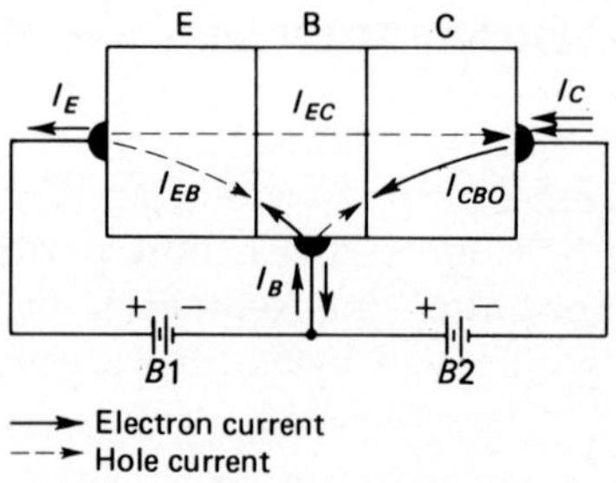

Fig. 2-59

→ Electron current
--→ Hole current

2-60
(I_{CBO}) Remember that the collector (C) of a PNP transistor contains acceptor ions, so that the majority carrier in the collector must be _____ rather than electrons.

2-61
(*holes*) Hence, the current I_{CBO} consisting of electrons as shown by the solid arrow must be a current due to the _____ carriers in the P-type germanium.

2-62
(*minority*) This is the small reverse current that flows because of the minority carriers, as we have seen in Fig. 2-59. Note that this electron current is met at the barrier between the base and collector by a current consisting of _____ as shown by the broken arrow.

2-63
(*holes*) Since the base is N-type germanium, its majority carriers are electrons and its minority carriers are _____.

46 A PROGRAMMED COURSE IN BASIC ELECTRONICS

2-64
(*holes*) These base holes recombine with the electrons composing I_{CBO}. As the holes are canceled, electrons from electron-pair bonds leave the base and return to the battery. These electrons are shown as the solid arrow going in a _____ direction out of the base.

2-65
(*downward*) Thus, in the base wire, some electrons are flowing upward to recombine with holes composing I_{EB}. This is indicated by the arrow pointing _____ alongside the base wire.

2-66
(*upward*) Also, in the base wire, some electrons are flowing _____ because of the entry of electrons from electron-pair bonds that are broken in the base when base holes recombine with electrons forming I_{CBO}.

2-67
(*downward*) The total base current I_B is therefore composed of 2 electron currents flowing in _____ directions.

2-68
(*opposite*) Hence, the net base current must be the difference between the upward and the _____ currents.

2-69
(*downward*) Now let us explain the double arrow symbolizing the electron current I_C (the collector current). One of these represents the electron current from the negative end of B2 that flows into the collector to recombine with the holes that reach this terminal due to current _____.

2-70
(I_{EC}) The second solid arrow at I_C represents the electrons that flow into the collector to form the minority carrier current which we have labeled _____.

2-71
(I_{CBO}) Thus, the total collector current I_C is the sum of I_{EC} and I _____.

2-71
(CBO)

CURRENTS IN AN NPN TRANSISTOR

2-72
The currents in an NPN transistor may be similarly analyzed with the aid of Fig. 2-72. We start with the emitter. Since this is N-type germanium, its majority carriers are _____.

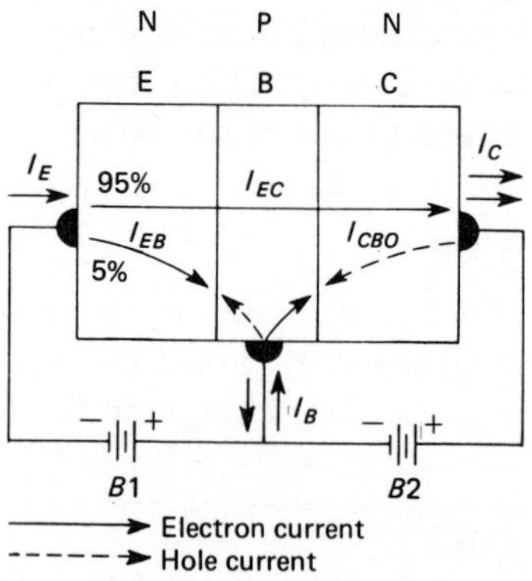

Fig. 2-72

→ Electron current
----→ Hole current

2-73
(*electrons*) The electron current due to the majority carriers flows across the EB barrier to the right (I_{EB}). It is met by a _____ current due to the majority carriers in the base, which is P-type germanium.

2-74
(*hole*) A few recombinations occur in the base. As the base holes are canceled, a small electron current flows back into the base battery $B1$. This small electron current is shown by the arrow alongside the base lead pointing _____.

2-75
(*downward*) Most of the electrons that cross the EB barrier, however, diffuse into the collector because the base is so narrow. This electron current is symbolized in Fig. 2-72 as _____.

2-76
(I_{EC}) These electrons leave the collector and flow back into the positive end of $B2$. Thus, we have symbolized I_{EC} outside the transistor by 1 of the 2 arrows labeled _____.

2-77
(I_C) A small reverse current flows from the righthand terminal of the transistor toward the left. The collector is N-type germanium; hence its minority carriers are _____.

2-78
(*holes*) The small reverse current flowing in the collector back toward the CB barrier therefore consists of holes and is labeled _____ in Fig. 2-72.

48 A PROGRAMMED COURSE IN BASIC ELECTRONICS

2-79
(I_{CBO}) I_{CBO} is a hole current. It is met by an _____ current flowing up out of the base terminal.

2-80
(*electron*) Recombinations of electrons and holes due to I_{CBO} therefore occur in the base. As electrons are canceled, more _____ enter the base through the base lead.

2-81
(*electrons*) These entering electrons are indicated by the arrow pointing _____ alongside the base lead.

2-82
(*upward*) Thus, the total base current I_B consists of the difference between I_{EB} and I _____.

2-83
(*CBO*) Also, the total collector current I_C consists of the _____ of I_{CBO} and I_{EC}.

2-84
(*sum*) Let us evaluate the order of magnitudes of these currents. First, I_{EB} is due to a very small number of recombinations in the base. Thus, I_{EB} must be a very _____ current.

2-85
(*small*) I_{CBO} is due to minority carriers. Minority carriers are very few in number compared with the majority carriers. Hence, I_{CBO} must be a very _____ current.

2-86
(*small*) Assume that I_{EB} and I_{CBO} are of the same order of magnitude. A small current subtracted from another small current will yield a very _____ difference.

2-87
(*small*) Thus, the net base current I_B must be a very _____ current indeed.

2-88
(*small*) This small current may result in a flow either in or out of the base terminal, depending upon whether I_{EB} is larger or _____ than I_{CBO}.

TRANSISTOR FUNDAMENTALS | 49

2-89
(*smaller*) As shown in Fig. 2-72, I_{EC} contains a total of _____ percent of all the electrons that enter the transistor from the left.

2-90
(95) Thus, I_C outside the transistor is the *sum* of a large current and a small current. The sum of a large current and a small current is also a _____ current.

2-91
(*large*) This large current is provided by the potential of B1 and _____ in series connection.

2-92
(B2) Thus, despite the fact that the CB junction is reverse-biased, the current flowing out of the collector terminal into the positive terminal of battery _____ may be relatively large.

2-93
(B2) Note: The potential of B1 is responsible for the current flow from E across the barrier to B. Since I_{EC} consists of the carriers that crossed the barrier due to B1, then I_{EC} is controlled by the potential of battery _____.

2-94
(B1) The total collector current I_C consists mostly of the carriers in I_{EC}. If I_{EC} is controlled by the emitter-base potential, then I _____ is also controlled by the emitter-base potential.

2-95
(C) This is true for both NPN and PNP transistors. Restating this important fact, we may say that the collector current of any transistor is controlled by the potential applied between the _____ and emitter of the transistor.

2-96
(*base*) Figure 2-96 illustrates a static characteristic for a typical transistor. Collector current, along the X axis, is labeled I_C. The voltage between the emitter and base, along the Y axis, is labeled _____.

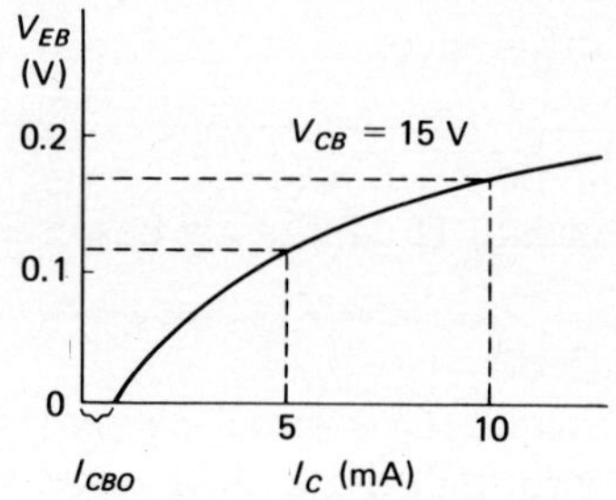

Fig. 2-96

2-97
(V_{EB}) The voltage between the collector and base, indicated as V_{CB}, was held constant at _____ V while the data for this curve was taken.

2-98
(15) A small collector current flows when $V_{EB} = 0$. This current, labeled I_{CBO}, is the current due to the _____ carriers in the collector according to the action previously described.

2-99
(*minority*) As V_{EB} is applied in increasing magnitude, the collector current I_C also _____ along the curve.

2-100
(*increases*) Note that the characteristic is nonlinear for very small values of V_{EB}. However, as V_{EB} increases, the _____ improves.

2-101
(*linearity*) The curve is reasonably linear between 5 and 10 mA of collector current. Along this portion of the curve, small changes of V_{EB} would produce proportional changes in _____ .

2-102
(I_C) We shall return to a study of static characteristics later. For the time being, it will be sufficient to recognize that the _____ current of any transistor may be varied by varying the emitter-to-base voltage.

2-102
(*collector*)

INPUT AND OUTPUT CIRCUITS

2-103
Refer to Fig. 2-103. To learn how a transistor amplifies, we must first see how the signal to be amplified is put into the transistor and how the amplified signal is taken out. The transistor type shown in Fig. 2-103 is a(n) _____ type.

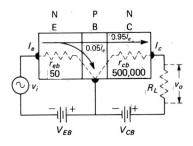

Fig. 2-103

2-104
(*NPN*) The input signal is represented by the generator v_i. This signal is applied between the base (assume that the battery impedance is negligible) and the _____ of the transistor.

2-105
(*emitter*) The output signal is v_o. This signal is taken out across the base (assume that the battery impedance is negligible) and the _____ of the transistor.

2-106
(*collector*) The input impedance is roughly the same as the emitter-to-base resistance r_{eb}. Since this junction is forward-biased, the input impedance is low. The value, given for a typical transistor in Fig. 2-103, is _____ Ω.

2-107
(*50*) The output impedance is roughly the same as the collector-to-base resistance r_{cb}. Since this junction is reverse-biased, the output impedance is high. The value for a typical transistor as given in Fig. 2-103 is _____ Ω.

2-108
(*500,000*) The internal currents are shown by the branching arrows. As we know, the emitter current inside the transistor may be taken as 5 percent toward the base and 95 percent toward the _____.

2-109
(*collector*) Thus, we may say that $I_c = 0.95 \times$ _____, ignoring the small current I_{CBO}, which is not shown in Fig. 2-103.

2-110
(I_e) We can now provide an approximate definition of the *current-amplification factor* for the transistor shown in Fig. 2-103. This factor, symbolized by α, is roughly the ratio of collector current to emitter current. In formula: $\alpha = I_c/$ _____.

2-111
(I_e) The symbol α is the Greek letter alpha. We can find alpha for the typical transistor in Fig. 2-103. It is: $\alpha = 0.95\ I_e/I_e$. Canceling out the I_e factors, we see that alpha for the typical transistor is _____.

2-112
(*0.95*) This merely tells us that, in going from the input to the output circuit of the transistor in the connection shown in Fig. 2-103, there is a current *loss* since the collector current is _____ percent of the emitter current.

52 A PROGRAMMED COURSE IN BASIC ELECTRONICS

2-113
(95) The configuration given in Fig. 2-103 is known as the *common-base* configuration because the _____ electrode is common to both the input and the output circuits.

2-114
(*base*) Thus, alpha in a common-base or CB configuration is almost always less than 1. This is so because there is always some recombination in the base, thus usually making the collector current _____ than the emitter current. (The exceptions are transistors with alpha greater than 1 because of a collector multiplication factor.)

2-115
(*less*) It is possible, as we shall see, to connect a transistor in such a manner that any of the 3 elements is common to input and output circuits. Thus, we shall have a common-emitter circuit (CE) and a common-collector circuit (CC), as well as the common-_____ circuit just described.

2-116
(*base*) To help distinguish one alpha from another in different configurations, we shall use standard symbology. Current-amplification factor is given by α. To this we then add the subscript $_{fb}$ where f stands for *forward* and b stands for _____ when the configuration is the common-base type.

2-117
(*base*) In the above, *forward* is used to show that this alpha expresses the effect of the input circuit on the _____ circuit; that is, the progression through the transistor is in a forward direction.

2-118
(*output*) Thus, the current-amplification factor in a common-base configuration which shows the forward effect that occurs is given by the symbol _____.

2-119
(α_{fb}) Summarizing: α_{fb} is almost always less than _____.

2-119
(1)

HOW A TRANSISTOR AMPLIFIES

2-120
To show the mechanism whereby a transistor amplifies, we shall use the simplified equivalent CB circuit in Fig. 2-120. In this circuit, v_i is the signal input voltage, while v_o is the signal _____ voltage.

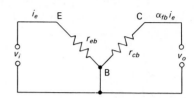

Fig. 2-120

2-121
(*output*) In Fig. 2-120, i_e is the signal current in the emitter (lowercase symbols are used for all signal values), while r_{eb} represents the input resistance. Similarly, r_{cb} represents the _____ resistance.

2-122
(*output*) We know that the collector current is equal to the emitter current times alpha. (Remember in our last example, alpha was 0.95 and the collector current was 0.95 I_e.) Thus, in Fig. 2-120, the collector current is symbolized by _____.

2-123
($\alpha_{fb}i_e$) Since voltage equals current times resistance, we may say that the input voltage equals the emitter current times the emitter-base resistance. In formula: $v_i = i_e \times$ _____.

2-124
(r_{eb}) We can do the same for the output voltage. The output voltage equals the collector current times the collector-to-base resistance. In formula: $v_o =$ _____ $\times r_{cb}$.

2-125
($\alpha_{fb}i_e$) Voltage amplification is defined as signal output voltage over signal input voltage. In equation form: voltage amplification = $VA =$ _____.

2-126
(v_o/v_i) We have seen that $v_i = i_e r_{eb}$ and that $v_o = \alpha_{fb} i_e r_{cb}$. Using these equalities instead of v_o and v_i in the definition of voltage amplification gives us $VA = \alpha_{fb} i_e r_{cb} / i_e r_{eb}$. Clearly, the i_e factors may be canceled so that $VA =$ _____ $\times r_{cb}/r_{eb}$.

2-127
(α_{fb}) This is our final result: the voltage amplification is equal to alpha times the ratio of output resistance to _____ resistance.

2-128
(*input*) The ratio r_{cb}/r_{eb} is sometimes referred to as the *resistance gain;* using this term, we can also define voltage amplification as being equal to the product of current-amplification factor and _____ gain.

2-129
(*resistance*) This concept of resistance gain assists us in obtaining a qualitative picture of the amplification mechanism in a transistor. From a simple statement of Ohm's law, we can say that voltage equals current times _____.

2-130
(*resistance*) In a transistor, the input signal is introduced into a circuit in which the resistance is low (r_{eb}). Thus, a relatively small voltage can produce a reasonably large _____ flowing through the low resistance.

2-131
(*current*) This current i_e is transferred with little change to a high-resistance output circuit r_{cb}. The current change is relatively small because 95 percent of the carriers pass through both junctions to the _____ electrode.

2-132
(*collector*) But now this current flows in a high-resistance circuit. For a given current, the output voltage will be greater if the resistance of the circuit is _____.

2-133
(*greater*) Hence, the output voltage may be considerably larger than the _____ voltage because of the resistance gain.

2-134
(*input*) Take a concrete example. In our typical transistor, the input resistance r_{eb} is only about 50 Ω, while the output _____ is of the order of 500,000 Ω.

2-135
(*resistance*) Thus, the resistance gain is r_{cb}/r_{eb} = 500,000/50 = _____.

2-136
(*10,000*) The current amplification is alpha. This is of the order of 0.95, as we have seen. Since $VA = \alpha \times$ resistance gain, then the VA of this transistor is $0.95 \times 10{,}000 =$ _____.

2-137
(*9,500*) Throughout this discussion, we have assumed that the load resistance is infinitely large. In practice, the load resistance may be from 10,000 to 25,000 Ω in a typical circuit. As we shall see, this reduces the actual *VA* severely. Hence, in practice the *VA* is considerably _____ than 9,500.

2-138
(*less*) For example, in an actual circuit using a 2N525 transistor, the input voltage was 0.01 V while the output voltage measured was 3.8 V. Thus, in this circuit, the actual voltage gain was $v_o/v_i = $ _____.

2-138
(*380*)

DESIGNATIONS AND SYMBOLS

2-139
Transistors are given a reference designation using the letter _____ as in Figs. 2-139*a* and 2-139*b*.

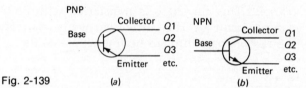

Fig. 2-139

2-140
(*Q*) The graphic symbol of a PNP transistor is given in Fig. 2-139*a*. Note that the arrow, which represents positive, or conventional current flow, points from the emitter toward the _____.

2-141
(*base* or *collector*) The graphic symbol of an NPN transistor is shown in Fig. 2-139*b*. In this symbol, the arrow points away from the base and toward the _____.

2-142
(*emitter*) Some convenient mnemonic devices are given in Figs. 2-142*a* and 2-142*b*. The first letter of the transistor type (NPN or PNP) indicates the polarity of the emitter with respect to the _____.

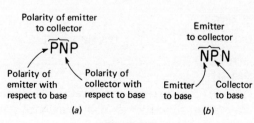

Fig. 2-142

2-143
(*base*) Thus, in a PNP transistor, the first letter is P, indicating that the emitter is _____ with respect to the base.

2-144
(*positive*) The upper diagram in Fig. 2-144 shows this relationship in a PNP transistor. The base is represented as zero (or reference level) while the emitter is shown as _____ with respect to the base.

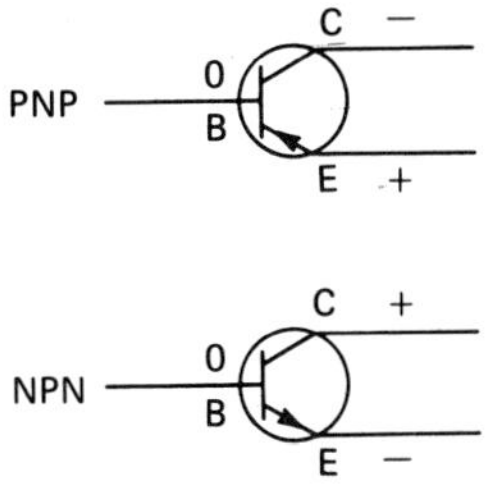

Fig. 2-144

2-145
(*positive*) The second letter of the type (*N* in a PNP, Fig. 2-142*a*) gives the relative polarity of the _____ with respect to the base.

2-146
(*collector*) Thus, in a PNP transistor, the collector is _____ with respect to the base.

2-147
(*negative*) In the upper diagram of Fig. 2-144, the base is represented as zero and the collector is shown to be _____ with respect to the base.

2-148
(*negative*) Also, in Fig. 2-142*a*, the first 2 letters of the type (*PN* in the PNP) give the relative polarity of the emitter with respect to the _____.

2-149
(*collector*) As is apparent from the upper diagram in Fig. 2-144, the emitter is Positive while the collector is relatively _____.

2-150
(*Negative*) Using Fig. 2-142*b* and the lower diagram in Fig. 2-144, we can trace the mnemonic device for the NPN transistor. The first letter *N* tells us the relative polarity, emitter to _____.

2-151
(*base*) The second letter *P* tells us the relative _____ of the collector with respect to the base.

TRANSISTOR FUNDAMENTALS | 57

2-152
(*polarity*) The first 2 letters together, *NP*, tell us that the emitter is negative and the _____ is positive, respectively.

2-153
(*collector*) In review: In either type of transistor, the base-emitter junction is always _____-biased.

2-154
(*forward*) In review: In either type of transistor, the collector-base junction is always _____-biased.

2-155
(*reverse*) In review: An input voltage that aids (increases) the forward bias always causes the emitter and collector currents to _____.

2-156
(*increase*) In review: An input voltage that opposes (decreases) the forward bias always causes the emitter and collector currents to _____.

2-156
(*decrease*)

TRANSISTOR CONFIGURATIONS

2-157
In the common-base (CB) configuration (Fig. 2-157), the input voltage is applied to the emitter-base circuit and the output voltage is taken from the _____-base circuit.

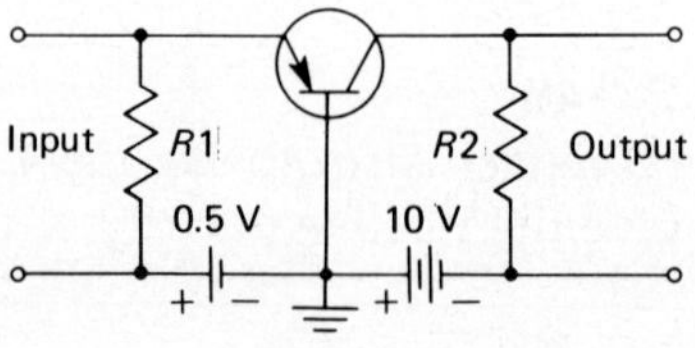

Fig. 2-157

2-158
(*collector*) The electrode that is common to both the input and output circuits is the _____ of the transistor.

2-159
(*base*) It is customary to "ground" the common element of a 3-element device. Since the base is the common element in the CB configuration, the base is shown _____ in Fig. 2-157.

2-160
(*grounded*) Thus, the common-base configuration is often referred to as the grounded-base arrangement. Since the arrow on the emitter element points toward the base in the diagrams of transistors in Figs. 2-157, 2-163, and 2-170, these transistors must be _____ types.

2-161
(*PNP*) As we have seen, in a PNP transistor the emitter is positive with respect to the base, the collector is negative with respect to the base, and the collector is _____ with respect to the emitter.

2-162
(*negative*) The positive voltage required for the emitter in Fig. 2-157 is supplied to this element through the resistor R1. Similarly, the negative voltage required for the collector is supplied through resistor _____.

2-163
(*R2*) A transistor circuit is given in Fig. 2-163. In this configuration, the _____ is the common or grounded element.

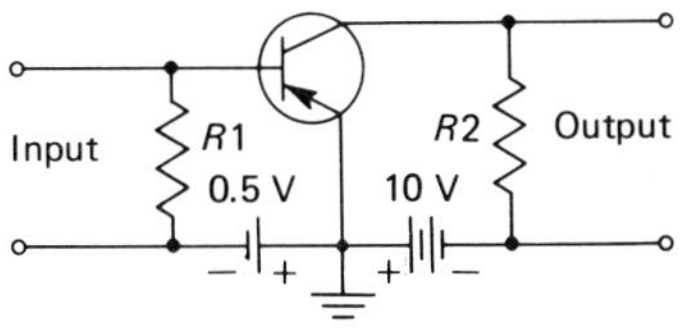

Fig. 2-163

2-164
(*emitter*) For this reason, this configuration is called the common-emitter or grounded-_____ configuration (CE).

2-165
(*emitter*) The input signal is applied in the emitter-base circuit, and the output signal is taken from the _____-emitter circuit.

2-166
(*collector*) Checking battery polarities in Fig. 2-163, we find that the emitter is positive with respect to the base and that the collector is negative with respect to the _____ since the base battery is smaller than the collector battery.

2-167
(*base*) Also, the *PN* mnemonic device tells us that the emitter should be positive with respect to the _____.

2-168
(*collector*) Checking the polarity of the collector battery we find that the positive end of the battery connects to the emitter, and the negative end connects to the _____, indicating thereby that the polarities adhere to the previously established requirements.

TRANSISTOR FUNDAMENTALS | 59

2-169
(*collector*) As we shall see, the common-emitter (CE) configuration is capable of greater power gain than any other arrangement and is therefore most widely used. Let us remember that the CE circuit is also often called a grounded-_____ connection.

2-170
(*emitter*) Now, the third possible configuration may be studied as given in Fig. 2-170. In this arrangement, the common or grounded element is the _____ of the transistor.

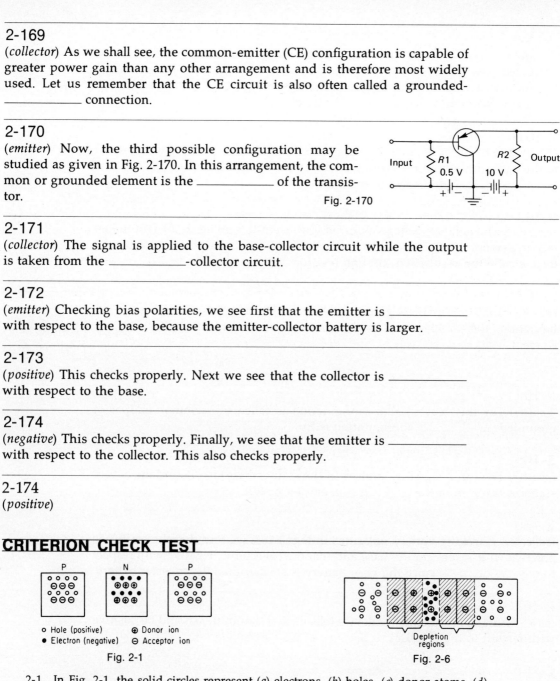

Fig. 2-170

2-171
(*collector*) The signal is applied to the base-collector circuit while the output is taken from the _____-collector circuit.

2-172
(*emitter*) Checking bias polarities, we see first that the emitter is _____ with respect to the base, because the emitter-collector battery is larger.

2-173
(*positive*) This checks properly. Next we see that the collector is _____ with respect to the base.

2-174
(*negative*) This checks properly. Finally, we see that the emitter is _____ with respect to the collector. This also checks properly.

2-174
(*positive*)

CRITERION CHECK TEST

Fig. 2-1

Fig. 2-6

_____ 2-1 In Fig. 2-1, the solid circles represent (*a*) electrons, (*b*) holes, (*c*) donor atoms, (*d*) acceptor atoms.

_____ 2-2 In Fig. 2-6, there are how many depletion regions? (*a*) 0, (*b*) 1, (*c*) 2, (*d*) 3.

_____ 2-3 A PNP transistor has (*a*) only acceptor ions, (*b*) only donor ions, (*c*) 2 P regions and 1 N region, (*d*) 3 PN junctions.

_____2-4 In Fig. 2-9, the circles with a minus sign represent (a) electrons, (b) Ge ions, (c) Si ions, (d) acceptor ions.

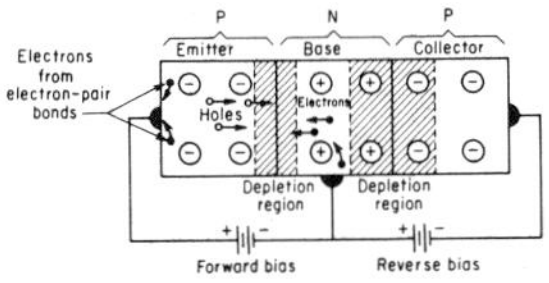

Fig. 2-9

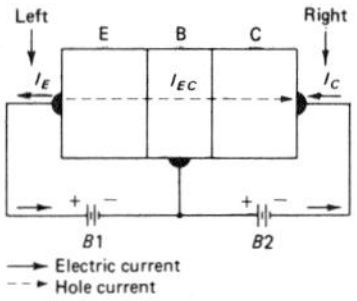

Fig. 2-44

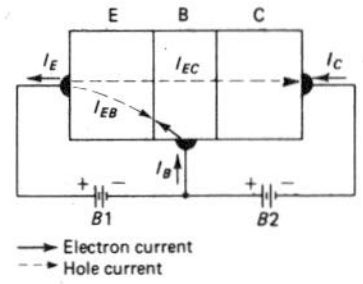

Fig. 2-52

_____2-5 In a PNP transistor, the N region is called the (a) base, (b) emitter, (c) collector, (d) cathode.

_____2-6 At the base-emitter junction of a transistor, one finds (a) reverse bias, (b) a wide depletion region, (c) high resistance, (d) low resistance.

_____2-7 In a junction transistor, (a) the collector region is made narrower than the other 2 regions, (b) the base region is made narrower than the other 2 regions, (c) the emitter region is thinner than the other 2 regions, (d) all 3 regions are the same thickness.

_____2-8 Most of the majority carriers from the emitter (a) recombine in the base, (b) recombine in the emitter, (c) pass through the base region to the collector, (d) are stopped by the junction barrier.

_____2-9 In the NPN transistors, (a) the current carrier is the hole, (b) the current carrier is the electron, (c) both junctions are forward-biased, (d) both junctions are reverse-biased.

_____2-10 In Fig. 2-44, the collector current is denoted by (a) I_E, (b) I_C, (c) I_B, (d) I_L.

_____2-11 In Fig. 2-52, the majority current flow from emitter to collector is denoted by (a) I_C, (b) I_E, (c) I_{EB}, (d) I_{EC}.

_____2-12 I_B denotes what kind of current? (a) electron current, (b) hole current, (c) acceptor ion current, (d) donor ion current.

_____2-13 The minority current flow in the collector region of Fig. 2-59 is denoted by (a) I_{EC}, (b) I_C, (c) I_E, (d) I_{CBO}.

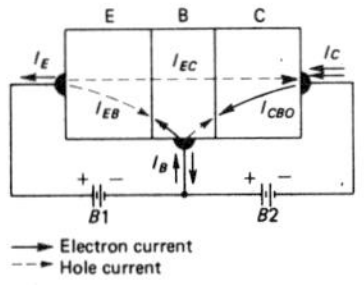

Fig. 2-59

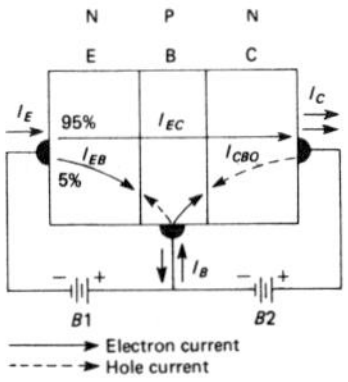

Fig. 2-72

The following 3 questions refer to Fig. 2-72.

_____2-14 In Fig. 2-72, I_{EB} represents a flow of (a) minority carriers, (b) majority carriers, (c) atoms, (d) ions.

_____2-15 In Fig. 2-72, the collector current I_C is made up of 2 currents: (a) I_R and I_{EB}, (b) I_E and I_{EC}, (c) I_{CBO} and I_{EB}, (d) I_{CBO} and I_{EC}.

_____2-16 In Fig. 2-72, the I_{EB} junction current is due to (a) battery B1, (b) battery B2, (c) diffusion, (d) recombination.

_____2-17 Figure 2-96 shows that the collector current is controlled by the (a) current I_{CBO}, (b) voltage V_{EB}, (c) voltage V_{CE}, (d) voltage V_{CB}.

Fig. 2-96

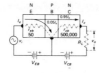

Fig. 2-103

Fig. 2-120

The following 3 questions refer to Fig. 2-103.

_____2-18 In Fig. 2-103, the signal source, v_i, is (a) in series with V_{EB}, (b) in parallel with V_{EB}, (c) connected to the collector lead, (d) connected to the positive plate of V_{CB}.

_____2-19 In Fig. 2-103, the output signal is denoted by (a) V_{CB}, (b) V_{EB}, (c) I_e, (d) v_o.

_____2-20 In the transistor of Fig. 2-103, the resistance r_{cb} is large because (a) I_{CBO} is small, (b) the CB junction is reverse-biased, (c) the BE junction is forward-biased, (d) r_{eb} is small.

_____2-21 The current gain in the common-base configuration is usually about (a) 0.15, (b) 0.95, (c) 1.35, (d) 15.

The next 2 questions refer to Fig. 2-120.

_____2-22 In Fig. 2-120, if $i_c = 0.1$ mA and r_{cb} is 400,000 Ω, compute v_o: (a) 0.4 V, (b) 4 V, (c) 40 V, (d) 400 V.

_____2-23 In Fig. 2-120, an expression for the voltage gain is (a) r_{cb}/r_{eb}, (b) r_{eb}/r_{cb}, (c) $r_{eb}/r_{cb}\alpha_{fb}$, (d) $r_{cb}\alpha_{fb}/r_{eb}$.

The next 2 questions refer to the figure below.

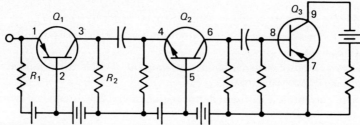

_____2-24 The PNP transistors are denoted by (a) Q1 and Q2, (b) Q1 only, (c) Q2 and Q3, (d) Q1 and Q3.

_____2-25 The emitter lead of Q2 is given by lead number (a) 1, (b) 4, (c) 5, (d) 6.

The following 2 questions refer to the figure below.

_____2-26 For transistor Q2, estimate the magnitude of B4: (a) 0.2, (b) 1.2, (c) 1.8, (d) 8.0.

_____2-27 The output resistor for Q2 is denoted by (a) R1, (b) R2, (c) R3, (d) R4.

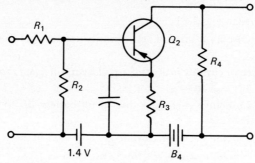

3 transistor fundamentals II

OBJECTIVES

(1) Describe fully the circuit and bias values for a single-battery CB configuration. (2) Compare the phase relationships of the output and input signals in a CB, CE, and CC amplifier configuration. (3) Explain why the JFET transistor has a high input impedance in comparison with the conventional transistor. (4) Using diagrams and curves, explain the electrical operation of a JFET. (5) Using diagrams and curves, explain the electrical operation of a MOSFET. (6) State the advantages of the IC over a conventional discrete circuit. (7) List the steps in fabricating a monolithic IC circuit, giving the dimensions of a typical chip. (8) Sketch and describe a thin-film IC resistor. (9) Sketch and describe a thin-film IC capacitor. (10) Describe how transistors are attached to a thin-film IC. (11) Compare the advantages and disadvantages of a monolithic IC and a thin-film IC with respect to each other.

SINGLE-BATTERY TRANSISTOR CIRCUITS

3-1
Regardless of the configuration, the dc bias produced by the batteries in a transistor amplifier must provide forward bias for the emitter-base circuit and _____ bias for the collector-base circuit.

3-2
(*reverse*) Up to now, we have used 2 separate batteries for bias purposes. The lower-voltage battery (refer back to Figs. 2-157, 2-163, and 2-170) supplies the emitter-_____ forward bias.

3-3
(*base*) The larger battery (Figs. 2-157, 2-163, 2-170) supplies the reverse bias for the _____-base circuit.

3-4

(*collector*) It is possible to operate any transistor amplifier with only 1 battery. Figure 3-4 shows a circuit in which a common-base transistor amplifier is operated from only _____ battery.

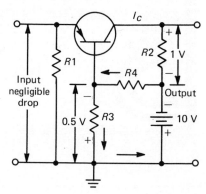

Fig. 3-4

3-5

(*1*) Since the arrow on the emitter symbol points toward the base, this transistor must be a(n) _____ type.

3-6

(*PNP*) In a PNP transistor, the emitter must be polarized so that it is _____ with respect to the base.

3-7

(*positive*) This polarization is achieved by means of the voltage drop caused by the flow of current from the battery through R4 and R3 in series (see arrows). The voltage drop across R3 is in such a direction as to make the top of R3 _____ with respect to the bottom.

3-8

(*negative*) The base of the transistor is connected directly to the "−" end of R3, while the _____ of the transistor is connected to the "+" end through resistor R1.

3-9

(*emitter*) Hence the emitter is positive with respect to the base, thus providing the forward bias required. The N in PNP tells us that the collector must be _____ in polarity with respect to the base.

3-10

(*negative*) The value of R3 is selected so that the voltage drop across it is, say, 0.5 V. The emitter resistor R1 is selected so that the voltage drop across it is so small as to be considered _____, as indicated in Fig. 3-4.

3-11
(*negligible*) Since the base is connected directly to the negative end of R3 and the emitter is connected through R1 to the positive end of R3, the voltage difference between the emitter and base, neglecting the drop across R1, is the same as the voltage drop across R3. Hence, the actual forward bias is _____ V.

3-12
(*0.5*) The collector resistor R2 is selected so that the voltage drop across it is 1 V. Thus, the voltage on the collector is _____ V negative with respect to ground.

3-13
(*9.0*) But the base is 0.5 V negative with respect to ground; thus, the collector is more negative than the base by _____ V.

3-14
(*8.5*) This meets the conditions required for reverse collector-to-base bias. The PN in PNP tells us that the emitter must be _____ with respect to the collector.

3-15
(*positive*) The emitter is at the same potential as ground since the drop in R1 is negligible. On the other hand, the collector is _____ V more negative than ground, as seen before.

3-16
(*9.0*) Hence, the emitter is _____ V more positive than the collector, thus meeting the PN requirement.

3-17
(*9.0*) If the transistor in Fig. 3-4 were NPN instead of PNP, correct bias voltages would be established merely by reversing the polarity of the basic voltage source. Thus, correct bias voltages are obtained merely by reversing the connections to the _____.

3-18
(*battery*) It should be noted that, in use, the bias point is determined principally by the _____ circuit, and not by the transistor itself.

3-18
(*bias*)

TRANSISTOR FUNDAMENTALS II **65**

PHASE RELATIONSHIPS IN A CB AMPLIFIER

3-19
In Fig. 3-19, the schematic symbol of the transistor indicates that we will be dealing with a(n) _____ type of transistor in this section.

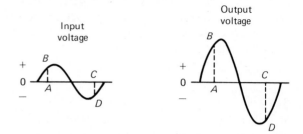

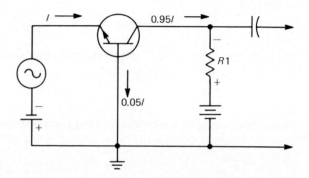

Fig. 3-19

3-20
(*NPN*) The current flowing in the emitter lead is symbolized by _____ in Fig. 3-19.

3-21
(*I*) As we know, this current divides into 2 parts. One part is the current flowing in the base lead and the other part is the current flowing in the _____ lead.

3-22
(*collector*) The current flowing in the collector lead is assumed to be 95 percent of the total emitter current, while the current in the base lead is assumed to be _____ percent of the total emitter current.

3-23
(*5*) The waveform at the left near the signal source is the input *voltage* waveform; the waveform at the right is the output _____ waveform.

3-24
(*voltage*) The output voltage waveform at the right represents the voltage that is being developed as a drop across the component identified as _____.

3-25
(*R1*) Consider voltage *AB* in the input waveform. Since *AB* is above the zero axis, its polarity is _____ with respect to common ground.

3-26
(*positive*) Since the battery polarity is such that the emitter is made negative by the battery, then a positive voltage such as *AB* opposes the bias. Thus, to find the net voltage applied by both emitter battery and input signal, we must _____ the input voltage from the battery voltage.

3-27
(*subtract*) The input signal *AB* therefore reduces the forward bias due to the emitter battery. This causes the current *I* to _____.

3-28
(*decrease*) Since both the collector and base currents are parts of the original emitter current, then both the base and collector currents must _____ correspondingly as the emitter current decreases.

3-29
(*decrease*) Since the collector current is the current flowing through *R1*, then the voltage drop across *R1* must _____ as the collector current decreases.

3-30
(*decrease*) If the voltage drop decreases, then the top of the resistor becomes less negative than before. Another way to say this is that the top of the resistor becomes more _____ than it was before.

3-31
(*positive*) But if the top of the resistor goes positive as a result of the input signal's going positive, then we can represent the output voltage as a positive voltage. This positive output voltage corresponding to input voltage *BA* is designated as _____ on the output waveform.

3-32
(*BA*) The same is true for the entire positive half of the input cycle. That is, whenever the input signal voltage goes positive, the output signal voltage also goes _____.

3-33
(*positive*) Now consider input voltage *CD* in Fig. 3-19. The fact that *CD* is below the zero axis indicates that *CD* has a _____ polarity with respect to ground.

TRANSISTOR FUNDAMENTALS II

3-34
(*negative*) Since the forward bias on the emitter is negative and the input voltage CD is negative, then input voltage _____ must aid the forward bias.

3-35
(*CD*) If the forward bias is increased, then the emitter current must increase; hence both the collector and _____ currents must also increase.

3-36
(*base*) An increase of collector current flowing through R1 must cause a(n) _____ in the voltage drop across R1.

3-37
(*increase*) An increase in the voltage drop across R1 must mean that the top of R1 becomes more _____ than it was before.

3-38
(*negative*) Hence, the output voltage CD on the output waveform must also go in a _____ direction.

3-39
(*negative*) Thus, a negative-going input signal gives rise to a negative-going _____ signal in the CB amplifier.

3-40
(*output*) Comparing the phase relationships between the input voltage waveform and the output waveform in Fig. 3-19, we see that the two waveforms are in the same _____.

3-41
(*phase*) Hence, in a CB amplifier there is no _____ reversal between input and output signals.

3-41
(*phase*)

68 A PROGRAMMED COURSE IN BASIC ELECTRONICS

PHASE RELATIONSHIPS IN A CE AMPLIFIER

3-42
In Fig. 3-42, the schematic symbol of the transistor shows that we will be discussing a(n) _____ type of transistor in this section.

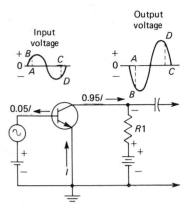

Fig. 3-42

3-43
(*NPN*) The emitter current is symbolized by *I*. The base current is again assumed to be only 5 percent of the emitter current. The collector current is assumed to be _____ percent of the emitter current.

3-44
(*95*) The input voltage shown at the left is the voltage developed by the signal generator in the base lead. The output voltage at the right is the voltage developed across the component identified as _____.

3-45
(*R1*) Consider voltage *BA* in the input voltage waveform. Since *BA* is above the zero axis, it is a positive voltage with respect to ground. The base battery is polarized to make the base _____ with respect to ground.

3-46
(*positive*) Since voltage *BA* and the battery voltage tend to make the base positive, then the total voltage applied by both the signal *BA* and the battery is the _____ of the 2 voltages.

3-47
(*sum*) Thus voltage *BA* aids the forward bias of the battery on the base. When voltage *BA* is applied, it causes the emitter current *I* to _____.

3-48
(*increase*) Since both the collector and base currents are parts of the original emitter current, then both the base and collector currents must _____ correspondingly as the emitter current increases.

TRANSISTOR FUNDAMENTALS II 69

3-49
(*increase*) This increased collector current flows through R1, causing an increased _____ drop across R1.

3-50
(*voltage*) If the voltage drop across R1 increases, then the top of the resistor becomes more negative than it was before. Thus, the output signal is _____-going when the input signal is positive-going.

3-51
(*negative*) The same is true for the entire positive half of the input cycle. That is, whenever the input signal voltage goes positive, the output signal voltage goes _____.

3-52
(*negative*) Now, consider input voltage CD. This is a negative-going voltage which opposes the _____ bias applied by the base-emitter battery.

3-53
(*forward*) As a result of this opposition to the forward bias, the emitter current must decrease. This causes a corresponding _____ in the collector current.

3-54
(*decrease*) The voltage drop across R1 then decreases. The top of R1, then, must become less negative or more _____.

3-55
(*positive*) This gives rise to a positive-going output signal. We see that this positive-going output voltage is the result of a _____-going input voltage.

3-56
(*negative*) Comparing the phase relationships between the input voltage waveform and the output voltage waveform in Fig. 3-42, it is evident that the 2 waveforms are out of phase by _____ degrees.

3-57
(*180*) Hence, in a CE amplifier there is a _____ reversal of 180° between input and output voltages.

3-57
(*phase*)

PHASE RELATIONSHIPS IN A CC AMPLIFIER

3-58
In Fig. 3-58, we are again dealing with an NPN transistor in which the emitter current is identified as I, the base current as $0.05I$, and the collector current as _____.

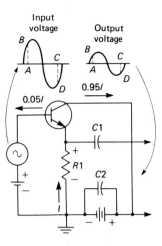

Fig. 3-58

3-59
($0.95I$) Capacitor $C2$ may be assumed to have a large capacitance; hence its reactance at the signal frequency may be considered quite _____.

3-60
(*small*) Resistor $R1$ may have a resistance of several thousand ohms. This resistor is in series with the emitter. It isolates the _____ from ground.

3-61
(*emitter*) Thus, the emitter is isolated from ground while capacitor $C2$ places the collector at the same signal potential as _____.

3-62
(*ground*) Since the input signal is applied between base and ground and the output is taken from emitter and ground while the collector is at signal ground potential, then the common element is the _____.

3-63
(*collector*) Hence, this circuit is a common-_____ configuration.

3-64
(*collector*) Consider voltage BA in the input voltage waveform. It is a positive voltage with respect to ground. Also, the battery in the base circuit is polarized to make the base _____ with respect to ground.

TRANSISTOR FUNDAMENTALS II **71**

3-65
(*positive*) Since voltage BA and the battery voltage tend to make the base positive, then the total voltage applied by both the signal BA and the battery is the _____ of the 2 voltages.

3-66
(*sum*) Thus, voltage BA aids the forward bias of the base battery. When voltage BA is applied, therefore, it causes the emitter current I to _____.

3-67
(*increase*) The increased current in R1, flowing upward through the resistor, causes the voltage drop across R1 to increase, thereby making the top of the resistor more _____ than it was before.

3-68
(*positive*) The positive-going voltage is the output voltage BA. Thus, a positive-going input voltage causes a positive-going output voltage. This is true for all parts of the first _____ of the input cycle.

3-69
(*half*) Similarly, when the input voltage is negative, as CD in the input waveform, then the input voltage acts as opposition to the forward bias, thus causing the emitter current to _____.

3-70
(*decrease*) As a result of the decreased I, the voltage drop across R1 must also _____.

3-71
(*decrease*) When this occurs, the top of the resistor R1 becomes more negative than it was previously, thereby producing a _____-going output voltage.

3-72
(*negative*) Hence, a _____-going input voltage produces a negative-going output voltage.

3-73
(*negative*) Since this is true for the entire second half of the input voltage waveform, it is clear that there is no _____ reversal of the signal amplified by a CC amplifier.

3-74
(*phase*) Reviewing: In a CB amplifier, a positive-going input produces a positive-going output and a negative-going input produces a negative-going output. Thus, the phase angle between input and output signals is _____°.

3-75
(*0*) Reviewing: In a CE amplifier, a positive-going input produces a negative-going output and a negative-going input produces a positive-going output. Thus, the phase angle between input and output signals is _____°.

3-76
(*180*) Reviewing: In a CC amplifier, the relationships between input and output phase are the same as they are for the CB amplifier. Hence, the phase angle between input and output waveforms in the CC amplifier is _____°.

3-77
(*0*) Thus, the only configuration that can be used for phase inversion is the _____-_____ configuration.

3-77
(*common-emitter* or *CE*)

FETS
3-78
Consider a design problem where a solar battery is to be used to trigger a relay. A solar battery is a device which produces voltage when activated by _____ light.

3-79
(*sun*) Suppose that the solar battery you are working with produces 0.2 V in bright light, and the relay to be triggered requires 5 V. Then you need an amplifier with a voltage amplification of 5/0.2 or _____.

3-80
(*25*) The amplifying device discussed so far in this text is the _____, so that it *appears* as though a _____ amplifier of voltage gain 25 is needed.

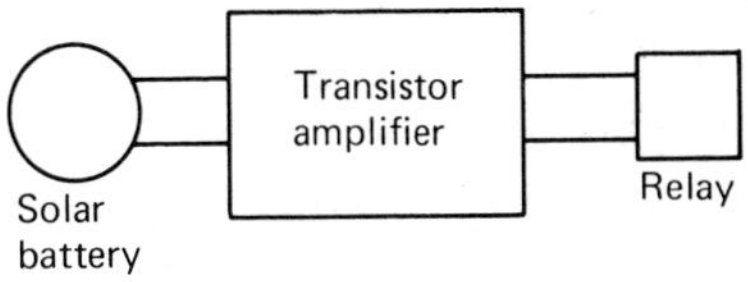

Fig. 3-80

TRANSISTOR FUNDAMENTALS II **73**

3-81

(*transistor, transistor*) Unfortunately, this simple argument is not valid. To see why this is so, consider Fig. 3-81 where we have represented the solar cell by an effective battery of 0.2 V, and an internal resistance, _____ .

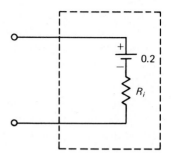

Fig. 3-81

3-82

(R_i) In fact, *all* physical sources have some internal resistance. For a typical solar cell, R_i is about 200 Ω. Further, recall that the input resistance of a typical transistor is about _____ Ω.

3-83

(*50*) Thus, if the solar cell is connected to the input of a transistor (Fig. 3-83), a voltage divider of resistance 200 Ω and _____ Ω is formed.

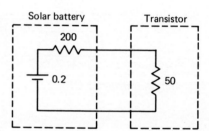

Fig. 3-83

3-84

(*50*) Hence, the actual voltage delivered to the input of the transistor is $[50/(200 + 50)] \times 0.2 =$ _____ V.

3-85

(*0.04*) Since the output of the transistor must still be 5 V, the true voltage amplification needed is _____ .

3-86

(*125*) Clearly, if we were involved with a source of higher internal resistance than the solar battery, then the transistor amplifier would need to have an even _____ gain.

3-87

(*higher* or *larger*) When you consider that there are sources with an internal resistance in the megohm range, then low input resistance amplifiers become quite _____ (practical, impractical) for such sources.

3-88
(*impractical*) The basic reason why the transistor has a low input resistance is that the BE junction is normally operated in the forward-biased region. A junction which is forward-biased has a _____ resistance, whereas a reverse-biased junction has a very _____ resistance.

3-89
(*low, high*) Thus, a device whose input is essentially a reverse-biased junction would provide a _____ input resistance, and therefore would be a good voltage amplifier for high-impedance sources. Such a reverse-biased input device has been developed and is called the *junction field-effect transistor* (JFET).

3-90
(*high*) Figure 3-90 is a schematic of a JFET. The device has 3 terminals, denoted by source, _____, and gate.

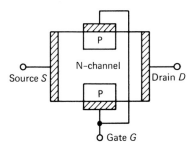

Fig. 3-90

3-91
(*drain*) The bulk of the device is _____-type semiconductor material which has 2 areas of P-type material. Charge flow from source to drain must pass between the channel formed by the 2 P-type areas.

3-92
(*N*) The basic idea of the device is to control source-to-drain current by varying the width of the channel. To see how this is done, examine Fig. 3-92. The voltage between drain and source is denoted by V_{DS}, and the voltage between gate and source is denoted by _____.

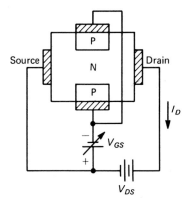

Fig. 3-92

3-93

(V_{GS}) Initially, set $V_{GS} = 0$. Then since the PN interface in region 1 has no applied voltage ($V_{GS} = 0$), it is neither forward- nor reverse-biased. However, in region 2, the PN interface is _____-biased because of battery V_{DS}.

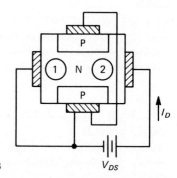

Fig. 3-93

3-94

(*reverse*) Because of the reverse bias in region 2, there will be a depletion region as shown in Fig. 3-94. In Fig. 3-94a, a small V_{DS} has caused a finite depletion region to spread out from each P-type region. In Fig. 3-94b, V_{DS} has been _____ to the point where the depletion regions just touch. The channel has reached the *pinchoff* condition.

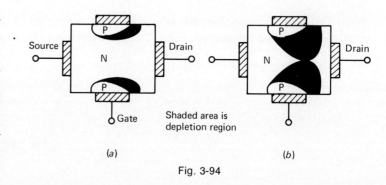

Fig. 3-94

3-95

(*increased*) Since the *pinchoff* condition means a closing of the channel, an increase in the voltage V_{DS} produces a *very* small increase in drain current (I_D). This is shown graphically in Fig. 3-95, where the drain current _____ until the pinchoff voltage is reached, after which the current remains _____.

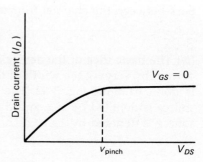

Fig. 3-95

3-96

(*increases, constant*) If now we change V_{GS} from zero to some negative value (Fig. 3-92), the PN interface in region 1 of Fig. 3-93 will also be _____-biased. Thus the depletion region of the channel will exist in both region 1 *and* region 2.

76 A PROGRAMMED COURSE IN BASIC ELECTRONICS

3-97
(*reverse*) Hence, the more negative V_{GS}, the sooner pinchoff voltage will be reached and the smaller the current at pinchoff. This is shown graphically by a series of curves in Fig. 3-97, where each curve is labeled by a value of _____.

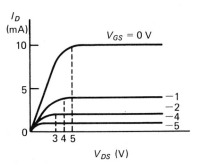

Fig. 3-97

3-98
(V_{GS}) When the gate-to-source voltage is 0 V, the pinchoff voltage is about 5 V; when the gate-to-source voltage is −2 V, the pinchoff voltage is reduced to _____ V.

3-99
(3) Further, by changing the input voltage V_{GS}, the output (drain) current is also changed. For example, at $V_{GS} = 0$, the current beyond pinchoff is 10 mA, while at $V_{GS} = -2$ V, the output current is reduced to about _____ mA.

3-100
(2) Thus, by varying the input voltage, _____, one can vary the output current I_D. Moreover, since V_{GS} is negative, the gate-to-source PN junction is reverse-biased, producing a large resistance looking into the gate-source terminals.

3-101
(V_{GS}) In later chapters, we shall show in detail how to use curves such as those shown in Fig. 3-97 in circuit design. In fact, you will see that the *output* characteristics of the PNP(NPN) transistor are quite similar to those of the JFET. Summarizing, the JFET is a 3-terminal device: source, _____, and gate.

3-102
(*drain*) Output current is controlled by depleting carriers in the source-to-_____ channel. This depletion region is controlled by a negative voltage on the gate terminal.

3-103
(*drain*) A high input resistance (as much as 10 megohms or MΩ) is achieved because of the reverse bias of the _____-to-source junction.

TRANSISTOR FUNDAMENTALS II 77

3-104

(*gate*) Another device has been developed which goes one step beyond the JFET. This is the *metal-oxide-semiconductor field-effect transistor* (MOSFET). The high input resistance of the JFET is achieved by maintaining a _____ bias on the gate; the MOSFET will yield high input resistance with *either* positive *or* negative bias on the gate.

3-105

(*negative* or *reverse*) To see how this is achieved, examine Fig. 3-105. As in the JFET, there are 3 terminals: drain, gate, and _____. Note that the gate electrode is separated from the bulk N material by an *insulating* oxide layer.

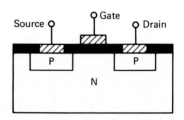

Fig. 3-105

Crosshatched areas are metal electrodes.
Shaded area is an oxide layer

3-106

(*source*) Because of this insulating layer, no gate current can flow whether the gate voltage is positive or negative. Hence the input (gate) resistance is extremely _____.

3-107

(*high* or *large*) However, even though the gate current is essentially zero, the gate *voltage* can still affect the drain-to-source current. Observe Fig. 3-107a, where a battery has been inserted into the source-to-_____ circuit. Since the two N regions are separated by a P region, the drain-to-source current, I_D, is zero.

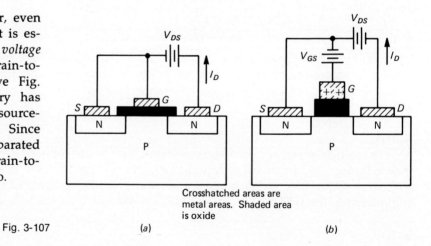

Crosshatched areas are metal areas. Shaded area is oxide

Fig. 3-107 (a) (b)

3-108

(*drain*) If now, as in Fig. 3-107b, a positive voltage is applied to the gate, the gate electrode and P channel will behave like the plates of a capacitor, with the oxide layer between the two. The positive voltage will deposit _____ charge on the top surface of the oxide layer and _____ charge on the bottom of the oxide layer in the P channel.

3-109

(*positive, negative*) This negative charge in the P channel will cause the P channel to look like N-type material. Current will then flow from drain to source through the *pseudo-N channel*. Clearly, as the positive voltage is increased, the drain current will _____.

3-110

(*increase*) If one measures the drain-to-source characteristics of a MOSFET, one will find characteristics quite similar to those for the JFET (Fig. 3-97). The drain current of the MOSFET rises with increasing drain voltage until the induced charges are swept out of the pseudo-N channel. Then the drain current levels off and becomes _____.

3-111

(*constant*) As an aid in applying field-effect transistors (FETs) to circuit design, a parallel between NPN(PNP) transistors and FETs is presented in Fig. 3-111. For example, an NPN common-emitter configuration would be analogous to a FET common-_____ configuration.

PNP(NPN) Transistor	FET
Collector	Drain
Emitter	Source
Base	Gate

Fig. 3-111

3-112

(*source*) Just as the 2 types of transistors, NPN and PNP, require 2 symbols, a FET can also have either a P channel or an _____ channel. Figure 3-112 shows the 2 symbols used for FETs.

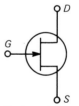

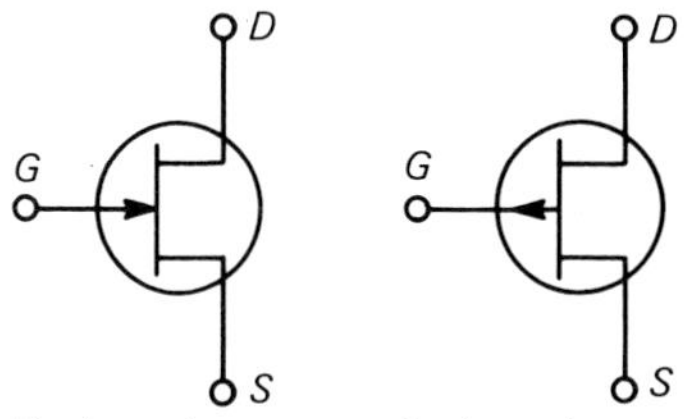

Fig. 3-112   N-channel type   P-channel type

3-112

(*N*)

TRANSISTOR FUNDAMENTALS II   79

INTEGRATED CIRCUITS

3-113
The circuit of Fig. 3-113 is a typical building block of a communication system. If we count the number of components, one sees 10 resistors, 6 transistors, and _____ capacitors.

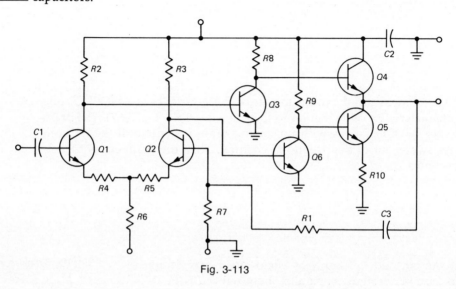

Fig. 3-113

3-114
(3) In Fig. 3-114, the circuit of Fig. 3-113 is shown as a printed circuit board configuration. If you count them, you will find that there are _____ soldered connections to the circuit board.

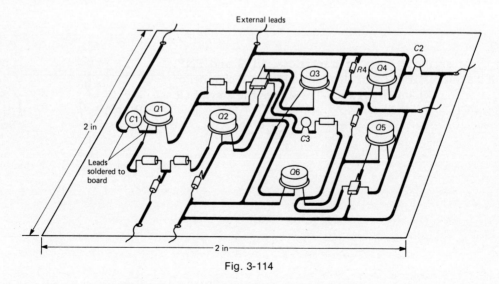

Fig. 3-114

80 A PROGRAMMED COURSE IN BASIC ELECTRONICS

3-115
(44) Further, from Fig. 3-113, the size of the printed circuit board is a square _____ inches (in) on each side. A guidance system or computer may have 1,000 such circuits.

3-116
(2) This would give a total of _____ soldered joints, posing a considerable problem of reliability. Further, 1,000 circuit boards could take up considerable space.

3-117
(44,000) To meet the two problems of reliability and space conservation, new techniques have been used to dramatically reduce the size of circuits. In Fig. 3-117, we see that new technology has permitted the circuit of Fig. 3-114 to be packaged in a case the size of a single conventional _____.

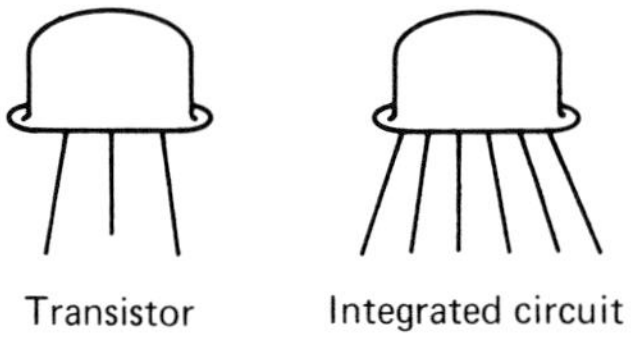

Transistor   Integrated circuit

Fig. 3-117

3-118
(*transistor*) To achieve this degree of miniaturization, the concept of discrete components soldered to a board must be discarded. Instead, circuit elements are now made as _____ circuits. (See Fig. 3-117.)

3-119
(*integrated*) The first class of integrated circuit (IC) to be discussed is the *monolithic* IC. In a monolithic IC, all circuit elements such as transistors, diodes, capacitors, and _____ are realized by semiconductor material (P or N) or PN junctions.

3-120

(*resistors*) In Fig. 3-120a, a simple monolithic IC is shown. From the schematic (Fig. 3-120b), we see that the circuit is to contain 2 transistors and _____ resistor.

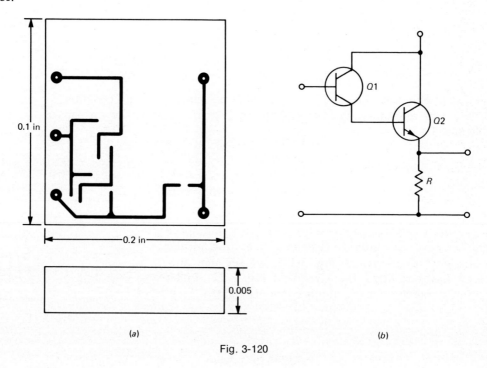

Fig. 3-120

3-121

(*1*) The first point to note about the IC is its extremely small size. The dimensions are given as 0.1 in by _____ in.

3-122

(*0.2*) The thickness of the IC *chip* is given as 0.005 in. The area of the IC is obtained by multiplying the length by the width, so that the Area = 0.2 × 0.1 = _____ in^2.

3-123

(*0.02*) The volume of the chip is the product of area times thickness, or Volume = 0.02 × 0.005 = _____ in^3.

3-124

(*0.0001*) Said otherwise, if each chip occupies one _____-thousandth of a cubic inch, then _____ thousand of these chips could be packed in 1 cubic inch. This is what is making the desk top computer feasible.

3-125

(10, 10) The second point to note about the chip of Fig. 3-120a is that in the end view *no* components are seen to protrude above the surface of the chip. This is because all components are formed *within* the chip. This tiny chip contains within itself 2 transistors and _____ resistor.

3-126

(1) To see how these elements are formed and connected in the chip, we shall examine step by step the fabrication of the IC of Fig. 3-120. In step 1, we see that the base of the IC consists of a substrate made of _____-type semiconductor material.

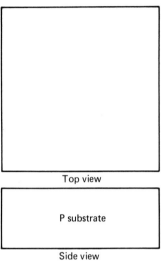

Fig. 3-126   Step 1

3-127

(P) All necessary components—transistors and resistors (or capacitors)—will be formed by diffusing semiconductor material into the substrate. In step 2, 3 pieces of _____-type semiconductor material have been diffused into the substrate.

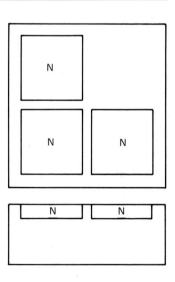

Fig. 3-127   Step 2

3-128

(*N*) In step 3, a P region has been diffused into each of the 3 _____ regions.

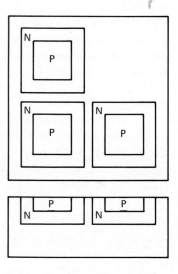

Fig. 3-128   Step 3

3-129

(*N*) In step 4, additional N regions have been diffused into the IC chip. However, only _____ of the 3 regions have received this second N region.

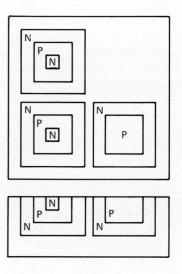

Fig. 3-129   Step 4

3-130
(2) The reason for this can be seen by comparing steps 4 and 4a. In step 4a, the sections of the areas on the substrate have been labeled. The two areas on the left of the substrate form 2 _____ _____ _____ transistors.

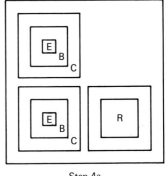

Fig. 3-130   Step 4a

3-131
(N P N) The center N area is labeled as the _____ region. The outer N area is labeled the collector region.

3-132
(emitter) The P region between the collector and emitter is labeled as the base region. The area to the right of the transistors is used to form the _____.

3-133
(resistor) The value of resistance depends on the physical dimensions of the P material (see Fig. 3-130). If a large value of resistance is needed, the P region is made long and narrow; if a small resistance is needed, the P region is short and _____.

3-134
(wide) Step 4 represents completion of the component fabrication. The final procedure is to connect the components. In step 5, an _____ layer has been formed over the surface of the IC chip.

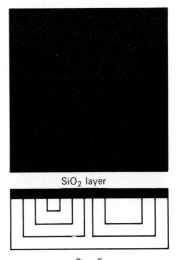

Fig. 3-134   Step 5

TRANSISTOR FUNDAMENTALS II

3-135

(*oxide*) The oxide layer has 2 functions: one is to prevent contamination of the chip; the other is to prevent contact between the interconnections and components passed over by the interconnections. In step 6, we see that parts of the oxide layer have been _____ away.

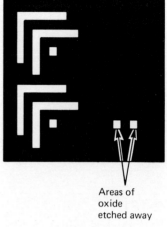

Fig. 3-135   Step 6

Areas of oxide etched away

3-136

(*etched*) The purpose of these etched areas is to make contact between the components and the interconnections. For this IC, there are 3 connections to each transistor and _____ to the resistor.

3-137

(*2*) The result of the last step is shown in step 7. The interconnections are made by depositing a thin film of _____ onto the oxide layer.

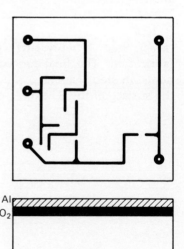

Fig. 3-137   Step 7

86 A PROGRAMMED COURSE IN BASIC ELECTRONICS

3-138
(*aluminum*) Because of the insulating properties of the oxide layer, the aluminum will make contact with the components only where the oxide is etched away. Thus, for example, the connection from the emitter of Q1 to the base of Q2 is prevented from touching the collector of Q2 by the _____ _____.

3-139
(*oxide layer*) To complete the IC, external leads (5 for the circuit of Fig. 3-126) are connected to the IC, and the IC is mounted in a package such as that of Fig. 3-117. In conclusion, all components of the monolithic IC are formed within a _____.

3-140
(*substrate*) A second type of integrated circuit is the thin-film IC. For the thin-film IC, passive elements such as resistors and _____ are formed of thin films of metals or insulators.

3-141
(*capacitors*) Figure 3-141a shows the schematic of a flip-flop logic circuit. The circuit has a total of 8 passive components: _____ resistors and _____ capacitors.

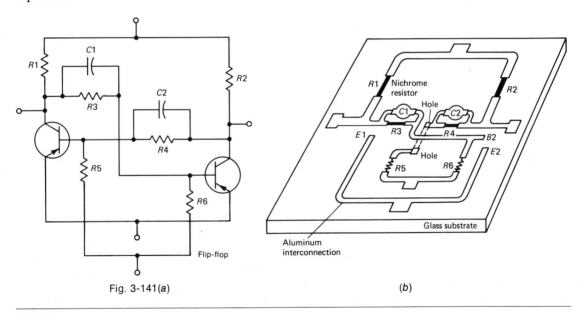

Fig. 3-141(a)   (b)

3-142
(*6, 2*) In Fig. 3-141b, the passive components have been realized as a thin-film IC. All the passive components are deposited on the surface of a _____.

3-143

(*glass substrate*) The interconnections among the components are made by depositing aluminum on the surface of the glass. The resistors are made of _____ metal deposited on the glass.

3-144

(*nichrome*) For proper circuit operation, resistors *R5* and *R6* must be large resistors. Extra resistance is obtained for these by using a winding deposit of nichrome instead of a simple straight line. Note that the connection between *R5* and *C2R4* is shown as a _____ line.

3-145

(*dotted*) This is to indicate that that interconnection is made on the *bottom* of the substrate. This must be done to prevent crossing of interconnections. To make connections between the top and bottom of the substrate, _____ are drilled in the substrate.

3-146

(*holes*) In Fig. 3-146, an exploded view of a thin-film capacitor is shown. The upper and lower plates are thin films of aluminum while the dielectric is a thin film of _____ material.

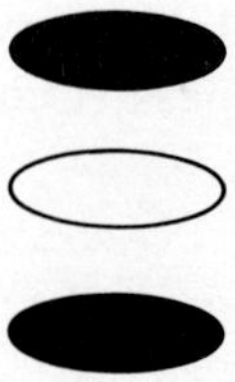

Fig. 3-146

C1 upper plate

Ceramic thin-film dielectric

C2 lower plate

3-147

(*ceramic*) After the passive components are deposited on the glass substrate, the last step is to attach the transistors. Conventional transistors (without the case) are _____ to the interconnections at the appropriate places on the circuit.

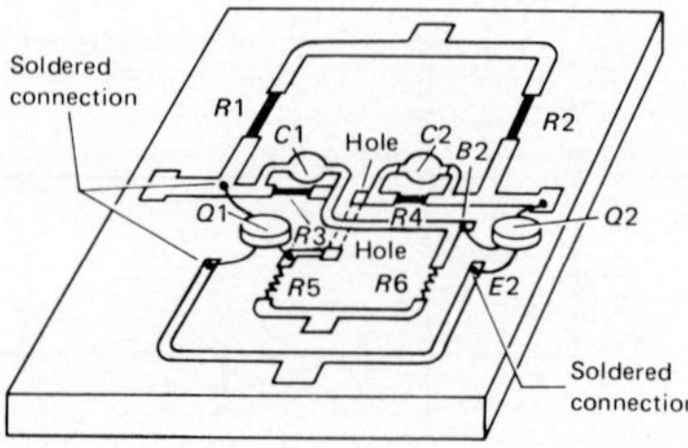

Fig. 3-147

3-148

(*soldered*) The monolithic IC has certain advantages and disadvantages as compared with the thin-film IC. First, since the components of a monolithic IC are formed _____ the IC chip, the monolithic circuit is more compact.

3-149
(*within*) Second, since the monolithic IC is made only of _____ materials, it is less expensive and easier to fabricate than the thin-film IC.

3-150
(*semiconductor*) The major disadvantage of the monolithic IC is that the range of resistance and capacitance values is limited to relatively low values. For example, a low-frequency phase shift oscillator which requires large capacitance and resistance values would probably be fabricated as a _____ (thin-film, monolithic) IC.

3-150
(*thin-film*)

CRITERION CHECK TEST

____3-1 For a junction transistor, the bias conditions are (*a*) forward bias on the BE junction, reverse on the CB junction, (*b*) forward bias on the BE junction, forward on the CB junction, (*c*) reverse bias on the BE junction, forward on the CB junction, (*d*) reverse bias on the BE junction, reverse on the CB junction.

____3-2 In Fig. 3-4, the dc voltage between collector and base is provided by the voltage divider consisting of (*a*) R1 and R2, (*b*) R1 and R3, (*c*) R3 and R4, (*d*) R2 and R4.

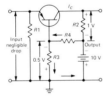

Fig. 3-4

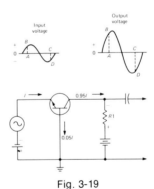

Fig. 3-19

____3-3 In Fig. 3-4, the base-emitter bias voltage is about (*a*) 0.5 V, (*b*) 9.5 V, (*c*) 9.0 V, (*d*) 10 V.

____3-4 In Fig. 3-19, the ac signal source is connected directly to the (*a*) resistor R1, (*b*) collector-base bias battery, (*c*) emitter-base bias battery, (*d*) collector lead.

____3-5 In Fig. 3-19, when the input voltage is *CD*, the (*a*) base current decreases, (*b*) emitter current decreases, (*c*) output voltage is becoming more negative, (*d*) output voltage is becoming less negative.

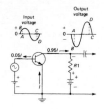

Fig. 3-42

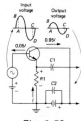

Fig. 3-58

Fig. 3-90

___3-6 In Fig. 3-42, the common or ground element is the (*a*) cathode, (*b*) collector, (*c*) emitter, (*d*) base.

___3-7 In Fig. 3-42, the output voltage is taken from the (*a*) base, (*b*) emitter, (*c*) collector, (*d*) anode.

___3-8 In a common-emitter amplifier, (*a*) both PN junctions are reverse-biased, (*b*) there is a 180° phase shift between input and output, (*c*) there is no phase shift between input and output, (*d*) there is a 90° phase shift between input and output.

___3-9 In Fig. 3-58, the function of C2 is to (*a*) pass dc, (*b*) block dc, (*c*) pass ac signals, (*d*) block ac signals.

___3-10 In a common-collector amplifier, (*a*) both PN junctions are forward-biased, (*b*) the output is taken from the collector, (*c*) there is a 0° phase shift between input and output, (*d*) there is a 180° phase shift between input and output.

___3-11 A junction transistor has a low input resistance because (*a*) it is made of semiconductor material, (*b*) of the impurity atoms, (*c*) the input is a forward-biased junction, (*d*) the base region is narrow.

___3-12 In Fig. 3-90, the charge carrier in the channel is the (*a*) hole, (*b*) electron, (*c*) Ge ion, (*d*) Si ion.

___3-13 For a JFET, above pinchoff voltage, the (*a*) drain current decreases, (*b*) drain current increases strongly, (*c*) drain current remains fairly constant, (*d*) depletion region becomes smaller.

___3-14 The input control parameter of an FET is the (*a*) source voltage, (*b*) drain voltage, (*c*) gate voltage, (*d*) gate current.

___3-15 A common-base configuration of a PNP transistor is analogous to an FET (*a*) common-drain configuration, (*b*) common-source configuration, (*c*) common-gate configuration, (*d*) common-emitter configuration.

___3-16 In the curves below for a JFET, what would V_{GS} for curve 3 be? (*a*) +1 V, (*b*) +2 V, (*c*) +3 V, (*d*) −2 V.

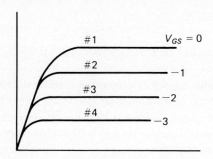

_____ 3-17 A single monolithic IC chip has a volume of about (a) 0.01 ft³, (b) 0.001 ft³, (c) 0.01 in³, (d) 0.0001 in³.

_____ 3-18 In monolithic ICs, resistors are formed from (a) nichrome wire, (b) ceramic material, (c) aluminum ribbon, (d) P-type material.

_____ 3-19 A function of the oxide layer on a monolithic IC is to (a) form the dielectric for the capacitors, (b) prevent shorting of elements by the interconnections, (c) prevent heat radiation from the substrate, (d) form the cathode of diodes.

_____ 3-20 Most computer circuits use monolithic ICs rather than thin-film ICs because (a) larger resistance values are possible with monolithic ICs, (b) larger capacitance values are possible with monolithic ICs, (c) monolithic ICs are more compact, (d) logic circuits cannot be built as thin-film ICs.

4 transistor parameters, equivalent circuits, and gain calculations

OBJECTIVES

(1) List and diagram the 5 parameters which can represent any linear circuit. **(2)** Draw the v-i characteristics of the resistor, the ideal voltage source, and the ideal current source. **(3)** Using static curves, sketch the curves for the input of a transistor in the CE configuration. **(4)** State the basic circuit element which best describes the input of a transistor circuit. **(5)** Give the range of input resistances found in transistors. **(6)** State a limitation on the use of hybrid parameters in ac analysis. **(7)** Derive a representation for the output of a junction transistor in the CE configuration, sketching a family of curves relating collector current to collector voltage. **(8)** State the basic circuit element which best describes the output of a transistor. **(9)** Define and compute the current amplification factor in the hybrid model. **(10)** Diagram the hybrid circuit showing input and output parameters. **(11)** List two methods for determining hybrid parameters. **(12)** Derive the current gain of a CE amplifier in terms of hybrid parameters. **(13)** Derive the voltage gain of a CE amplifier in terms of hybrid parameters and external circuit parameters. **(14)** Evaluate the current and voltage gain for a typical CE amplifier using a single-stage diagram showing typical component values. **(15)** Compute the power gain of any given CE amplifier. **(16)** Diagram the complete 4-parameter hybrid equivalent circuit. **(17)** Compare the simple hybrid circuit to the complete hybrid circuit, explaining the reasons for neglecting h_{re} and h_{oe}. **(18)** State the convention for expressing the hybrid parameter symbols in the CB and CC configurations. **(19)** Draw a table showing typical values of the hybrid parameters for the three configurations and compare their values. **(20)** Detail the procedure for obtaining the hybrid parameters of one configuration from the parameters of a different configuration. **(21)** Relate the symbols β and α to the h parameters.

GENERAL INFORMATION—EQUIVALENT CIRCUITS

4-1
The relationships that exist between 2 or more variable quantities can be expressed in the form of a *mathematical equation*. For example, Ohm's law may be written: $I = V/R$. When Ohm's law is expressed in this manner, it takes the form of a mathematical _____.

4-2
(*equation*) The circuit behavior of *any* electrical element or device can be expressed in terms of one or more _____ equations.

4-3
(*mathematical*) In electricity, mathematical equations deal specifically with voltages and currents, in addition to elements such as resistance, inductance, and capacitance. Such elements are present in electric circuits to govern and control voltages and _____.

4-4
(*currents*) Voltages and currents, together with their controlling elements, are called the dimensions or *parameters* of a circuit. Thus, in any circuit containing a resistance element, the resistance element is called a _____ of the circuit.

4-5
(*parameter*) In a circuit in which the current is controlled by a resistance, an inductance, and a capacitance, then all the factors—voltage, current, resistance, capacitance, and inductance—are called _____ of the circuit.

4-6
(*parameters*) A circuit element such as a voltage generator or a current generator, as well as a resistance, inductance, or capacitance, exhibits a single characteristic and therefore may be represented in an equation by a single parameter. For instance, the pure-resistance parameter is symbolized by the letter _____ in any circuit.

4-7
(R or r) In a similar manner, the voltage parameter appears as V or v in equations, and the current parameter appears as I or _____ in equations.

4-8

(*i*) The basic circuit parameters are shown diagramatically in Fig. 4-8. The symbols for the parameters *R, L,* and *C* are familiar. We shall use the symbol adjacent to *v* (circle with sine cycle) for a constant voltage generator. We shall here use the two interlocking circles adjacent to *i* as a constant _____ generator; this latter symbol is not a standard one in the general literature but is convenient for use here.

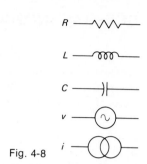

Fig. 4-8

4-9

(*current*) For later comparison with transistors and vacuum tubes, Fig. 4-9 presents the *v-i* characteristic of a resistor. We see that there is a linear relation between the current and _____.

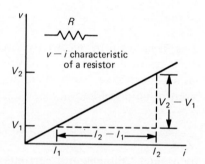

Fig. 4-9

4-10

(*voltage*) In fact, from Ohm's law, $V = IR$. Since the equation of a straight line is $y = mx$, where *m* is the slope, we can see that the slope of the *v-i* characteristic for a resistor is _____.

4-11

(*R*) Graphically, from Fig. 4-9, the slope of the straight line is given by $(V_2 - V_1)/(I_2 - I_1)$. Hence, $R = (V_2 - V_1)/$_____. We shall use this fact when discussing the transistor and vacuum tube.

4-12

($I_2 - I_1$) Figure 4-12 depicts the *v-i* characteristic of a voltage source. It can be seen from Fig. 4-12 that the voltage of the source _____ (does, does not) depend on the current.

Fig. 4-12

4-13
(does not) The voltage will be *V* no matter how much current is drawn. Finally, Fig. 4-13 shows the *v-i* characteristic of a _____ source.

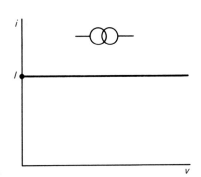

Fig. 4-13

4-14
(current) The current supplied by the source will be _____ no matter what voltage is across the current source.

4-15
(I) It is a fact that any linear device or circuit, regardless of its complexity, may be discussed mathematically in terms of only 5 parameters: voltage, current, resistance, inductance, and _____.

4-16
(capacitance) A simple transistor amplifier is drawn in Fig. 4-16. In this circuit, the emitter is common to the input and output circuits; hence this is a _____-_____ configuration.

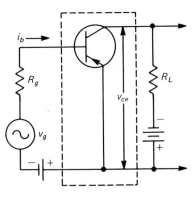

Fig. 4-16

4-17
(common-emitter) The source of input signal is a voltage generator identified symbolically as _____.

4-18
(v_g) The signal source has a definite resistance. This is normally called the source impedance and is considered to be in series with the source itself. In Fig. 4-16, the source impedance is symbolized by _____.

4-19
(R_g) In Fig. 4-16, the base current is i_b, the load resistance is R_L, and the voltage from emitter to collector is _____.

4-20
(v_{ce}) An equivalent circuit of a CE transistor amplifier is shown in Fig. 4-20. It is seen that the transistor is represented by a resistor and a _____ _____.

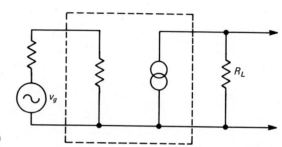

Fig. 4-20

4-21
(*current source*) The reasons for each of these parameters and their connections will be developed in the next section. This equivalent circuit will simplify the development of equations for voltage gain, current _____, and power gain.

4-21
(*gain*)

HYBRID EQUIVALENT CIRCUIT
4-22
In this section, we shall derive a simple ac equivalent circuit for a transistor. To begin, Fig. 4-22 shows the important quantities for the input and _____ of a transistor.

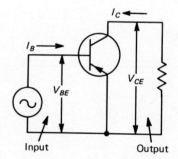

Fig. 4-22

4-23
(*output*) The function of the equivalent circuit is to take into account the behavior of the input and output of a transistor. Hence, we must examine curves which describe the _____ and _____ properties.

4-24
(*input, output*) Figure 4-24 shows a set of curves which describes the input of a transistor. These are static curves in which dc base voltage V_{BE} is plotted against dc base _____ I_B.

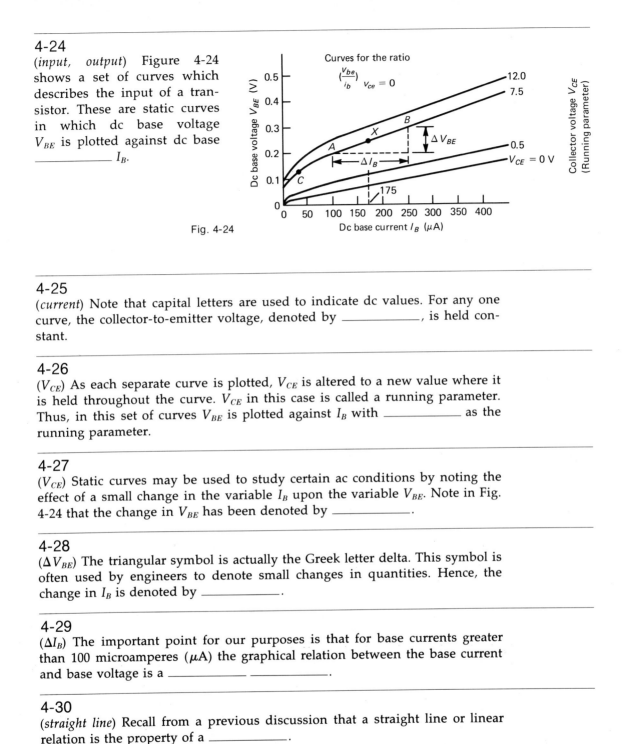

Fig. 4-24

4-25
(*current*) Note that capital letters are used to indicate dc values. For any one curve, the collector-to-emitter voltage, denoted by _____, is held constant.

4-26
(V_{CE}) As each separate curve is plotted, V_{CE} is altered to a new value where it is held throughout the curve. V_{CE} in this case is called a running parameter. Thus, in this set of curves V_{BE} is plotted against I_B with _____ as the running parameter.

4-27
(V_{CE}) Static curves may be used to study certain ac conditions by noting the effect of a small change in the variable I_B upon the variable V_{BE}. Note in Fig. 4-24 that the change in V_{BE} has been denoted by _____.

4-28
(ΔV_{BE}) The triangular symbol is actually the Greek letter delta. This symbol is often used by engineers to denote small changes in quantities. Hence, the change in I_B is denoted by _____.

4-29
(ΔI_B) The important point for our purposes is that for base currents greater than 100 microamperes (μA) the graphical relation between the base current and base voltage is a _____.

4-30
(*straight line*) Recall from a previous discussion that a straight line or linear relation is the property of a _____.

4-31
(*resistor*) Thus the input of a transistor looks like a simple resistor. Further, recall that the slope of the *v-i* characteristic of a resistor is the value of _____.

4-32
(*resistance*) By inspection of Fig. 4-24, it is seen that the slope of the straight line is given by $\Delta V_{BE}/$_____.

4-33
(ΔI_B) Since the slope of the *v-i* characteristic of a resistor is the value of resistance, the input resistance of a transistor is simply _____/_____.

4-34
($\Delta V_{BE}, \Delta I_B$) Using Fig. 4-24, we can compute a value for the input resistance. For the $V_{CE} = 7.5$-V curve, in going from point A to point B, I_B changes from 100 μA to 250 μA. Hence, $\Delta I_B = 250\ \mu\text{A} - 100\ \mu\text{A} = $ _____.

4-35
(150 μA) Similarly, from A to B, V_{BE} goes from 0.2 V to 0.3 V. Hence, $\Delta V_{BE} = $ _____.

4-36
(0.1) This merely states that when the base current is changed 150 μA, then the _____ voltage changes 0.1 V.

4-37
(*base*) Note that the collector voltage (V_{CE}) has been kept _____ throughout this process.

4-38
(*constant*) Substituting these values in the equation for input resistance gives

$$\left.\frac{\Delta V_{BE}}{\Delta I_B}\right|_{V_{CE}=7.5\text{ V}} = \frac{0.1\text{ V}}{150 \times 10^{-6}\text{ A}} = \underline{\qquad}\ \Omega$$

4-39
(666.7) Note that volts divided by amperes gives ohms, the unit of resistance. In the hybrid model, this input resistance is denoted by h_{ie}. The letter *h* stands for hybrid, *i* stands for input, and *e* tells us that this is a common-_____ configuration.

4-40
(*emitter*) Thus, when we subsequently see h_{ie} in equations we shall recognize it as the input resistance of a common-emitter configuration, obtained by noting the effect of I_B upon V_{BE} with V_{CE} held _____.

4-41
(*constant*) Depending upon the transistor, h_{ie} may run from 300–800 Ω. Thus, the transistor in our example is quite typical because its _____ resistance is 666.7 Ω.

4-42
(*input*) Before examining the output circuit of the transistor, we shall show that the derivation just given for the input resistance has an important limitation. In Fig. 4-24, we see that situated midway between points A and B on the v-i characteristic is a point denoted by the letter _____.

4-43
(*X*) Assume that X represents the dc operating point of the transistor; i.e., the direct current fed to the base is _____ μA.

4-44
(*175*) If an ac signal is now fed to the base, the net base current will vary around the dc point, X. As long as the ac signal is small enough so that the base current stays within points A and _____, the operation stays in a linear region.

4-45
(*B*) However, if the base current swing is large enough to reach a point such as C, then operation is in a nonlinear region. Hence, the assumption that the input resistance is of constant value _____ (is, is not) valid.

4-46
(*is not*) Summarizing: The hybrid circuit is only valid for small ac signals operating in a _____ region.

4-47
(*linear*) To complete the hybrid equivalent circuit, we next examine the output curves of a transistor. Figure 4-47 is a graph of a family of curves in which the collector current, I_C, is the ordinate, and the _____ voltage is the abscissa.

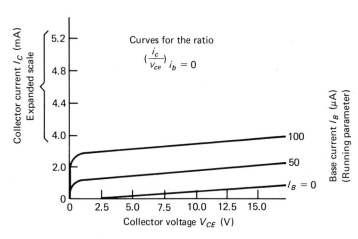

Fig. 4-47

4-48
(*collector*) Each curve is labeled by a different value of _____ current.

4-49
(*base*) For example, the first curve is for a base current of 0 µA, the second curve for a base current of 50 µA, the third curve for a base current of _____ µA.

4-50
(*100*) The important aspect of Fig. 4-47 is that each curve is almost a straight, horizontal line. As the collector voltage increases, the collector current varies very _____ (little, much).

4-51
(*little*) From a previous discussion, recall that a current source has a *v-i* characteristic which is a horizontal line. Hence, the output of a transistor behaves like a _____ source.

4-52
(*current*) Since the basic function of a transistor is to be an amplifying device, it is useful to express the value of this output current source in terms of the _____ current.

4-53
(*input*) By definition, the input current in this configuration is the base current (refer to Fig. 4-22). Thus, we wish to find the ratio of the collector current to the _____ current.

4-54
(*base*) Since the signals to be amplified are small ac signals, we wish to compute the ratio $\Delta I_C/\Delta I_B$. Using Fig. 4-47, we see that at a fixed collector voltage of 7.5 V, a base-current change of $(100 - 50) = 50$ µA produces a collector-current change of $(3.3 - 1.8) =$ _____ mA.

4-55
(*1.5*) The ratio of $\Delta I_C/\Delta I_B$ is then easily computed to be

$$\frac{\Delta I_C}{\Delta I_B} = \frac{1.5 \times 10^{-3}}{50 \times 10^{-6}} = \underline{\qquad}$$

4-56
(*30*) Thus, as long as the signal voltages are small, for this transistor the collector ac will be _____ times the base current. Since the collector current is larger than the base current, the ratio $\Delta I_C/\Delta I_B$ is called the *forward-current-amplification factor*.

4-57
(30) In the hybrid equivalent model, the forward-current-amplification factor is denoted by h_{fe}, where h stands for _____, f stands for forward, and e indicates common-emitter configuration.

4-58
(*hybrid*) Using the identity $\Delta I_C = (\Delta I_C/\Delta I_B) \Delta I_B$, one has $\Delta I_C = h_{fe} \Delta I_B$. This relation again states that the collector current is h_{fe} times the _____ current.

4-59
(*base*) Figure 4-59 summarizes the simple hybrid equivalent circuit. The input circuit consists of a resistor, h_{ie}. The output circuit consists of a _____ source of value $h_{fe} \times i_b$.

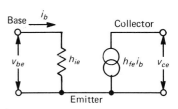

Fig. 4-59

4-60
(*current*) Note in Fig. 4-59 that the delta signs have been omitted. This can be done because, by convention, lowercase letters such as i_b and v_{ce} denote signal or ac values in which periodic changes are occurring. Since _____ means *a change in*, this symbol is no longer required.

4-61
(Δ) In practice, the hybrid parameters are either measured using an instrument such as a curve tracer, or they can be obtained from a transistor handbook or specification sheet. In subsequent sections, this simple model will be used to calculate the ac gain of a transistor in the common-_____ configuration.

4-61
(*emitter*)

CURRENT AND VOLTAGE GAIN

4-62
The equations that enable us to find the current or voltage gain of a transistor may be derived from the hybrid equivalent circuit (Fig. 4-62). Current gain is conventionally symbolized by A_i and is defined as the ratio of load current (ac) to input _____ (ac).

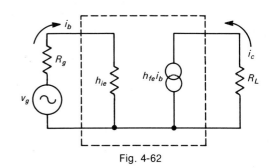

Fig. 4-62

4-63
(*current*) In the circuit of Fig. 4-62, the input current is the base current, denoted by i_b. The output current is the collector current, denoted by _____.

4-64
(i_c) Since the output circuit consists of a constant current source in series with a resistor, the output (or load) current is given by $h_{fe} \times$ _____.

4-65
(i_b) We can now compute the current gain, A_i:

$$A_i = \frac{i_c}{i_b} = \frac{h_{fe} \times i_b}{i_b} = \underline{\qquad}$$

4-66
(h_{fe}) Summarizing: For this simple common-emitter amplifier the current gain is h_{fe}, where h stands for hybrid, f for forward, and e indicates the common-_____ configuration.

4-67
(*emitter*) Next, we shall use the hybrid equivalent circuit (Fig. 4-62) to compute the voltage gain. The voltage gain, A_v, is defined as the output voltage divided by the _____ voltage.

4-68
(*input*) In Fig. 4-62, the output voltage is the voltage across the load resistor R_L while the input voltage is the source voltage, denoted by _____.

4-69
(v_g) The voltage across R_L can be obtained by Ohm's law. The current through R_L is i_c, so by Ohm's law the output voltage (voltage across R_L) is given by $v_{out} = i_c \times$ _____.

4-70
(R_L) Because of the constant current generator, $i_c = h_{fe} \times i_b$. This gives $v_{out} = h_{fe} \times$ _____ $\times R_L$.

4-71
(i_b) Finally, i_b can be expressed in terms of v_g by another application of Ohm's law. Examining the input circuit of Fig. 4-62, we see that $v_g = i_b \times (R_g +$ _____ $)$.

4-72
(h_{ie}) We may solve the equation $v_g = i_b \times (R_g + h_{ie})$ for i_b. The result is $i_b = v_g/(\underline{\qquad} + h_{ie})$.

4-73
(R_g) This can be substituted in the equation for v_{out} as follows:

$$v_{\text{out}} = h_{fe} \times i_b \times R_L = h_{fe} \times \frac{v_g}{(R_g + h_{ie})} \times R_L$$

Dividing both sides of this equation by v_g gives

$$A_v = \frac{v_{\text{out}}}{v_g} = \frac{h_{fe} \times ?}{R_g + h_{ie}}$$

4-74
(R_L) Figure 4-74 summarizes the steps in the derivation of the voltage gain. Note that as the source resistance, R_g, becomes larger, the voltage gain becomes _____.

(a) $v_{\text{out}} = i_c \times R_L$ (Ohm's law)
(b) $i_c = h_{fe} \times i_b$ (definition of h_{fe})
(c) $i_b = \dfrac{v_g}{(R_g + h_{ie})}$ (Ohm's law)
(d) $A_v = h_{fe} \times \dfrac{R_L}{(R_g + h_{ie})}$

Fig. 4-74

4-75
(*smaller*) This means that a single transistor stage will provide little voltage amplification for a source which has a _____ internal impedance.

4-76
(*high* or *large*) Reviewing: It has been shown that for a simple common-emitter stage, the current and voltage gains are given by $A_i = h_{fe}$ and $A_v = h_{fe} \times R_L/(R_g + h_{ie})$, where h_{fe} is the current-amplification factor and h_{ie} is the hybrid input _____.

4-76
(*resistance*)

TYPICAL CIRCUIT FOR NUMERICAL EVALUATION OF GAIN

4-77
Figure 4-77 illustrates a typical CE amplifier including all required components. (Note: The functions of many of these will be discussed later.) The diagram shows a signal source identified as _____ with an internal resistance, R_g, of 1,500 Ω.

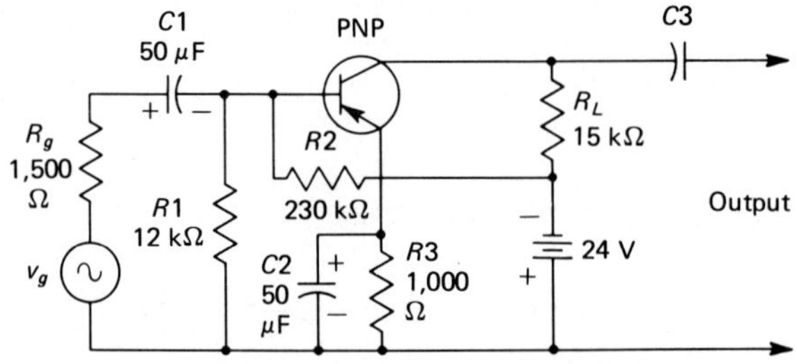

Fig. 4-77

4-78
(v_g) The component which prevents dc voltages from reaching the signal source and couples the ac signal to the base of the transistor is _____ in Fig. 4-77.

4-79
(C1) A single battery is used to establish the correct values of both base-emitter forward bias and collector-emitter _____ bias.

4-80
(*reverse*) The base-emitter forward bias is adjusted for magnitude by the use of a voltage divider consisting of 2 resistors, both of which terminate at the base lead. These resistors are $R1$ and _____.

4-81
($R2$) Resistor $R3$ is used for temperature-stabilization purposes (to be discussed later). To prevent signal degeneration, resistor $R3$ is bypassed by _____ in Fig. 4-77.

4-82
(C2) The load resistance into which the transistor works is _____ kilohms (kΩ).

4-83
(15) The component labeled _____ blocks dc voltages from, and couples the output signal to, the following stage.

---

104 A PROGRAMMED COURSE IN BASIC ELECTRONICS

4-84
(C3) Figure 4-84 shows the equivalent circuit of the CE amplifier with all parts that do not directly affect the calculation eliminated. Clearly, the only components (other than the transistor) that do influence the numerical results for amplification are the internal resistance of the generator R_g and the _____ resistance.

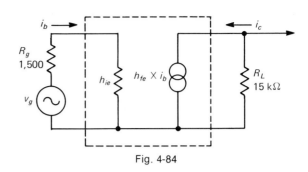

Fig. 4-84

4-85
(*load*) Now imagine that this circuit is actually set up with its operating point such that the emitter current (dc) I_E equals 0.1 mA and its dc collector voltage V_{CE} equals _____ as in Fig. 4-85.

$I_E = 0.1$ mA
$V_{CE} = 8.0$ V
Establish operating point

$h_{ie} = 1{,}500\ \Omega$
$h_{fe} = 50$

Fig. 4-85

Factors obtained by measurement and from static characteristics.

4-86
(*8.0 V*) All the necessary static curves are then assumed to be drawn in order to obtain values for the essential factors. To obtain the current amplification A_i it is only necessary to know _____ .

4-87
(h_{fe}) This factor is given in Fig. 4-85. Thus, $A_i = h_{fe} =$ _____ .

4-88
(*50*) Thus the current gain is 50. This answer indicates that an input current of any value within operating limits would be multiplied by 50 in the _____ .

4-89
(*output*) To obtain the value of voltage gain, we can use the formula of Fig. 4-74: $A_v = h_{fe} \times$ _____ $/(R_g + h_{ie})$.

4-90
(R_L) With the values of Fig. 4-85 substituted,

$$A_v = \frac{? \times 15{,}000}{1{,}500 + 1{,}500}$$

4-91
(*50*) When multiplied out, the numerator may be written as $75 \times 10^?$.

4-92
(*4*) The denominator becomes _____ .

4-93
(*3,000*) Dividing the numerator by the denominator gives a final result for voltage gain equal to _____ .

4-94
(*250*) This result shows that this transistor in the common-emitter configuration will provide a voltage gain of 250. Further, recall from a previous discussion (Frame 3-57) that the output voltage has a phase of _____ ° with respect to the input voltage.

4-94
(*180*)

THE POWER-GAIN EQUATION AND NUMERICAL EXAMPLE
4-95
The *power input* to any device is equal to the power that is transferred from the generator or other input device to the _____ circuit of the device.

4-96
(*input*) The *power output* from any device is the power transferred from the output circuit of the device to the _____ resistance or impedance.

4-97
(*load*) The *power gain* of any device is defined as the ratio of power output to power _____ .

4-98
(*input*) Power can be defined as the product of current squared and resistance; it can be defined as the voltage squared divided by the resistance; and it can be defined as the current times the _____ in a purely resistive circuit.

4-99
(*voltage*) We will use the last definition given above. The input circuit of a transistor in the CE configuration is the base circuit. The _____ input, therefore, is the product of i_b and v_g.

4-100
(*power*) In equation form, the power input to a transistor can therefore be written as $P_i =$ _____.

4-101
($i_b v_g$) The output circuit of a transistor in the CE configuration is the collector-emitter circuit. The power output, therefore, is the power delivered to the load, or the load current times the load _____.

4-102
(*voltage*) The equivalent circuit shows that the load current is the same as the collector current. That is, $i_L = i_c$. Similarly, the load voltage equals the collector-to-emitter voltage. That is, $v_L =$ _____.

4-103
(v_{ce}) Hence, the output power is the product of collector current and collector-to-emitter voltage. In equation form, $P_o =$ _____.

4-104
($i_c v_{ce}$) Since power gain is the ratio of power output to power input, then the equation for power gain in terms of base and collector currents and voltages is

$$G = \frac{i_c v_{ce}}{?}$$

4-105
($i_b v_g$) Writing the equation, we have $G = i_c v_{ce} / i_b v_g$. If this equation is rearranged thus

$$G = \frac{i_c}{i_b} \times \frac{v_{ce}}{v_g}$$

We can recognize the first term as the current gain and the second term as the _____ gain.

4-106
(*voltage*) Thus, power gain is merely the product of _____ gain and voltage gain.

4-107
(*current*) Using the symbols introduced in previous sections, we can write the equation for power gain as: $G = A_i \times$ _____.

4-108
(A_v) Thus, power gain of a transistor in the CE configuration is: $G = A_i A_v$. In the CE amplifier of Fig. 4-77, we found the current gain to be 50 and the voltage gain to be 250. The power gain is, therefore, $G =$ _____ $\times 250$.

4-109
(*50*) The power gain in this circuit as obtained from this product is _____.

4-110
(*12,500*) Power gain is normally expressed in decibels (dB). The equation is: $G_{dB} = 10 \log G$, where G is the actual power gain. If the log of 12,500 is 4.1, then $10 \log G =$ _____.

4-111
(*41.0*) Thus, the power gain is 41.0 dB. This shows that the power gain in decibels can always be obtained directly from G (actual power gain) by multiplying the _____ of G by 10.

4-111
(*log*)

COMPLETE HYBRID CIRCUIT
4-112
So far we have derived a simple hybrid circuit which involves 2 hybrid parameters, h_{ie} and _____. Figure 4-112 reviews the simple hybrid circuit.

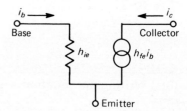

Fig. 4-112

4-113
(h_{fe}) This model is quite adequate for many transistor applications. However, most transistor specification sheets list 2 additional hybrid parameters which have not been mentioned. Figure 4-113 depicts the complete hybrid model containing a total of _____ hybrid parameters.

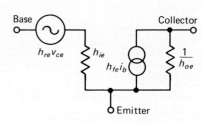

Fig. 4-113

4-114
(4) The new component of the input circuit is a voltage source whose value is _____ × v_{ce}.

4-115
(h_{re}) The parameter h_{re} is called the reverse-voltage-amplification factor. The h stands for hybrid, the r denotes reverse, and the e denotes the common-_____ configuration.

4-116
(emitter) What was the justification for neglecting this voltage source in the simple hybrid model? A typical value of h_{re} is 0.000575 or 5.75 × 10$^?$.

4-117
(−4) The fact that h_{re} is very small, of the order of 10^{-4}, means that it is often omitted from transistor calculations. Next, let us consider the "new" parameter in the output circuit, a resistor denoted by 1/_____.

4-118
(h_{oe}) Since the unit of resistance is ohms, and the resistance is given as $1/h_{oe}$, the units of h_{oe} should be reciprocal ohms, or _____.

4-119
(mhos) A typical value of h_{oe} is 24.5 × 10^{-6} mho. To convert to a resistance in ohms, we must find 1/_____.

4-120
(h_{oe}) The conversion is done as follows:

$$\frac{1}{h_{oe}} = \frac{1}{24.5 \times 10^{-6} \text{ mho}} = \frac{10^6}{24.5 \text{ mho}} = 0.041 \times 10^6 \underline{\qquad}$$

4-121
(Ω) This result can also be expressed as 41 × 10^3 or 41,000 Ω. Recall that when 2 resistors are in parallel, current will take the path of least resistance, so that more of the current will go through the _____ (smaller, larger) of the 2.

4-122
(smaller) Figure 4-122 depicts only the output circuit of a transistor. The load resistor, R_L, has a value of _____ Ω.

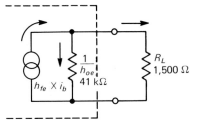

Fig. 4-122

4-123
(*1,500*) The current source has a value of $h_{fe} \times i_b$. This current will split, some going through the 41-kΩ internal resistance of the transistor, some going through _____.

4-124
(R_L) Since $1/h_{oe}$ is about 25 times the size of R_L, only about 1/25 of the total current will go through the resistor _____.

4-125
($1/h_{oe}$) Therefore, the amount of current fed to the load resistor R_L is very little changed by the presence of the internal resistor, $1/$_____.

4-126
(h_{oe}) Summarizing: The complete hybrid model has 4 parameters: h_{ie}, h_{fe}, h_{re}, and h_{oe}. These 4 are usually listed on transistor specification sheets. However, because of the small size of _____ and _____, the simple hybrid model is often adequate.

4-126
(h_{re}, h_{oe})

CB AND CC CONVERSIONS
4-127
All equations and formulas thus far presented have been based upon the CE configuration. Similar equations, however, are available for the CB and _____ configurations.

4-128
(*CC* or *common-collector*) We shall not derive these relationships; we should study their connections with the CE equations, however. Note the parameter symbols in Fig. 4-128. For each parameter symbol used in the CE configuration, there is an equivalent symbol for the _____ and CC configurations.

	CC	CB	CC
h_f	h_{fe}	h_{fb}	h_{fc}
h_r	h_{re}	h_{rb}	h_{rc}
h_i	h_{ie}	h_{ib}	h_{ic}
h_o	h_{oe}	h_{ob}	h_{oc}

Fig. 4-128

4-129
(*CB*) For example, the forward-current-amplification factor may be written in general form as h_f. To indicate that the particular h_f we are discussing is a parameter for the common-emitter configuration, we add a(n) _____ to the symbol.

4-130
(*e*) To show that the particular h_f we are discussing is a parameter for the common-collector configuration, we add the letter _____ to the symbol.

4-131
(*c*) Thus, the symbol for forward-current-amplification factor in a common-base circuit is _____ .

4-132
(h_{fb}) Similarly, the general symbol for the reverse-voltage-amplification factor is h_r. To show that we are dealing with the h_r for the common-emitter circuit, we have been adding the letter _____ to the symbol.

4-133
(*e*) Thus, the symbol for reverse-voltage-amplification factor when this factor is a parameter for the common-base circuit is h_{rb}. Hence, when h_r is a parameter for the common-collector circuit, the symbol is _____ .

4-134
(h_{rc}) In the same way, the input resistance for common emitter is symbolized by h_{ie}; for common base the symbol is h_{ib}; and for common collector it is _____ .

4-135
(h_{ic}) Also, the output conductance for common base is h_{ob} and the output conductance for common collector is _____ .

4-136
(h_{ic}) The typical values of the various parameters in different configurations are quite different from each other. For example, h_{fe} has a typical value of 49 (see Fig. 4-136), while h_{fc} has a typical value of _____ .

	CE	CB	CC
h_f	$h_{fe} = 49$	$h_{fb} = -0.98$	$h_{fc} = -50$
h_r	$h_{re} = 575 \times 10^{-6}$	$h_{rb} = 380 \times 10^{-6}$	$h_{rc} = 1$
h_i	$h_{ie} = 1,950\ \Omega$	$h_{ib} = 39\ \Omega$	$h_{ic} = 1,950\ \Omega$
h_o	$h_{oe} = 24.5 \times 10^{-6}$ mho	$h_{ob} = 0.49 \times 10^{-6}$ mho	$h_{oc} = 24.5 \times 10^{-6}$ mho

Fig. 4-136

4-137
(−50) Now, comparing the current-amplification factor for the common-base configuration with the other typical values, we find that this particular parameter has a typical value of _____ .

4-138

(-0.98) The same thing is true of all the other parameters as indicated in Fig. 4-136. For example, h_{rc} has a typical value of 1 while the same parameter in the CB configuration has a typical value of _____.

4-139

(380×10^{-6}) Similarly, the parameter h_i has its smallest typical value in the _____ configuration.

4-140

(*CB*) Both h_i and h_o have exactly the same values in the CE and _____ configurations.

4-141

(*CC*) Manufacturers normally provide only 1 (or at best 2) sets of parameters: one set for CE and/or CB. Figure 4-141 provides a set of simple conversions which permit you to obtain any CB from its equivalent CE parameter, any CC from its CE parameter, any CE from its CB parameter, and any _____ from its CB parameter.

From CE to CB	From CE to CC	From CB to CE	From CB to CC
$h_{ib} = \dfrac{h_{ie}}{1 + h_{fe}}$	$h_{ic} = h_{ie}$	$h_{ie} = \dfrac{h_{ib}}{1 + h_{fb}}$	$h_{ic} = \dfrac{h_{ib}}{1 + h_{fb}}$
A	E	I	M
$h_{rb} = \dfrac{h_{ie} h_{oe}}{1 + h_{fe}} - h_{re}$	$h_{rc} = 1 - h_{re} \cong 1$	$h_{re} = \dfrac{h_{ib} h_{ob}}{1 + h_{fb}} - h_{rb}$	$h_{rc} = \dfrac{1 - h_{ib} h_{ob}}{1 + h_{fb}} + h_{rb} \cong 1$
B	F	J	N
$h_{fb} = \dfrac{-h_{fe}}{1 + h_{fe}}$	$h_{fc} = -(1 + h_{fe})$	$h_{fe} = \dfrac{-h_{fb}}{1 + h_{fb}}$	$h_{fc} = \dfrac{-1}{1 + h_{fb}}$
C	G	K	O
$h_{ob} = \dfrac{h_{oe}}{1 + h_{fe}}$	$h_{oc} = h_{oe}$	$h_{oe} = \dfrac{h_{ob}}{1 + h_{fb}}$	$h_{oc} = \dfrac{h_{ob}}{1 + h_{fb}}$
D	H	L	P

Fig. 4-141

4-142

(*CC*) For example, suppose a manufacturer gave the CB parameters for transistor Q, and you wanted to find its current and voltage amplifications in a given circuit in the CC configuration. You would obtain the CC parameters from the equations lettered M, N, O, and _____ in Fig. 4-141.

4-143
(*P*) Or if the manufacturer provided the parameters for the CE configuration and you wanted to work with a CB circuit, you would use the conversion equations lettered *D, C, B,* and _____.

4-144
(*A*) Consider a numerical example. Suppose you are given the typical CB figures in Fig. 4-136 for a certain circuit; suppose further that you want to find the CE parameters for this circuit. This calls for the use of equations _____, _____, _____, _____ in Fig. 4-141.

4-145
(*I, J, K, L*) Thus, to find h_{ie}, we use equation *I*.

$$h_{ie} = \frac{h_{ib}}{1 + h_{fb}} = \frac{39}{1 + (-0.98)} = \underline{\qquad} \, \Omega$$

4-146
(*1,950*) Note that this checks nicely with the value of h_{ie} given in the first column of Fig. 4-136. To find h_{re} we use equation *J* thus

$$h_{re} = \frac{h_{ib}h_{ob}}{1 + h_{fb}} - h_{rb} = \underline{\qquad} \text{ (no unit)}$$

4-147
(575×10^{-6}) Note that this checks nicely with the value of h_{re} in the first column of Fig. 4-136. To find h_{fe}, we use equation *K* thus

$$h_{fe} = \frac{-h_{fb}}{1 + h_{fb}} = \underline{\qquad} \text{ (no units)}$$

4-148
(*49*) Thus, the conversion process merely requires that you find the manufacturer's parameters and convert these (if necessary) to the parameters for the configuration you wish to work with, using the appropriate conversion equations extracted from the proper vertical row in Fig. _____.

4-149
(*4-141*) It has become almost universal to list the hybrid parameters on transistor specification sheets. However, in the engineering literature, one often finds several other notations. For example, the hybrid notation for the current gain in the CE configuration is _____, but this same parameter is often called beta.

4-150

(h_{fe}) Thus, an engineer who says that a transistor with a beta (β) of 50 is needed means that the hybrid parameter _____ should be 50.

4-151

(h_{fe}) Another notation used instead of h is the symbol α (alpha), which is most commonly used in conjunction with the current gain in the CB configuration. Whereas the current gain in the CB configuration is denoted in the h system by _____, engineers often denote this same quantity by α_{fb}, or simply α.

4-151

(h_{fb})

CRITERION CHECK TEST

___4-1 Typical parameters of an electric circuit are (a) length and weight, (b) voltage and resistance, (c) speed and acceleration, (d) mass and specific gravity.

The following 3 questions refer to the figure below.

___4-2 The symbol which represents a resistance is number (a) 1, (b) 2, (c) 3, (d) 4.
___4-3 The symbol which represents an inductor is number (a) 1, (b) 2, (c) 3, (d) 4.
___4-4 The symbol which represents a current source is number (a) 1, (b) 2, (c) 3, (d) 4.
___4-5 The symbol for dc base-emitter voltage is (a) V_B, (b) V_E, (c) V_{be}, (d) V_{BE}.
___4-6 Another way of writing 150 μA is (a) 0.00150 A, (b) 15×10^{-2} A, (c) 150×10^{-6} A, (d) 0.000150 mA.
___4-7 A typical ac input resistance in the CE configuration is (a) 10 Ω, (b) 100 Ω, (c) 450 Ω, (d) 40,000 Ω.
___4-8 The parameter which best describes the output of a transistor is a (a) voltage source, (b) resistor, (c) capacitor, (d) current source.

____4-9 The hybrid symbol h_{fe} is given the name (a) forward voltage amplification, (b) output resistance, (c) output capacitance, (d) forward current amplification.

____4-10 Hybrid parameters are found (a) in the *Handbook of Chemistry and Physics*, (b) stamped on the transistor case, (c) in a transistor handbook, (d) in an electron tube handbook.

____4-11 In Fig. 4-62, the generator voltage is denoted by (a) v_o, (b) v_L, (c) v_g, (d) v_c.

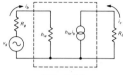

Fig. 4-62

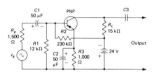

Fig. 4-77

____4-12 In Fig. 4-62, the base current is given by (a) $v_g R_g$, (b) v_g/R_g, (c) $v_g(R_g + h_{ie})$, (d) $v_g/(R_g + h_{ie})$.

The following 5 questions refer to Fig. 4-77 ($h_{fe} = 50$).

____4-13 Base-emitter bias is provided by the voltage divider (a) $R3R1$, (b) $R1R_g$, (c) $R2R1$, (d) $R_L R2$.

____4-14 The load resistor has a value of (a) 1,000 kΩ, (b) 230 kΩ, (c) 12 kΩ, (d) 15 kΩ.

____4-15 The 2 coupling capacitors are (a) $C1C3$, (b) $C1C2$, (c) $C2C3$, (d) $C2C4$.

____4-16 If the base ac is 500 μA, then the collector ac is (a) 5 mA, (b) 10 mA, (c) 25 mA, (d) 50 mA.

____4-17 If R_g were changed from 1,500 Ω to 3,000 Ω, the voltage gain would then be (a) 100, (b) 133, (c) 167, (d) 200.

____4-18 A typical value of h_{fb} is (a) -4.6×10^{-4}, (b) -4.6×10^{-1}, (c) -0.99, (d) -460.

____4-19 Which of the following parameters refer to the common-emitter configuration? (a) h_{ob}, (b) h_{ie}, (c) h_{fc}, (d) h_{oc}.

The following 3 questions refer to Fig. 4-141 ($h_{ie} = 500$ Ω, $h_{fe} = 80$).

From CE to CB	From CE to CC	From CB to CE	From CB to CC
$h_{ib} = \dfrac{h_{ie}}{1 + h_{fe}}$ A	$h_{ic} = h_{ie}$ E	$h_{ie} = \dfrac{h_{ib}}{1 + h_{fb}}$ I	$h_{ic} = \dfrac{h_{ib}}{1 + h_{fb}}$ M
$h_{rb} = \dfrac{h_{ie}h_{oe}}{1 + h_{fe}} - h_{re}$ B	$h_{rc} = 1 - h_{re} \approx 1$ F	$h_{re} = \dfrac{h_{ib}h_{ob}}{1 + h_{fb}} - h_{rb}$ J	$h_{rc} = \dfrac{1 - h_{ib}h_{ob}}{1 + h_{fb}} + h_{rb} \approx 1$ N
$h_{fb} = \dfrac{-h_{fe}}{1 + h_{fe}}$ C	$h_{fc} = -(1 + h_{fe})$ G	$h_{fe} = \dfrac{-h_{fb}}{1 + h_{fb}}$ K	$h_{fc} = \dfrac{-1}{1 + h_{fb}}$ O
$h_{ob} = \dfrac{h_{oe}}{1 + h_{fe}}$ D	$h_{oc} = h_{oe}$ H	$h_{oe} = \dfrac{h_{ob}}{1 + h_{fb}}$ L	$h_{oc} = \dfrac{h_{ob}}{1 + h_{fb}}$ P

Fig. 4-141

____4-20 Compute h_{fb}: (a) -0.25, (b) -0.99, (c) -1.6, (d) -3.4.

____4-21 Compute h_{fc}: (a) -10, (b) -41, (c) -81, (d) -146.

____4-22 Compute h_{ib}: (a) 2.5 Ω, (b) 50 Ω, (c) 6.17 Ω, (d) 12.4 Ω.

5 transistor bias stability

OBJECTIVES

(1) Define and discuss the *operating point* and *stable operating point* of a transistor. **(2)** Sketch and discuss curves showing the relation of a collector current to temperature. **(3)** State a general method used to control the collector direct current. **(4)** Describe the effect on bias stability of interchanging transistors. **(5)** Trace the temperature-induced shift in collector current and base-emitter resistance and state the relation between base current and collector current. **(6)** Describe the temperature-stabilizing effect of the emitter-swamping resistor and its effect on base-emitter resistance changes. **(7)** Sketch a family of curves showing the effect of the emitter resistor on bias stability. **(8)** State the effect of base circuit resistance on bias stability. **(9)** Describe the effect of the emitter resistor in minimizing transistor differences. **(10)** Sketch a transistor circuit with bias supply, collector supply, and emitter resistor, and specify the sources of bias voltage. **(11)** State the effects on the bias voltage when transistors are interchanged. **(12)** Describe the action of the collector-base resistor in stabilizing the operating point. **(13)** Define feedback, negative feedback, and positive feedback. **(14)** Draw the general transistor bias circuit, labeling the appropriate resistors and batteries. **(15)** Reduce the general bias network to the CB, CE, and CC configurations and to a single supply battery network. **(16)** Sketch and discuss a circuit which uses a single diode to compensate for the variation in base-emitter junction resistance. **(17)** Describe a stabilization method used for integrated circuits. **(18)** Draw a circuit using a zener diode, demonstrating that it can be used as a voltage regulator. **(19)** State the behavior of the zener diode resistance as a function of temperature. **(20)** Sketch and describe a two-diode temperature-compensating circuit for a zener diode. **(21)** Draw and analyze a zener-diode-stabilized transistor circuit. **(22)** Analyze a circuit using a shunt limiting diode. **(23)** Draw a CE circuit with transformer coupling on the input and output. **(24)** Discuss the effect of transformer transients on the collector of the transistor.

STABILITY OF OPERATING POINT

5-1
Figure 5-1 shows a graph of certain transistor "static" characteristics. In drawing these characteristics, the independent variable was the quiescent base direct current I_B and the dependent variable was the quiescent collector direct current _____.

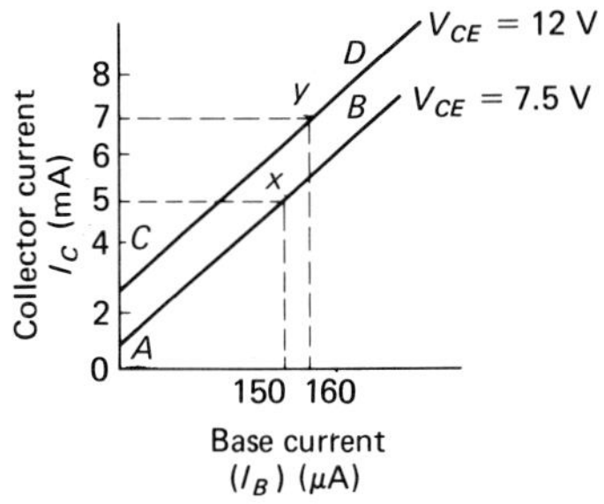

Fig. 5-1

5-2
(I_C) For one of these curves, that is, AB, the dc quiescent collector voltage was held constant at 7.5 V. For the other curve, that is, CD, the dc quiescent collector voltage was held constant at _____ V.

5-3
(12) As the base current is varied with the collector voltage held constant at, say, 7.5 V, the _____ current varies in a linear fashion since the characteristic curves are straight lines.

5-4
(collector) For example, if the base current is adjusted to be 150 μA with the collector voltage equal to 7.5 V, the collector current is _____ mA.

5-5
(5) On the other hand, if the base current is made 160 μA while the collector voltage is held at 12 V, then the collector current is _____ mA.

5-6
(7) Point X is called the *operating point* of the transistor when the collector voltage V_{CE} is 7.5 V and when the base current I_B is _____ μA.

5-7
(150) Point Y is called the operating point of the transistor when the collector voltage V_{CE} is _____ V and when the base current I_B is 160 μA.

5-8
(12) In other words, the _____ point of a transistor is established by selecting a specific value of dc quiescent collector voltage V_{CE} and a specific value of quiescent _____ direct current I_B.

TRANSISTOR BIAS STABILITY 117

5-9
(*operating, base*) In Fig. 5-9, the operating point X has been specified by using a base current of 150 μA and a collector voltage of _____ V.

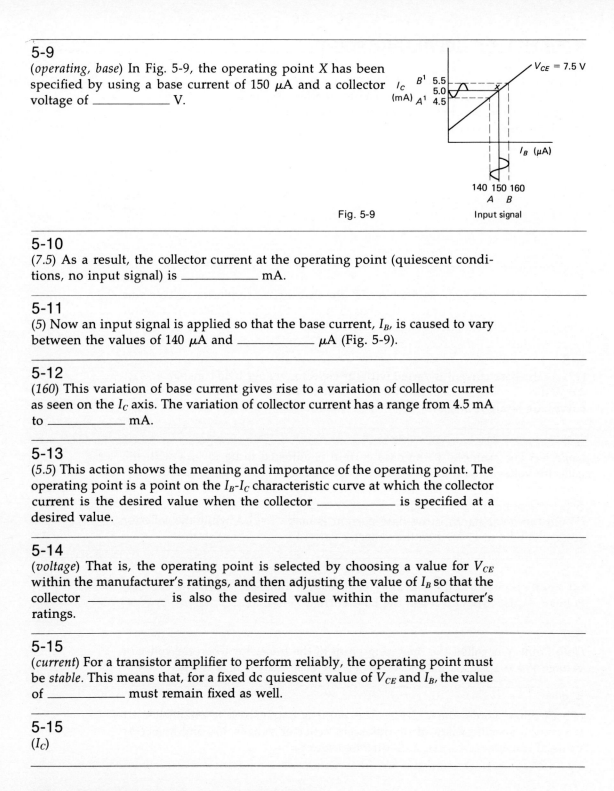

Fig. 5-9

5-10
(*7.5*) As a result, the collector current at the operating point (quiescent conditions, no input signal) is _____ mA.

5-11
(*5*) Now an input signal is applied so that the base current, I_B, is caused to vary between the values of 140 μA and _____ μA (Fig. 5-9).

5-12
(*160*) This variation of base current gives rise to a variation of collector current as seen on the I_C axis. The variation of collector current has a range from 4.5 mA to _____ mA.

5-13
(*5.5*) This action shows the meaning and importance of the operating point. The operating point is a point on the I_B-I_C characteristic curve at which the collector current is the desired value when the collector _____ is specified at a desired value.

5-14
(*voltage*) That is, the operating point is selected by choosing a value for V_{CE} within the manufacturer's ratings, and then adjusting the value of I_B so that the collector _____ is also the desired value within the manufacturer's ratings.

5-15
(*current*) For a transistor amplifier to perform reliably, the operating point must be *stable*. This means that, for a fixed dc quiescent value of V_{CE} and I_B, the value of _____ must remain fixed as well.

5-15
(I_C)

CAUSES OF BIAS INSTABILITY

5-16
Figure 5-16 is a graph that indicates the variation of collector _____ with temperature for a germanium transistor. (Silicon units also show temperature variations, but to a lesser extent.)

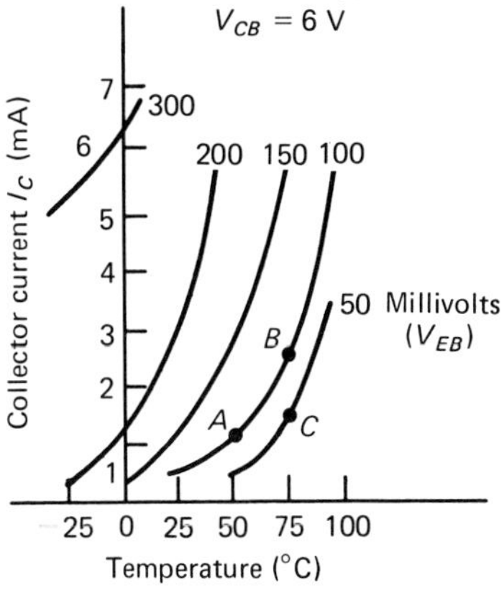

Fig. 5-16

5-17
(*current*) The figure shows a family of curves. In this family, each curve has been plotted with a fixed collector-base voltage V_{CB} and a fixed _____-_____ voltage V_{EB}.

5-18
(*emitter-base*) As we study the curves in Fig. 5-16, it is apparent that there is a definite variation of I_C with temperature. For example, the collector current is _____ mA when the base-emitter voltage is 100 mV and the temperature is 50°C (point A).

5-19
(*1*) If the base-emitter voltage is kept fixed, and the temperature increases to 75°C, then the collector current increases to about _____ mA.

5-20
(*2.6*) This is very poor bias stability. The method generally used to correct this instability is to reduce the bias. For example, at point B (Fig. 5-16), the base-emitter bias voltage is _____ mV.

5-21
(*100*) If the bias voltage is reduced to 50 mV, then the collector current will be reduced to about _____ mA (point C).

5-22
(*1.2*) A collector current of 1.2 mA represents only a small change from the original collector current of 1.0 mA. Thus, reducing the bias voltage compensates for changes caused by the increase of _____.

TRANSISTOR BIAS STABILITY **119**

5-23
(*temperature*) A second important cause of bias instability is variation among transistors. Figure 5-23 shows static curves for collector current as a function of _____ current, with the collector-emitter voltage fixed at 10 V.

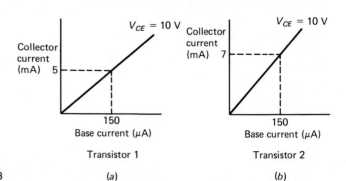

Fig. 5-23   (a)   (b)

5-24
(*base*) Figure 5-23a is a static curve for a type 2N3903 transistor. When the base current is 150 μA, the collector current is _____ mA.

5-25
(5) Therefore, the collector current amplifies the base current by a factor of

$$\frac{5 \text{ mA}}{150 \text{ μA}} = \frac{5 \times 10^{-3}}{150 \times 10^{-6}} = 0.033 \times 10^3 = \underline{\qquad}$$

5-26
(33) The dc collector current is 33 times the dc base current. Figure 5.23b is a static curve for another sample of a type 2N3903 transistor. For this sample, a base current of 150 μA produces a collector current of _____ mA.

5-27
(7) The collector direct current of this transistor is then greater than the base current by a factor of

$$\frac{7 \text{ mA}}{150 \text{ μA}} = \frac{7 \times 10^{-3}}{150 \times 10^{-6}} = \underline{\qquad}$$

5-28
(47) The 2 samples have different dc gains: 47 versus _____. In fact, if 10 samples of a given transistor type were examined, a large range of dc gains may be found.

5-29
(33) Let us suppose a transistor radio is built using transistor 1 of Fig. 5-23. Assume the radio designer sets the base bias of this transistor at 150 μA, so that the collector dc will be _____ mA.

5-30
(5) If this particular transistor requires replacement because it has burnt out, or for some other reason, it would be preferable to insert another type 2N _____.

5-31
(*3903*) But note what happens if, for example, transistor 2 is inserted to replace transistor 1. With a base drive of 150 µA, transistor 2 will give an operating collector current of _____ mA.

5-32
(*7*) The operating current has now increased from 5 mA to _____ mA. This represents poor operating point stability.

5-33
(*7*) Summarizing: The 2 most important factors which affect the operating point of transistors are _____ variations and variations among transistors.

5-34
(*temperature*) The circuit engineer *must* take these factors into account when designing a radio, TV, etc. Fortunately, several techniques which compensate for temperature changes also minimize effects of _____ among transistors.

5-35
(*variations*) In the following sections we shall study several methods of stabilizing the _____ point.

5-35
(*operating*)

RESISTOR BIAS STABILIZATION
5-36
Figure 5-36*a* depicts a single-stage transistor circuit. The collector supply is denoted by V_{CC} and the base bias voltage is denoted by _____.

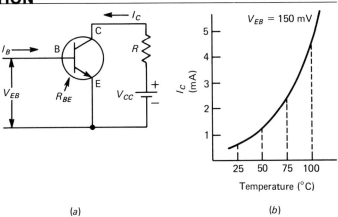

Fig. 5-36   (a)   (b)

TRANSISTOR BIAS STABILITY

5-37

(V_{EB}) As we have already seen, changes in temperature can produce changes in the collector current. Figure 5-36b depicts the dependence of collector current on _____.

5-38

(*temperature*) Note that the base-emitter voltage V_{EB} is held constant at _____ mV.

5-39

(*150*) Recall that the collector current is proportional to the base current (Fig. 5-1). Since the collector current is increasing with temperature, the base current must be _____ with increasing temperature.

5-40

(*increasing*) If we apply Ohm's law to the base circuit of Fig. 5-36a, we find $V_{EB} = I_B R_{BE}$, where I_B is the base current and R_{BE} is the dc resistance looking into the _____ - _____ circuit.

5-41

(*base-emitter*) We can solve this equation for R_{BE} to obtain $R_{BE} = V_{EB}/$_____.

5-42

(I_B) Since V_{EB} is held constant and I_B is increasing with temperature, R_{BE} must be _____ with increasing temperature.

5-43

(*decreasing*) Summarizing: As the temperature of a transistor goes up, the base-emitter resistance goes _____.

5-44

(*down*) Summarizing: The decreased base-emitter resistance causes more current to flow in the _____ circuit.

5-45

(*base*) Summarizing: The increasing base current causes an increase in _____ current.

5-46

(*collector*) Summarizing: The increase in collector current causes a shift in the dc _____ point.

5-47

(*operating*) One way to reduce the effect of the variable base-emitter resistance is to place a large-valued resistor in the emitter lead. To see how this works, we use an analogy (Fig. 5-47). In this simple dc circuit, the maximum current flows when R1 is reduced to zero. Then, $I_{max} = 10/1$ = _____ A.

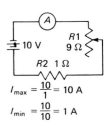

$I_{max} = \frac{10}{1} = 10$ A

$I_{min} = \frac{10}{10} = 1$ A

Fig. 5-47   Range = 10 − 1 = 9 A

5-48

(*10*) The minimum current flows when R1 is made 9 Ω. Then $I_{min} = E/(R_1 + R_2)$ = 10/10 = _____ A.

5-49

(*1*) Thus, the range of variation caused by changing the circuit resistance by 9 Ω is altogether _____ A.

5-50

(*9*) This is a large current variation. Now, however, we change the value of R2 to 11 Ω as in Fig. 5-50. The maximum current that can now flow when R1 is reduced to zero is _____ A.

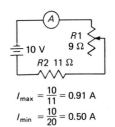

$I_{max} = \frac{10}{11} = 0.91$ A

$I_{min} = \frac{10}{20} = 0.50$ A

Fig. 5-50   Range = 0.91 − 0.50 = 0.41 A

5-51

(*0.91*) When R1 is made 9 Ω, the current drops to _____ A. Thus, I_{min} = _____ A.

5-52

(*0.50, 0.50*) The total current variation due to a resistance variation of 9 Ω now comes to _____ A.

5-53

(*0.41*) Thus, when the external resistance R2 is small, the current variation caused by changing the internal resistance of 9 Ω covers a large range; when the external resistance is increased, however, the same internal resistance variation now permits the range of current variation to become much _____.

TRANSISTOR BIAS STABILITY

5-54

(*smaller*) This occurs because the addition of a large external resistance causes the internal resistance to become a smaller *percentage* of the total resistance. This means that the current variation produced by varying the internal resistance becomes a smaller _____ of the total current.

5-55

(*percentage*) Placing a large-valued resistor in the emitter lead has a similar effect. This resistor causes the variation of the emitter-base junction resistance to be a _____ percentage of the *total* resistance in the circuit.

5-56

(*small*) The external resistor, shown in Fig. 5-56, is called a "swamping" resistor because it swamps or overcomes the junction resistance. Obviously, the swamping resistance must be large enough in value to overcome the junction resistance but it must not be so _____ as to introduce undesirable electrical effects in the circuit.

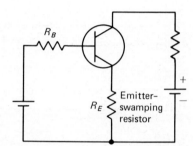

Fig. 5-56

5-57

(*large*) The group of curves in Fig. 5-57 shows how _____ current varies with changes of temperature for 3 different pairs of values of R_E and R_B.

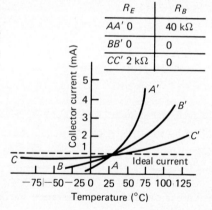

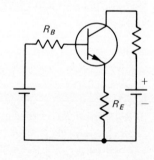

Fig. 5-57

5-58

(*collector*) The dashed line labeled *ideal current* shows that the variation of collector current for an ideal transistor circuit would be zero regardless of the changes in _____ from one extreme to the other.

5-59
(temperature) The curve that most nearly approaches the ideal curve is the one labeled _____.

5-60
(CC′) The legend in Fig. 5-57 states that curve CC′ was obtained by using a base resistor of 0 Ω and an emitter- _____ resistor of 2 kΩ.

5-61
(swamping) The worst stability of all is shown in curve _____, obtained by using no resistance in the emitter circuit and 40 kΩ in the base circuit.

5-62
(AA′) These curves substantiate the conclusion that the best stability is obtained with a large emitter resistor. Also, they illustrate a new point, that the base resistance R_B should be as _____ as possible.

5-63
(small) In any configuration, the best current stability is realized when an emitter-swamping resistor is included in the circuit and the base resistance is made as _____ as possible.

5-64
(small) Poor current stability results when the _____-_____ resistor is omitted, even though R_B may be made very small.

5-65
(emitter-swamping) It was mentioned in a previous section that there are 2 factors affecting the operating point: temperature changes and transistor variability. We will show now that the emitter-_____ resistor minimizes changes due to transistor variability as well as temperature changes.

5-66
(swamping) Figure 5-66 depicts a transistor circuit with bias battery V_{BB} and supply battery _____.

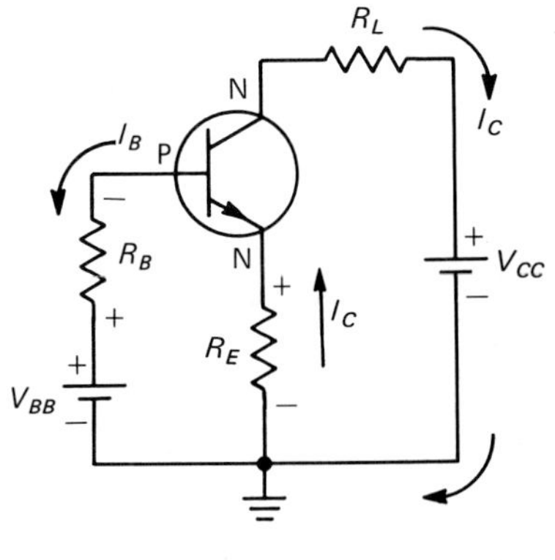

Fig. 5-66

TRANSISTOR BIAS STABILITY **125**

5-67

(V_{CC}) Examine first the collector circuit. The collector current flows out of the negative plate of the battery, up through resistor _____, into the emitter of the transistor, out of the collector, through resistor R_L, and back to the positive plate of the battery.

5-68

(R_E) Since the current in R_E is flowing up, the _____ of the resistor is positive with respect to the _____ of the resistor.

5-69

(*top, bottom*) Examine next the base circuit. The net voltage as seen by the base-emitter junction of the transistor consists of the bias battery V_{BB}, the ohmic drop across _____, and the ohmic drop across R_E.

5-70

(R_B) The current flowing through R_B is the base current, whereas the current through R_E is the _____ current.

5-71

(*emitter*) Since the emitter current is much larger than the base current, the ohmic drop across _____ will be much larger than the ohmic drop across _____.

5-72

(R_E, R_B) V_{BB} is providing forward bias for the base circuit of the NPN transistor in Fig. 5-66 since the positive plate of V_{BB} is connected to the "____" of the PN junction.

5-73

(*P*) However, the ohmic drop across R_E tends to *reverse*-bias the transistor since the positive end of R_E is connected to the "____" of the PN junction.

5-74

(*N*) The net bias of the base, then, consists of the forward bias of V_{BB} and the _____ bias of the ohmic drop of R_E.

5-75

(*reverse*) If the transistor of Fig. 5-66 is replaced by another transistor whose current amplification is greater, then _____ (more, less) collector current will *tend* to flow.

5-76
(*more*) But, the increased collector current will cause a greater ohmic drop across the emitter resistor _____ .

5-77
(R_E) This means a greater reverse bias caused by R_E. Since V_{BB} is a fixed voltage, the *net* forward bias will be _____ (increased, decreased).

5-78
(*decreased*) The reduced forward bias tends to cancel the increased collector current. Thus, the emitter-swamping resistor helps to _____ the collector current when different transistors are used in a given circuit.

5-79
(*stabilize* or *hold constant*) Figure 5-79 illustrates another method of using a resistor to stabilize the operating point. Note first that R_F in this circuit is connected directly from the _____ of the transistor to the base of this PNP transistor.

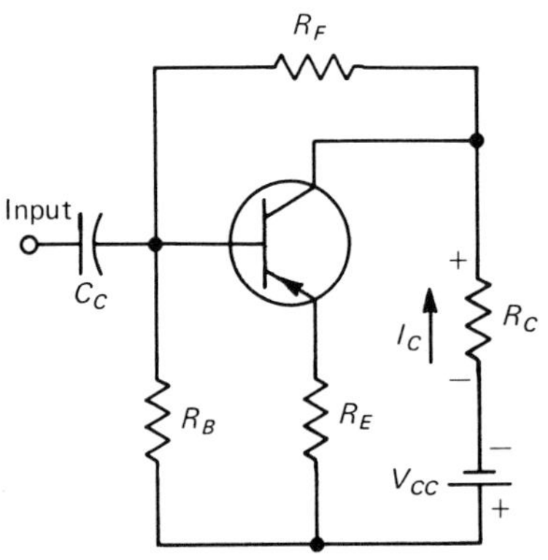

Fig. 5-79

5-80
(*collector*) To see how this circuit improves current stability, consider what happens if the collector current should increase because of a temperature or transistor change. The collector current flows through resistor _____ from the battery.

5-81
(R_C) When the collector current is small, the voltage drop across R_C is small. When the collector current increases, the voltage drop across R_C _____ .

5-82
(*increases*) When the voltage drop across R_C increases, the collector must become less negative than it was previously. Thus, the base, which is connected to the collector via R_F, must also become less _____ .

TRANSISTOR BIAS STABILITY

5-83
(*negative*) But the base-emitter forward bias in a PNP transistor calls for a negative base with respect to the emitter. Hence, when the base becomes less negative, the forward _____ is reduced.

5-84
(*bias*) A reduction in forward bias automatically tends to reduce the collector current. Thus, when the collector current increases, it *feeds back* a voltage to the base which tends to cancel this _____ of current.

5-85
(*increase*) A comparison of the emitter-swamping resistor with the resistor R_F will show a similarity. In both cases, the output current (or voltage) is made to react back on the _____ current (or voltage).

5-86
(*input*) This feature of the output affecting the input is an extremely important topic and is given the name *feedback*. In the case of the emitter-swamping resistor and R_F, an increase in the output causes a(n) _____ in the input.

5-87
(*decrease*) This kind of feedback is called *negative feedback*. Conversely, when an increase in the output causes an increase in the input, the term used is _____ feedback.

5-88
(*positive*) In the case of the emitter-swamping resistor, we say it stabilizes the bias point by _____ feedback. In subsequent chapters of this book, we shall see several more examples of positive and negative feedback.

5-88
(*negative*)

GENERAL BIAS CIRCUIT
5-89
The circuit of Fig. 5-89 is called the *general bias circuit*. By properly selecting input and output points and resistor sizes, this circuit may be transformed into any one of the 3 configurations: common-base, common-emitter, or _____-_____.

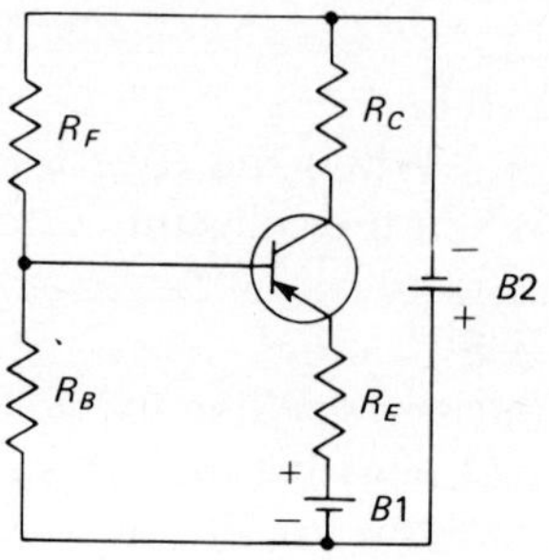

Fig. 5-89

128 A PROGRAMMED COURSE IN BASIC ELECTRONICS

5-90
(*common-collector*) Let us start with common-base. To transform the general bias circuit to CB, we first must recognize that the base is common to output and input circuits (base shown grounded) and that the input is applied between the _____ and ground (Fig. 5-90).

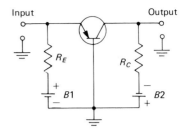

Fig. 5-90

5-91
(*emitter*) The output in this circuit (Fig. 5-90) is taken from the _____ electrode and ground. Note that the *common* element is always grounded; hence ground is always common to both input and output.

5-92
(*collector*) In the CB circuit, resistor R_F is omitted entirely; hence the resistance of R_F is taken to be "infinite." That is, by calling R_F infinite, we are saying that this resistor has been _____-circuited (Fig. 5-92).

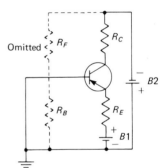

Fig. 5-92

5-93
(*open*) Note also the short-circuiting bar across R_B in Fig. 5-92. Thus, for the CB configuration, the resistance of R_B is short-circuited; hence R_B has _____ resistance.

5-94
(*zero*) Figure 5-90 shows how Fig. 5-92 is normally drawn. In this method of presentation, it is easy to see that the input signal is applied between emitter and ground, and that the output is taken from across the _____ and ground.

5-95
(*collector*) The general bias circuit may be converted to the CE configuration by again omitting resistor _____ as a first step (Fig. 5-95).

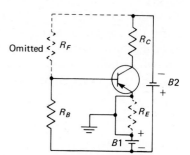

Fig. 5-95

5-96
(R_F) Thus, R_F is infinite in value. The second change is that resistor _____ has been short-circuited out so that the top of B1 goes right to the emitter.

5-97
(R_E) The grounded electrode in the CE circuit is the _____.

5-98
(*emitter*) When the circuit of Fig. 5-95 is redrawn in its standard form, it has the appearance of Fig. 5-98. In this mode of presentation, it is easier to see that the common element to input and output circuits is the _____.

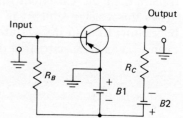

Fig. 5-98

5-99
(*emitter*) The drawing of Fig. 5-98 also shows clearly that the input is applied between the base and ground, while the output is taken from across the _____ and ground.

5-100
(*collector*) The general bias circuit may be converted to the CC configuration by again omitting resistor R_F and by short-circuiting resistor _____ as shown in Fig. 5-100.

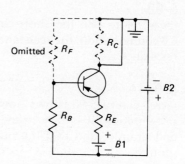

Fig. 5-100

130 A PROGRAMMED COURSE IN BASIC ELECTRONICS

5-101
(R_C) The grounded electrode in the CC circuit is the _____.

5-102
(*collector*) When the CC configuration is redrawn in standard form it appears as in Fig. 5-102. In this diagram, it is easier to see that the element common to both input and output is the _____.

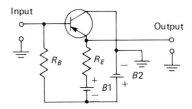

Fig. 5-102

5-103
(*collector*) The drawing of Fig. 5-102 also shows clearly that the input is applied between the _____ and ground.

5-104
(*base*) The same drawing also indicates that the output is taken from between the _____ and ground.

5-105
(*emitter*) In all 3 configurations, the ground terminal is *always* common to both input and output circuits. Turning this about, we may say that the grounded electrode (emitter, base, or collector) is always the electrode that is _____ to both input and output circuits.

5-106
(*common*) It is usually desirable to use a single battery rather than 2 bias batteries. The circuit of Fig. 5-106 shows how this is done. Base-emitter bias is obtained by means of a voltage divider across the battery. The voltage divider consists of R_B and _____.

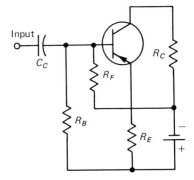

Fig. 5-106

5-107
(R_F) Since the transistor illustrated in Fig. 5-106 is a PNP type, the polarity of the base must be made _____ with respect to the emitter.

TRANSISTOR BIAS STABILITY **131**

5-108
(*negative*) Since the bottom of R_F is connected to the "—" terminal of the collector battery and the top of R_F is connected to the _____ of the transistor, the polarity of the base will be negative with respect to the emitter, which is returned to the positive terminal of the battery through R_E (Fig. 5-106).

5-108
(*base*)

DIODE STABILIZING CIRCUITS
5-109
Bias instability of a silicon transistor due to variation of collector current with temperature is caused by changes in junction _____.

5-110
(*resistance*) Such resistance variations are natural consequences of the structure of a PN junction. PN junctions are present in transistors; also, _____ junctions are present in junction diodes.

5-111
(*PN*) Whether the junction is part of a transistor or a diode, the behavior of the junction is similar. That is, variations of junction resistance follow the same pattern in the junction transistor and in the junction _____.

5-112
(*diode*) Furthermore, the junction diode can be made of the same materials as the transistor. When this is done, we find that the temperature changes of the transistor and diode are also the _____.

5-113
(*same*) This condition permits a more constant collector current over a wide range of _____ because the chances for good tracking are improved.

5-114
(*temperature*) Junction diodes show a decrease in resistance with increasing temperature whether they are forward- or _____-biased.

5-115
(*reverse*) Let us determine whether the diode in Fig. 5-115 is forward- or reverse-biased. The conduction direction of a diode is always indicated by the schematic symbol. Electron _____ always flows from the flat of the symbol toward the point as shown by the arrow.

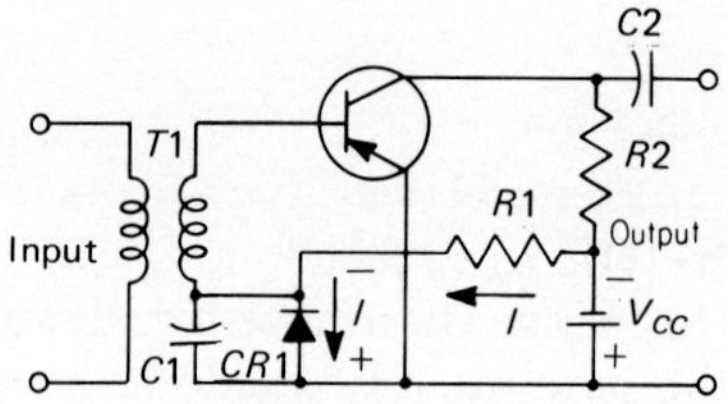

Fig. 5-115

132 A PROGRAMMED COURSE IN BASIC ELECTRONICS

5-116
(*current*) The battery V_{CC} is polarized with respect to the diode so that electrons are fed to the flat; hence the connections are such as to produce conduction. Thus, the diode is _____-biased.

5-117
(*forward*) Compensation is produced by the action of the voltage divider $R1CR1$. When the temperature rises in an uncompensated circuit, the collector current tends to _____.

5-118
(*rise* or *increase*) In the compensated circuit, a rise of temperature also changes the behavior of the diode. Its resistance tends to _____ with a rise of temperature.

5-119
(*decrease*) This encourages the current flowing through the voltage divider to _____.

5-120
(*increase*) As a result of the increased current through $R1$, the voltage drop across $R1$ _____.

5-121
(*increases*) This leaves less voltage across $CR1$. The voltage drop across $CR1$ applies the required _____ bias to the base-emitter circuit.

5-122
(*forward*) But when the drop across $CR1$ falls, there is less forward bias. This reduction in forward bias tends to cause the collector current to _____.

5-123
(*decrease*) Hence, the diode tends to stabilize the bias current, since a rising temperature which would normally cause an increasing collector current is compensated by the _____ forward bias produced by the variation of diode resistance with temperature.

TRANSISTOR BIAS STABILITY 133

5-124
(*decreasing*) The closer the characteristics of the compensating diode and the transistor, the better the stabilization. Figure 5-124 depicts a circuit where the transistor Q1 is being used to compensate the PN junction of transistor _____.

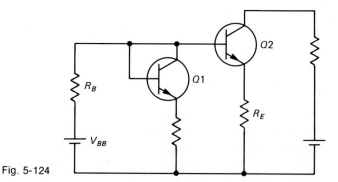

Fig. 5-124

5-125
(Q2) Transistor Q1 is made into a diode by attaching the collector directly to the _____.

5-126
(*base*) To obtain the best stabilization, transistors Q1 and Q2 are made the same type. Then the junction characteristics of the _____ diode are almost identical to those of the transistor being stabilized.

5-127
(*compensating*) This method of stabilization is very common for integrated circuits. Integrated circuit technology makes transistors less expensive than components such as _____, so that using extra transistors is in fact quite feasible.

5-127
(*resistors* or *capacitors*)

THE ZENER DIODE
5-128
Refer to Fig. 5-128. As we have pointed out previously, the symbol of a junction diode contains implicit instructions as to the normal conduction direction. Current flows easily from the flat toward the point of the symbol. Thus, in Fig. 5-128, the normal conduction direction is from A to _____.

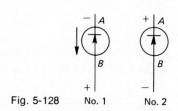

Fig. 5-128   No. 1   No. 2

5-129
(B) The diode in drawing 1 of Fig. 5-128 is connected for normal conduction. Since the top of the diode is negative and bottom positive, current flows easily from the "−" toward the _____ in the diode.

5-130
("+") We can say that the bias on the junction diode in drawing 1, Fig. 5-128, is forward bias. However, the diode in drawing 2, Fig. 5-128, is polarized in the nonconduction direction and thus is _____-biased.

5-131
(*reverse*) Even when a diode is reverse-biased, as in drawing 2, Fig. 5-128, a small current flows because of minority carriers. You will recall that this current is referred to as _____-bias current. It is also commonly called *reverse saturation current*.

5-132
(*reverse*) Now refer to Fig. 5-132. Note the polarity of the diode. The voltage applied to it has the same direction as that shown in drawing 2 of Fig. 5-128. Hence, the diode in Fig. 5-132 is _____-biased.

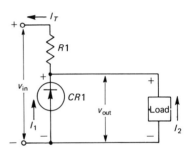

Fig. 5-132

5-133
(*reverse*) Thus, I_1 must represent the reverse-bias current flowing in $CR1$ (Fig. 5-132). In this figure, V_{in} is a voltage source that may be raised continuously from zero upward. The voltage across the diode for any condition of the input voltage is symbolized by _____.

5-134
(V_{out}) The graph in Fig. 5-134 shows how the reverse-bias current through the diode varies with the voltage across it. When $V_{out} = 0$, the reverse-bias current is equal to _____.

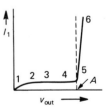

Fig. 5-134

5-135
(*zero*) As V_{out} is slowly increased, the current also _____ as shown from point 1 to point 2 on the curve of Fig. 5-134.

5-136
(*rises*) However, after the initial small current increase (from 1 to 2), the current remains _____ from point 2 to 3 to 4.

5-137
(*constant*) However, at a certain voltage the current very suddenly increases as shown from point 5 to point 6 in Fig. 5-134. The voltage line at which this sudden rise of current occurs is identified by the letter _____ in this figure.

5-138
(*A*) The voltage at which this occurs is called the *breakdown* or *zener* voltage, after the scientist who first described it. Thus, the letter *A* in Fig. 5-134 identifies the breakdown or _____ voltage for the diode under discussion.

5-139
(*zener*) The long, almost horizontal slope of the curve in Fig. 5-134 (1–2–3–4) indicates a condition in which the voltage is rising, but the _____ remains virtually constant.

5-140
(*current*) However, the long vertical slope (5–6) indicates a condition in which the current is increasing rapidly, but the _____ remains almost constant.

5-141
(*voltage*) When the reverse-biased junction diode is used to take advantage of this characteristic, it is called a breakdown or zener diode. The symbol of a zener diode as used here is given in Fig. 5-141. It should be noted that this symbol is different from that of an ordinary diode in that the letter _____ appears inside the circle. (It should be pointed out, however, that different authors use different symbols for a zener diode.)

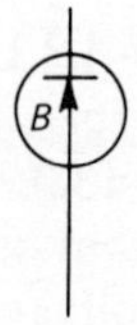

Fig. 5-141

5-142
(*B*) The constant-_____ characteristic of a zener diode may be used to advantage in voltage regulator circuits.

5-143
(*voltage*) Consider first the possibility of a resistance change in the load in Fig. 5-132. If the load resistance should decrease, the current through the load I_2 would tend to _____.

5-144
(*increase*) Because of the practically vertical aspect of the breakdown voltage curve (Fig. 5-134), the voltage output across *CR*1 tends to remain _____.

5-145
(*constant*) Thus, the increase of current through the load is not obtained at the expense of increasing I_T. Rather, the current in CR1 *decreases*, allowing the current through the load to _____ without changing the total current I_T.

5-146
(*increase*) Similarly, if the resistance of the load should rise for some reason, the current in the load would tend to _____ .

5-147
(*decrease*) Since the voltage across CR1 remains constant, the current in CR1 would tend to increase, thus keeping I_T at a _____ value.

5-148
(*constant*) Let us see what happens when the *source voltage* changes. Assume that V_{in} rises. This would cause the current I_T to _____ as well.

5-149
(*increase* or *rise*) When this occurs, the current through R1 must increase, causing the voltage drop across R1 to _____ .

5-150
(*increase*) Since the voltage across CR1 tends to remain constant, the voltage drop across R1 is just large enough to absorb the increase that occurred in the source voltage. Thus, the voltage applied to the load, and consequently the current through the load, both remain _____ .

5-151
(*constant*) This voltage regulator, within the limits prescribed by its design, is capable of maintaining a constant output voltage despite variations that occur in the load current or variations in the source _____ .

5-152
(*voltage*) The magnitude of the breakdown voltage is controlled during manufacture of each specific diode type. Special zener diodes used for voltage regulators may be obtained with _____ voltages ranging from 2 to over 400 V.

5-153
(*breakdown* or *zener*) A normal reverse-biased junction diode *below* the zener voltage exhibits a decrease in resistance with an increase in temperature. Above the zener voltage, the reaction of the diode takes on the opposite aspect. That is, above the zener voltage the resistance of the diode _____ with an increase in temperature.

5-154
(*increases*) Unless compensation is effected, this resistance-temperature dependence would militate against good voltage regulation. Thus, an uncompensated breakdown diode will serve as an effective voltage regulator only if the temperature remains _____.

5-155
(*constant*) One method of compensation is shown in Fig. 5-155. *CR1* and *CR2* are junction diodes connected in _____ to the breakdown diode *CR3*.

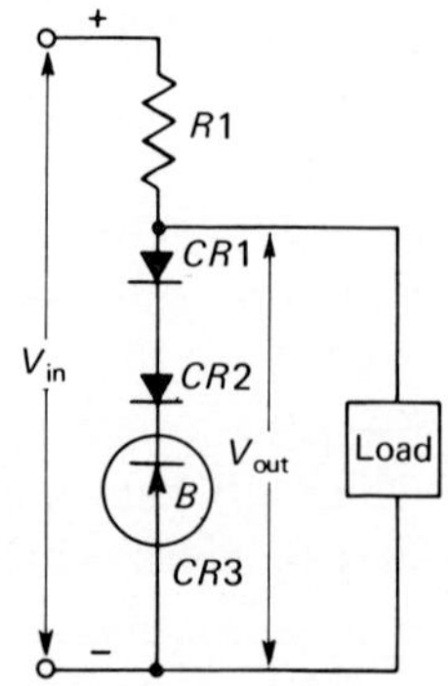

Fig. 5-155

5-156
(*series*) The zener diode *CR3* is reverse-biased. Note, however, that *CR1* and *CR2* are reversed from *CR3* in polarity; hence these 2 diodes are _____-biased.

5-157
(*forward*) With increasing temperature, each of the normal diodes (*CR1* and *CR2*) experiences a resistance decrease which, in magnitude, is equal to about half the resistance change of the zener diode. Thus, taken together, the negative resistance change of the forward-biased diodes is about _____ in magnitude to the positive resistance change of the zener diode.

5-158
(*equal*) In this way, the total resistance of the 3 diodes in series remains constant over a wide range of _____.

5-159
(*temperature*) The complete result of the circuit of Fig. 5-155 is a constant voltage output despite the fact that temperature, input voltage V_{in}, and load _____ may vary.

5-160
(*current*) Breakdown diodes may be used to stabilize collector voltage in a transistor amplifier (Fig. 5-160). To start the analysis, consider the current I_2. This current is provided by the source identified as _____.

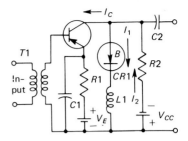

Fig. 5-160

5-161
(V_{CC}) One part of I_2 flows downward through the breakdown diode and is identified as _____.

5-162
(I_1) The remaining portion of I_2 is the current that flows in the _____ circuit of the transistor and is identified as I_C.

5-163
(*collector*) Assume that the temperature of the transistor rises. This would cause the collector current of the transistor I_C to _____.

5-164
(*increase*) With diode CR1 operating in the breakdown region, an *increase* of I_C would cause I_1 to *decrease* for reasons previously given. Thus, the total current I_2 would not _____.

5-165
(*change* or *vary*) If I_2 is constant, then the voltage drop across R_2 must also be _____.

5-166
(*constant*) The voltage supplied to the collector of the transistor is the battery voltage V_{CC} less the voltage drop across $R2$. Thus, the voltage supplied to the collector must also remain _____.

5-167
(*constant*) The reasoning presented for the action of *CR1* in the foregoing discussion is valid only if the temperature of *CR1* is constant. To overcome the resistance variation of *CR1* due to possible temperature variation, this diode may be stabilized by connecting forward-biased diodes in _____ with it, as in Fig. 5-155.

TRANSISTOR BIAS STABILITY

5-168
(*series*) The ac resistance of a breakdown diode may be quite low. This would tend to short-circuit the ac load circuit. To prevent this from occurring, a large reactance for the ac is connected in series with the zener diode. This reactance is identified as _____ in Fig. 5-160.

5-169
($L1$) The functions of other components in the circuit of Fig. 5-160 are as follows: input coupling is handled by the component identified as _____.

5-170
($T1$) The emitter-swamping resistor is the component identified as _____.

5-171
($R1$) Signal degeneration due to the emitter-swamping resistor is prevented by _____.

5-172
($C1$) Emitter-base bias voltage and current is provided by _____.

5-173
(V_E) The signal is coupled to the next stage without dc interaction through the component identified as _____.

5-173
($C2$)

SHUNT LIMITING ACTION OF A DIODE
5-174
Certain situations exist in transformer-coupled amplifiers which may endanger the transistor. The circuit shown in Fig. 5-174 uses _____ coupling in its input and output circuits.

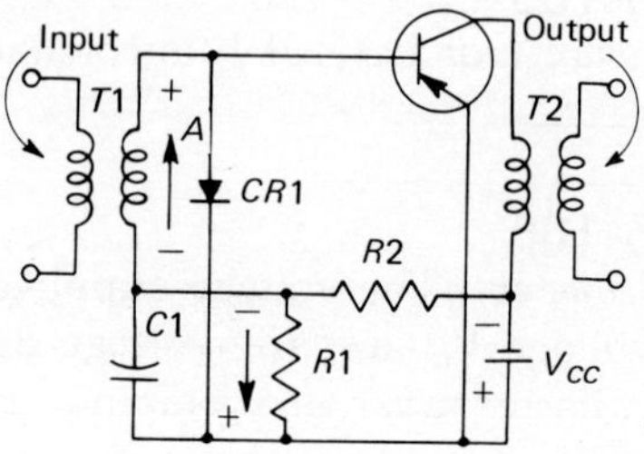

Fig. 5-174

5-175
(*transformer*) In the typical CE amplifier, as illustrated in Fig. 5-174, the collector circuit is reverse-biased and the base-emitter circuit is _____-biased.

5-176
(*forward*) The necessary negative voltage for the base of the PNP common-emitter amplifier in Fig. 5-174 is obtained from the battery V_{CC} through the voltage divider consisting of $R1$ and _____.

5-177
(*R2*) While the forward bias in the base-emitter circuit is normal, the collector current is normal. The load for the collector circuit is the _____ winding of transformer $T2$.

5-178
(*primary*) If the signal from the previous stage is suddenly terminated, it is possible for a surge of relatively high voltage to appear across the secondary of $T1$ having the polarity indicated by the arrow A. This polarity is such as to make the base positive with respect to the emitter. Hence, this surge applies _____ bias to the base.

5-179
(*reverse*) If $CR1$ is *not* present, the reverse bias thus applied may cut off the collector current suddenly. If this occurs, the magnetic field in the core of $T2$ will suddenly _____.

5-180
(*collapse*) As the field collapses, it may induce a very high _____ across the terminals of the primary winding of the transformer.

5-181
(*voltage*) If this high voltage is applied to the collector, there is danger of serious damage to the transistor, especially since the base-emitter circuit is reverse-biased at this instant and the collector is cut off so that collector _____ cannot flow.

5-182
(*current*) This situation is avoided by including $CR1$ in the circuit. Normally, the base is negative with respect to the emitter, making the top of $CR1$ negative with respect to the bottom. According to the symbol convention, this makes $CR1$ _____-biased.

5-183
(*reverse*) With $CR1$ reverse-biased it cannot conduct. When a circuit element is nonconducting, it may be considered to have a very high resistance, or, to behave like an open circuit. Thus, during normal operation $CR1$ behaves like an _____ circuit.

5-184
(*open*) The applicable portion of the circuit has been redrawn more simply in Fig. 5-184. The voltage normally applied as forward bias to the base is shown by arrow _____.

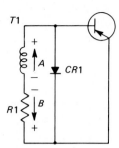

Fig. 5-184

5-185
(*B*) However, when a surge of opposite polarity appears across the secondary of T1 (arrow A), it tends to cancel the voltage across R1. If the surge voltage is large enough to overcome the drop across R1, the base-emitter bias would *tend* to reverse and would *tend* to cut off the _____ current.

5-186
(*collector*) However, the instant the surge voltage becomes larger than the drop across R1, the diode CR1 becomes *forward-biased* and therefore starts to _____.

5-187
(*conduct*) The moment CR1 begins to conduct, only a very small and insignificant voltage can appear across it. Thus, the base is prevented from going positive and is prevented from _____ off the collector current.

5-188
(*cutting*) Thus, CR1 is a protective device that prevents excessive voltage from appearing at the collector. In this application, the diode is said to be a _____ (series, shunt) limiter.

5-188
(*shunt*)

CRITERION CHECK TEST

The following 2 questions refer to Fig. 5-1.

___ 5-1 If the base current is 150 μA and $V_{CE} = 7.5$ V, then the operating point is denoted by the letter (*a*) B, (*b*) D, (*c*) x, (*d*) y.

___ 5-2 If $V_{CE} = 7.5$ V and $I_B = 300$ μA, the collector current is (*a*) 4 mA, (*b*) 5 mA, (*c*) 7.5 mA, (*d*) 10 mA.

___ 5-3 In Fig. 5-16, if the base-emitter voltage is 200 mV, and the temperature goes from 0°C to 25°C, the change in collector current is (*a*) 1 mA, (*b*) 2.5 mA, (*c*) 3.5 mA, (*d*) 5 mA.

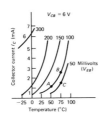 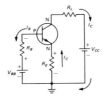

Fig. 5-1   Fig. 5-16   Fig. 5-66   Fig. 5-132

_____5-4 Which of the statements is true? (*a*) The type 2N3903 is a FET; (*b*) all type 2N3903 transistors have the same current gain; (*c*) the type 2N3903 is a point-contact transistor; (*d*) the type 2N3903 is a junction transistor.

_____5-5 As the temperature rises, the base-emitter resistance (*a*) goes up for a PNP transistor, (*b*) goes up for an NPN transistor, (*c*) stays the same, (*d*) goes down.

_____5-6 The purpose of the emitter-swamping resistor is to (*a*) limit the maximum emitter current, (*b*) limit the changes in emitter current, (*c*) provide collector-base bias, (*d*) provide base-emitter bias.

Refer to Fig. 5-66 for the next 3 questions.

_____5-7 The base bias battery is denoted by (*a*) V_{BB}, (*b*) V_{CC}, (*c*) V_{RR}, (*d*) V_{EE}.

_____5-8 The forward bias of V_{BB} is opposed by (*a*) V_{CC}, (*b*) V_{CE}, (*c*) the ohmic drop across R_E, (*d*) V_{R_L}.

_____5-9 If a transistor with less current gain were put in the circuit (all other components remaining the same), then the (*a*) forward bias voltage of R_E would tend to increase, (*b*) forward bias voltage of R_E would tend to decrease, (*c*) reverse bias voltage of R_E would tend to increase, (*d*) reverse bias voltage of R_E would tend to decrease.

_____5-10 When the output signal tends to reduce the input signal, we use the term (*a*) negative oscillation, (*b*) recombination, (*c*) negative gain, (*d*) negative feedback.

_____5-11 For the common-collector configuration, (*a*) the input is the base, the output the collector, (*b*) the input is the collector, the output the emitter, (*c*) the input is the base, the output the emitter, (*d*) the input is the emitter, the output the base.

_____5-12 The resistance changes of a junction transistor can be matched closely by a (*a*) MOSFET, (*b*) point-contact diode, (*c*) junction diode, (*d*) pure piece of germanium.

_____5-13 In IC technology, the most common element used to temperature-compensate a transistor is (*a*) a resistor, (*b*) a diode, (*c*) an identical transistor, (*c*) a thermistor.

_____5-14 In Fig. 5-132, the load is connected (*a*) in series with R1, (*b*) in parallel with v_{in}, (*c*) in parallel with the zener diode, (*d*) in series with the zener diode.

_____5-15 In Fig. 5-132, the sum of the zener current and the load current is equal to (*a*) I_1, (*b*) I_2, (*c*) I_T, (*d*) V_{out}.

_____5-16 In Fig. 5-132, the zener diode is denoted by (*a*) CR1, (*b*) R1, (*c*) v_{out}, (*d*) Load.

_____5-17 In Fig. 5-134, current flow in region 1–2–3–4 is due to (*a*) majority current flow,

Fig. 5-134

(b) minority current flow, (c) breakdown current flow, (d) forward bias current flow.

_____5-18 In Fig. 5-134, point A represents (a) the leakage current region, (b) the saturation current region, (c) zener breakdown region, (d) forward bias region.

_____5-19 In Fig. 5-132, if the resistance of the load increases, then the (a) voltage of CR1 changes greatly, (b) current in the load increases, (c) current in CR1 goes up, (d) current in R1 changes greatly.

_____5-20 In Fig. 5-132, if v_{in} decreases, then the (a) voltage of R1 increases, (b) current in the load increases, (c) current in CR1 decreases, (d) voltage of CR1 decreases.

The following 4 questions refer to Fig. 5-160.

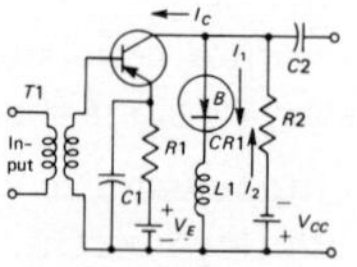

Fig. 5-160

_____5-21 The function of CR1 is to (a) provide BE bias, (b) maintain a constant collector voltage, (c) limit the amount of ac collector current, (d) couple the input to output.

_____5-22 The function of L1 is to (a) block dc from CR1, (b) block ac from CR1, (c) limit dc in CR1, (d) couple the collector signal to the next stage.

_____5-23 The function of C1 is to (a) block dc from R1, (b) couple the collector to the next stage, (c) bypass R1 to prevent signal degeneration, (d) tune L1.

_____5-24 The function of R2 is to act as (a) the collector load resistance, (b) a base bias resistor, (c) a frequency compensator, (d) the emitter-swamping resistor.

6 using transistor characteristic curves and charts

OBJECTIVES

(1) State the equation of and draw the dc load line for a transistor circuit. **(2)** Define the *cutoff* condition. **(3)** Locate the operating point of a transistor using the load line and characteristic curves. **(4)** Locate the operating point of a common-emitter amplifier by sketching the collector characteristics of a transistor, drawing the load line by using two appropriate intercepts, and combining the load line and characteristic curve into one graph. **(5)** Define and compute current, voltage, and power gains from given input and output voltage and current waveforms. **(6)** Define and contrast a *static* and a *dynamic transfer curve*. **(7)** Construct a dynamic transfer curve for a transistor using given operating points. **(8)** Sketch, on the graph of dynamic transfer, characteristic input and output waveforms demonstrating distortionless amplification, a clipped output waveform, and a nonlinear output waveform. **(9)** Sketch the output waveforms for class A, class AB, class B, and class C operation. **(10)** Diagram and describe the operation of a transistor push-pull amplifier. **(11)** Describe the use of the constant-power-dissipation curve in locating the load line. **(12)** Indicate, with reference to the constant-power-dissipation line, the acceptable and unacceptable regions of operation, and the positioning of the load line to obtain maximum safe power output. **(13)** Indicate the factors which influence the collector-base capacitance. **(14)** Describe the effects of the collector-base, emitter-base, and collector-emitter capacitances on high-frequency performance, including the feedback effect of the collector-base capacitance. **(15)** For each of the three transistor configurations, draw a graph showing input resistance versus load resistance, indicating which gives the highest input resistance, and a graph showing output resistance versus source resistance, indicating which gives the lowest output resistance. **(16)** Draw graphs showing voltage, current, and power gains versus the load resistance for the common-base, common-emitter, and common-collector configurations. **(17)** Draw a simple emitter-follower circuit and list several advantages of the emitter follower over the transformer as an impedance-matching device.

THE LOAD LINE

6-1
In order to determine the operating point of a transistor or vacuum tube and graphically compute gains, we must first learn how to draw a *load line*. Referring to Fig. 6-1, we start by noting that a transistor and its load resistor may be considered to be connected in _____ with each other.

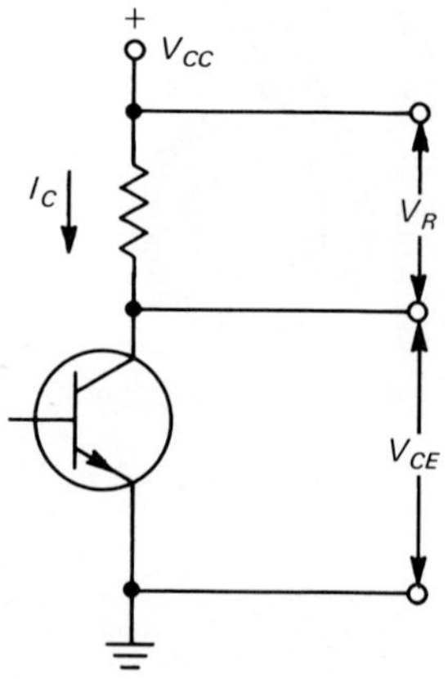

Fig. 6-1

6-2
(*series*) The total voltage applied across this series circuit is that of the supply battery V_{CC}. The individual voltage drops are shown as V_{CE} (the collector-emitter voltage) and V_R (the drop across R_L). From Kirchhoff's law, the total voltage V_{CC} is equal to the _____ of V_R and V_{CE}.

6-3
(*sum*) In equation form, this relationship may be written as $V_{CC} = V_{CE} + $ _____.

6-4
(V_R) The collector current of the transistor is symbolized as I_C. Since this current flows through R_L, the voltage drop across the load resistor may be given as $I_C R_L$. This product may be substituted in place of _____ in the above equation.

6-5
(V_R) If this substitution is used, the equation may be written $V_{CC} = $ _____.

6-6
($I_C R_L + V_{CE}$) In any transistor circuit, we can consider both the V_{CC} value and the load resistor R_L as known factors. This leaves 2 unknown factors: I_C and _____.

6-7
(V_{CE}) Let us "play" with this equation a bit. Suppose we imagine that the base is given a negative bias (note that the transistor in Fig. 6-1 is an NPN type), so as to cut off collector current flow. For this condition, I_C will have a value of _____.

6-8
(zero) Now apply this to the equation $V_{CC} = I_C R_L + V_{CE}$. When the transistor is cut off and the collector current is zero, the term $I_C R_L$ drops out completely, since any value of R_L multiplied by zero is zero. Hence, the equation reduces to $V_{CC} = $ _____.

6-9
(V_{CE}) Stated in words, when a transistor is cut off and its collector current is zero, the voltage on the collector of the transistor is equal to the _____ voltage (Fig. 6-9).

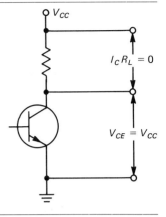

Fig. 6-9

6-10
(supply or V_{CC}) Let us try another condition. Imagine that the base of the transistor has been forward-biased to the point where the *entire voltage* of the source V_{CC} appears as a drop across R_L. For this condition, we could write that $I_C R_L$ equals the value of the _____ voltage.

6-11
(supply or V_{CC}) Now let us find the value of V_{CE} for the condition just described, that is, where $I_C R_L = V_{CC}$. To do this, we will first solve the equation $V_{CC} = I_C R_L + V_{CE}$ for V_{CE} as the unknown. This becomes $V_{CE} = V_{CC} - I_C R_L$. Now, since $I_C R_L$ and V_{CC} are the same quantities, we can replace the first with the second in the equation, and we see at once that the value of V_{CE} becomes _____.

6-12
(zero) This can be seen clearly when the physical facts are considered. If I_C is made to grow to the point where the entire V_{CC} voltage appears across R_L, there can be no voltage left for the collector. Thus, V_{CE} must be _____ when this condition is brought about (Fig. 6-12).

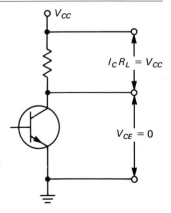

Fig. 6-12

6-13
(*zero*) In summary, we must remember that (1) when the transistor is cut off and $I_C = 0$, then _____ is equal to the supply voltage.

6-14
(V_{CE}) And (2) when the voltage drop across the load resistor is equal to the source voltage, then $V_{CE} =$ _____.

6-15
(*zero*) We are now ready to compute the points needed to draw the load line for the transistor. Assume that the supply voltage is 16 V and $R_L = 4$ kΩ. Start with the assumption that the transistor is cut off ($I_C = 0$). For this condition, $V_{CE} = V_{CC} = 16$ V when $I_C = 0$. This gives us point _____ on the diagonal line in Fig. 6-15.

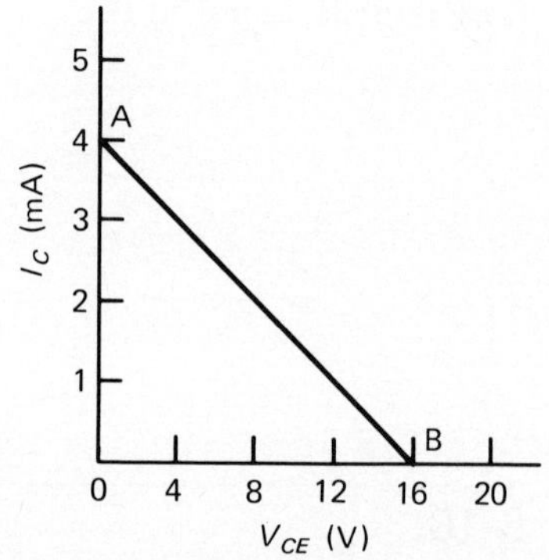

Fig. 6-15

6-16
(*B*) Next, assume that the entire V_{CC} voltage appears as a drop across R_L. For this condition, $V_{CE} = 0$, and we write $I_C R_L = V_{CC}$. To find the collector current, we solve for I_C and obtain $I_C = V_{CC}/$_____.

6-17
(R_L) Substituting 16 V for V_{CC} and 4,000 Ω for R_L, we find that the collector current $I_C =$ _____ mA.

6-18
(*4*) This gives us point _____ on the diagonal line in Fig. 6-15, since for this point $V_{CE} = 0$ and $I_C = 4$ mA.

6-19
(*A*) The line *AB* in Fig. 6-15 is called the *load line*. Since this line was drawn on the basis of a load resistance of 4,000 Ω and a source voltage V_{CC} of 16 V, this line is correct only when $R_L =$ _____ Ω and $V_{CC} = 16$ V.

6-20
(*4,000*) Whatever type of transistor is used in the circuit of Fig. 6-1, its operating point must lie somewhere on the _____ _____.

148 A PROGRAMMED COURSE IN BASIC ELECTRONICS

6-21
(*load line*) To illustrate the use of the load line in finding the operating point, we have superimposed the load line of Fig. 6-15 on the characteristic curves of the type _____ transistor (Fig. 6-21).

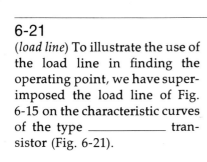

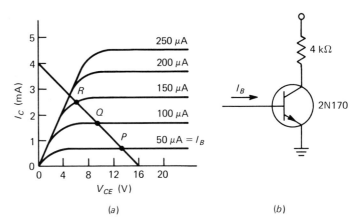

Fig. 6-21   (a)   (b)

6-22
(*2N170*) The curves in Fig. 6-21a are obtained by varying V_{CE} and measuring I_C. During the measurements, the base current, _____, is held constant.

6-23
(I_B) By feeding in different values of I_B, a set, or family, of curves is obtained. In Fig. 6-21a, the family has _____ distinct curves.

6-24
(*5*) The operating point of the transistor is determined both by the characteristic curves of the transistor and the load line. For example, if the base current is 100 μA, the intersection of the load line and characteristic curve is the point _____.

6-25
(*Q*) Point Q then represents the operating point of the 2N170 transistor of Fig. 6-21b when the base current is _____ μA.

6-26
(*100*) If the base current is changed to 150 μA, then the operating point is denoted by point _____ in Fig. 6-21a.

6-27
(*R*) And finally, if the base current is 50 μA, the operating point is denoted by point _____.

USING TRANSISTOR CHARACTERISTIC CURVES AND CHARTS **149**

6-28
(*P*) In the next section we shall see how the load line can be used to compute graphically the gain of a transistor. The reader should remember that whatever the device—transistor or tube—the operating point must be on the _____ _____.

6-28
(*load line*)

OUTPUT CURVES AND CALCULATION OF GAIN

6-29
The curves in Fig. 6-29b are the output static characteristics for a common-emitter amplifier as illustrated in Fig. 6-29a. In these curves, the independent variable is the collector voltage V_{CE} and the dependent variable is the _____ _____.

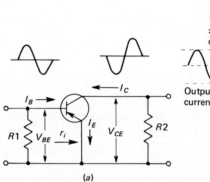
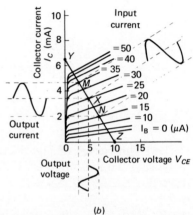

Fig. 6-29   (a)   (b)

6-30
(*collector current, I_C*) In this set of curves, the base current is the running parameter. This is a family of curves that shows how I_C varies with V_{CE} over a range of base currents that extends from $I_B = 0$ μA to $I_B =$ _____ μA.

6-31
(*50*) Let us now assume that the factors listed in Fig. 6-31 are given and we are asked to find the voltage, current, and power gains of the amplifier. Looking over the conditions given, we find that the $V_{CC} = 10$ V, $R2 = 1{,}500$ Ω, $r_1 = 500$ Ω, and that the operating point is specified as _____ μA of I_B.

Collector supply voltage	10 V
Load resistance (R2)	1,500 Ω
Emitter-base input resistance (r_1)	500 Ω
Operating point	25 μA of base current

Fig. 6-31

6-32
(25) We must first locate the load line on the output curves. First we assume the collector current as *zero*. If $I_C = 0$, the voltage drop in R2 is _____.

6-33
(0) If the voltage drop in R2 is zero, then the actual voltage between the collector and emitter must be equal to V_{CC}. That is, $V_{CE} = $ _____ V since V_{CC} is given as 10 V.

6-34
(10) This establishes one end of the load line. The point established is that for $I_C = 0$ and $V_{CE} = 10$ V. This point is identified as _____ in Fig. 6-29b.

6-35
(Z) Next, we assume that the collector voltage V_{CE} is zero. For this condition, the entire collector supply voltage must appear as a voltage drop across resistor _____.

6-36
(R2) That is, the supply voltage of 10 V appears as a drop across a resistance of 1,500 Ω. From Ohm's law, the collector current for this condition would have to be $I_C = 10 \text{ V}/1{,}500 \text{ Ω} = $ _____ mA. (Note that the answer asks for milliamperes, *not* amperes!)

6-37
(6.6) Thus, when the collector voltage is zero ($V_{CE} = 0$), the collector current is 6.6 mA ($I_C = 6.6$ mA). This establishes the second point for the load line. This point is identified as _____ in Fig. 6-29b.

6-38
(Y) We now draw in load line YZ. Since we are told that the operating point is to lie on the 25-μA bias line, we note that the load line crosses this line at a specific point. This point (i.e., where the load line intersects the 25-μA bias line) is identified as _____ in Fig. 6-29b.

6-39
(X) We can now determine the collector-voltage operating point by dropping a vertical line from point X until it intersects the horizontal axis. We find from this that when the base current is 25 μA, _____ is 4.8 V.

6-40
(V_{CE}) Now suppose we are told that the peak-to-peak input current is to be 20 μA. This tells us that the *deviation* of base current above and below the operating point will be half of the peak-to-peak current, or _____ μA.

USING TRANSISTOR CHARACTERISTIC CURVES AND CHARTS

6-41
(10) Adding 10 μA to the operating current of 25 μA, we find that the input current will reach a positive peak of 35 μA. Subtracting 10 μA from 25 μA, we find that the input current will reach a negative peak of _____ μA.

6-42
(15) The point of intersection of the positive peak current with the load line is identified as *M*, while the point of intersection of the negative peak current with the load line is identified as _____.

6-43
(N) We now can establish the waveform of the *input current* by extending 3 lines perpendicular to the load line, one from the operating point *X*, one from the positive deviation point _____, and one from the negative deviation point *N*.

6-44
(M) The waveform of the output current is next established by drawing 3 lines from the points *X*, *M*, and *N*. These 3 lines are perpendicular to the collector _____ axis.

6-45
(current) Finally, the waveform of the output voltage is established by drawing 3 lines from points *X*, *M*, and *N*; this time the lines are perpendicular to the collector _____ axis.

6-46
(voltage) We are now prepared to find the current gain. Current gain A_i is defined as the ratio of a change in collector current to a change in base current. Thus, $A_i = \Delta I_C / \Delta$ _____.

6-47
(I_B) The change of collector current is the difference between peaks of the output current, or $I_{C_{max}} - I_{C_{min}}$. Similarly, the change of base current is $I_{B_{max}} -$ _____.

6-48
($I_{B_{min}}$) The ratio of change of collector current to change of base current is therefore

$$\frac{I_{C_{max}} - I_{C_{min}}}{I_{B_{max}} - I_{B_{min}}}$$

Substituting the known quantities, this ratio becomes: 4.7 mA $-$ 2.1 mA divided by 35 μA $-$ _____ _____.

152 A PROGRAMMED COURSE IN BASIC ELECTRONICS

6-49
(15 μA) Thus $A_i = 2.6$ mA/20 μA. Converting the denominator to milliamperes as required for proper division, we have $A_i = 2.6$ mA/0.02 mA = _____ .

6-50
(130) Hence, the current amplification of this circuit is 130. Next, we solve for voltage gain. For this circuit, voltage gain is defined as the ratio of a change in collector voltage to the change in base voltage. That is, $A_v = \Delta V_{CE}/\Delta$ _____ .

6-51
(V_{BE}) Let us first obtain ΔV_{BE}. The change of input voltage, from Ohm's law, is the change in input current times the *input impedance*. Thus, $\Delta V_{BE} = \Delta$ _____ $\times r$.

6-52
(I_B) As we found previously, the change in input current is 20 μA and the input impedance is 500 Ω. Thus, the product of these 2 factors, after converting the change in input current to *amperes*, is 0.00002 A $\times$ 500 Ω = _____ V.

6-53
(0.01) The change of V_{CE} needed for the voltage gain equation is obtained by referring to the points of intersection of the vertical dashed lines in Fig. 6-29b with the horizontal axis. The lower intersection is 2.7 V, and the higher intersection is 6.7 V. Thus ΔV_{CE} is the difference between these, or _____ V.

6-54
(4) Thus, the input voltage change is 0.01 V and the output voltage change is 4 V. The ratio of ΔV_{CE} to ΔV_{BE} is the voltage gain. Hence, $A_v = 4/0.01 =$ _____ .

6-55
(400) Hence, the voltage gain of this transistor amplifier is 400. Next we shall find the power gain. It will be recalled that $G = A_v A_i$. In words, this expression may be read as: power gain equals the product of voltage gain and _____ .

6-56
(*current gain*) Since we have found the current gain to be 130 and the voltage gain to be 400, then the power gain is _____ .

6-57
(52,000) Expressed in decibels, the power gain is: $G = 10 \log 52{,}000$. The logarithm of 52,000 is 4.7. Therefore the power gain is _____ dB.

6-57
(47)

DYNAMIC TRANSFER CHARACTERISTIC
6-58
Refer first to the *right-hand* set of curves. These curves are the output characteristics discussed in the last section. The axis for the independent variable is labeled V_{CE}. Although the ordinate axis is unlabeled, we know from the previous study that this axis is collector _____ I_C in milliamperes.

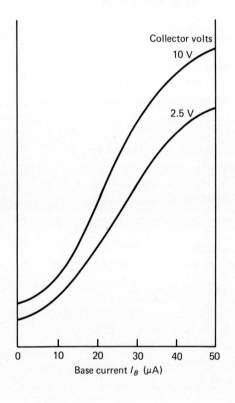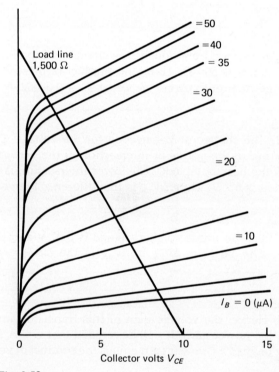

Fig. 6-58

6-59
(*current*) On the left, using exactly the same vertical axis (collector current I_C) as the output characteristics, are 2 *static transfer* characteristics shown as curved lines. One of these is labeled "Collector volts, 10 V" while the other is labeled _____ V.

154 A PROGRAMMED COURSE IN BASIC ELECTRONICS

6-60
(2.5) The independent variable for the curves on the left is shown along the horizontal axis and is the _____ _____ I_B in μA.

6-61
(*base current*) A static transfer characteristic relates an input parameter such as base current to an output parameter such as _____ _____.

6-62
(*collector current*) This explains the use of the word *transfer*. It is a descriptive term used to inform the reader that the curves show how variation of one *input* parameter can affect a dependent _____ parameter.

6-63
(*output*) The word *static* indicates that the second independent variable—collector volts in this case—is being held constant. For one static curve, the collector voltage is constant at 10 V. For the other static curve, the collector voltage is constant at _____ V.

6-64
(2.5) In an actual operating amplifier in which an input signal causes the base current to vary, however, the collector current also varies, thereby causing the voltage drop across the output load resistor to _____ in accordance.

6-65
(*vary*) Since the *IR* drop across the load resistor varies, the collector-to-emitter voltage also varies. Thus, in a *dynamic* system in actual operation, the collector voltage cannot remain _____.

6-66
(*constant*) The static transfer characteristics (left-hand curves) do not, therefore, present the true picture of events occurring dynamically in an actual amplifier. To correct this, we must plot the _____ transfer characteristic for the particular load resistor used.

6-67

(*dynamic*) We start this procedure by placing the 2 curves side by side, as shown, so that the collector current axes of both sets of curves are vertical and have the same scale. This may be done because the _____ current axis is *common to both sets of curves.*

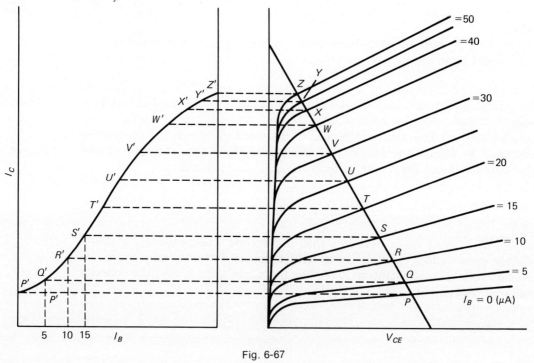

Fig. 6-67

6-68

(*collector*) Consider point P on the output characteristics. This point represents a condition at which the magnitude of the base current is _____.

6-69

(*zero*) We now draw a horizontal line to the left until it intersects the ordinate representing the condition of base current of zero on the left-hand set of curves. This ordinate is the vertical axis of the left-hand set of curves; hence we label the intersection point _____.

6-70

(P') We move now to point Q on the output load line. This point represents a base current of 5 μA. Therefore, we extend another line to the left until it intersects the 5-μA ordinate in the left coordinate system. This is Q'. Note that Q' is located directly above the _____-μA point on the horizontal I_B axis.

6-71
(5) Now consider point R. This point lies on the load line and also on the _____-µA base current curve.

6-72
(10) Drawing a line to the left from point R, we stop this line at R' because R' is immediately above the _____-µA point on the I_B axis in the left coordinate system.

6-73
(10) Point S is the intersection of the load line with the 15-µA base current curve. Hence, to plot the dynamic point that corresponds to S, that is S', we stop the horizontal line above the _____-µA coordinate in the transfer system.

6-74
(15) The same procedure is continued until we have finally plotted Z' in the transfer coordinate system. When P' through Z' are then connected sequentially by a heavy curve, we have plotted the complete _____ transfer characteristic.

6-75
(*dynamic*) This curve shows the way the collector current will actually vary under signal conditions when the signal is applied so that it causes the _____ current to vary.

6-76
(*base*) The dynamic transfer characteristic takes into account the fact that the collector voltage also varies as a result of the passage of the collector current through the 1,500-Ω _____ resistor.

6-77
(*load*) The dynamic transfer characteristic is useful in signal analysis. Figure 6-77 illustrates the conditions needed for distortionless amplification. The operating point, indicated as point _____, is selected at the center of the linear portion of the characteristic curve.

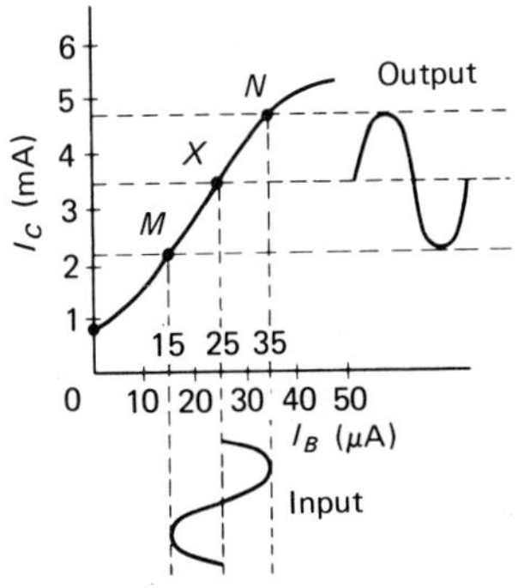

Fig. 6-77

6-78
(X) In this case, the base current flowing when the transistor is at its operating point with zero signal input is _____ µA.

6-79
(25) The input signal is applied along the vertical projected axis. The input signal is allowed to vary the base current from 15 µA to _____ µA.

6-80
(35) The collector current at the operating point with zero signal input is approximately _____ mA.

6-81
(3.5) When the base current is varied over the range of 20 µA (from 15 to 35 µA as noted previously) the collector current varies from about 2.2 mA to about _____ mA.

6-82
(4.7) Throughout this variation, all action occurs along the linear portion of the dynamic transfer characteristic. Therefore, the output signal is an excellent reproduction of the _____ signal.

6-83
(input) Now let us look at Fig. 6-83. In this arrangement, the operating point is again indicated by point X. It is the same operating point as in the previous case because the base current with zero signal input is again _____ µA.

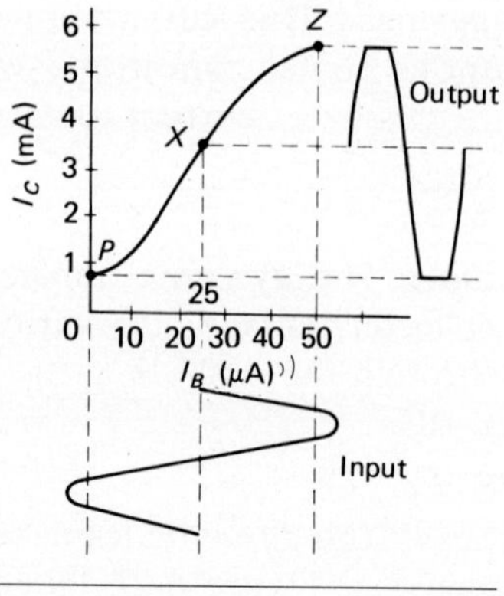

Fig. 6-83

6-84
(25) This time, however, the input signal is very much _____ than in the previous case. Note how it swings above 50 µA of base current.

6-85
(larger) This condition is known as "overdriving" because the input signal drives the base current into the nonlinear portion of the lower end of the characteristic at P and into the nonlinear portion of the upper end of the characteristic at _____.

6-86
(Z) The output signal is distorted as shown, because its positive peak is clipped and its _____ peak is also clipped.

6-87
(*negative*) Another form of distortion is introduced when the operating point is incorrectly chosen. In Fig. 6-87 the operating point around which the base current swings with the signal is labeled point _____.

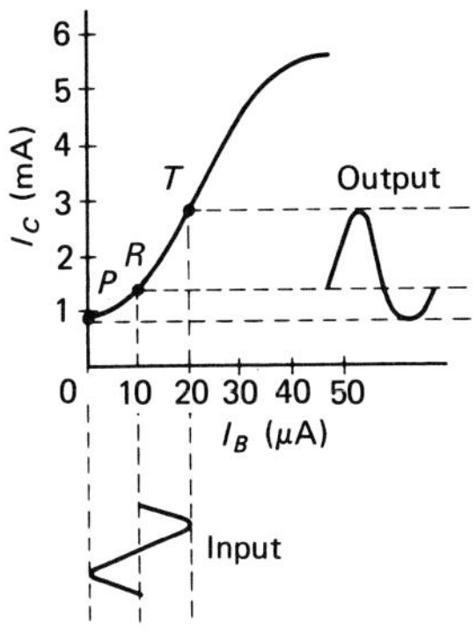

Fig. 6-87

6-88
(R) Note that this operating point is no longer at the _____ of the linear portion of the dynamic transfer characteristic.

6-89
(*center*) Thus, even a small negative-going base current will cause the collector current to go into the _____ linear portion of the characteristic.

6-90
(*non*) This occurs between R and P in Fig. 6-87. Because of the curvature of the characteristic, the negative peak of the output waveform is _____.

6-91
(*clipped*) Between R and T, however, the characteristic is relatively linear, giving rise to an output half-cycle which is not clipped. Thus, the output wave is asymmetrical and is not an exact duplicate of the _____ waveform.

6-91
(*input*)

USING TRANSISTOR CHARACTERISTIC CURVES AND CHARTS **159**

AMPLIFIER CLASSES

6-92
In Fig. 6-92, we have drawn the dynamic transfer characteristic of a transistor. Both the input current I_B and the output current _____ waveforms are shown.

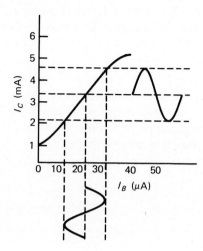

Fig. 6-92

6-93
(I_C) We note that the output waveform is a faithful reproduction of the input, and further, that the base current is such that the base is always _____-biased; i.e., the collector current flows for the complete cycle of input current.

6-94
(*forward*) When an amplifier, whether transistor or vacuum tube, is biased so that output current flows for the complete (360°) input signal, then that amplifier is said to be a class A amplifier. The amplifier depicted in Fig. 6-92 is an example of a class _____ amplifier.

6-95
(*A*) However, amplifiers are *not* always operated such that the output current flows for a complete cycle of input. In fact, by convention, one speaks about 3 other classes of amplifiers: AB, B, and C. Therefore, circuit designers recognize a total of _____ classes of amplifiers.

160 A PROGRAMMED COURSE IN BASIC ELECTRONICS

6-96
(4) Figure 6-96 shows a transistor operated in class AB. Note that the output current flows for less than a complete cycle (360°), but for more than a half-cycle (180°). For the paths A–B–C–D and F–G, current flows; but for path D–_____–_____, no output flows.

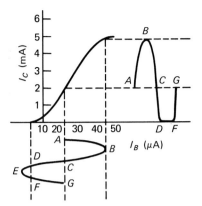

Fig. 6-96

6-97
(E, F) Hence, if an amplifier has output current for more than 180° but less than 360°, then it is in class _____ operation.

6-98
(AB) Figure 6-98 depicts the operation of a transistor in the class B mode. When the base current follows the path A–B–C, collector current flows, but when the base current follows the path _____–_____–_____, no collector current flows.

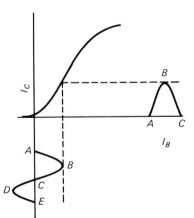

Fig. 6-98

6-99
(*C, D, E*) Thus, when a transistor or vacuum tube is operated so that output current flows for half of the input cycle, then that amplifier is in class B. The last class, C, is depicted in Fig. 6–99. In this case, output current flows for paths *B–C–D*, but not for paths *A–B* and *D*–_____–_____–_____.

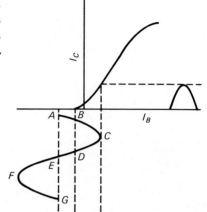

Fig. 6-99

6-100
(*E, F, G*) Hence, an amplifier whose output current flows for less than half a cycle is operating as a class _____ amplifier. Since class A gives distortionless amplification (Frames 6-77, 6-93), why would classes AB, B, and C be used?

6-101
(*C*) To answer this question, refer back to Fig. 6-77. The quiescent point (defined as the zero-signal operating point) is identified by the letter X. At that point, the collector current is about _____ mA.

6-102
(*3.5*) Hence, even when *no* input signal is present, considerable power is being used (wasted). In fact, it can be shown that a class A amplifier can achieve at most 25 percent power efficiency. A class A amplifier, then, has the advantage of _____ distortion, but the disadvantage of wasted power.

6-103
(*low*) This wasted power in the output stages of an amplifier is especially critical in battery-powered equipment. It can be shown that class B or C amplifiers can achieve power efficiencies of up to 80 percent. Hence, to avoid wasted power, the output stage of amplifiers is often *not* class _____, but rather class B or C.

6-104
(A) Since a single transistor or vacuum tube in class B only reproduces half the input, it alone is not sufficient. Figure 6-104 shows a circuit with 2 _____.

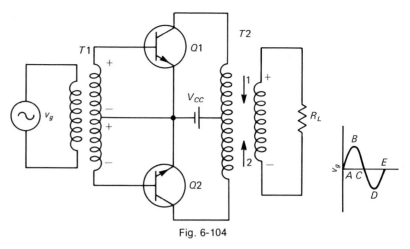

Fig. 6-104

6-105
(*transistors*) The collector supply for the transistors is denoted by V_{CC}. In order that the transistors be class B, they must be biased so that they are just *cut off* with *no* input signal. This is simply accomplished for transistors by providing a base current of _____ mA.

6-106
(*zero or no*) Consider feeding a complete cycle of input into this amplifier. During the path A–B–C, the polarities of the center-tapped transformer T1 are shown in Fig. 6-104. Q1 receives a _____ voltage and conducts, while Q2 receives a _____ voltage and does not conduct.

6-107
(*positive, negative*) Q1 sends current *down* through the primary of T2, causing a positive voltage in the secondary. When the input voltage reaches the path C–D–E, the secondary polarities of T1 reverse (not shown in Fig. 6-104), and transistor _____ conducts, while transistor _____ is cut off.

6-108
(Q2, Q1) Q2 sends current *up* through the primary of T2, causing a negative voltage in the secondary of T2. This arrangement whereby one transistor conducts while the other transistor is _____ _____, is called a *push-pull* amplifier.

6-109
(*cut off*) Just as with the class A amplifier, the class B push-pull amplifier reproduces the *entire* input signal. Further, the class B push-pull amplifier has the advantage of consuming power *only* when the input signal is present. This is a result of both transistors (or tubes) being _____ _____ in the quiescent condition.

6-110
(*cut off*) In conclusion, in low-power amplifier stages, class A is almost always used, but in power output stages class _____ amplifiers are very common.

6-110
(*B*)

CONSTANT-POWER-DISSIPATION LINE
6-111
Each transistor has a maximum collector *power* rating that should not be exceeded if damage to the transistor is to be avoided. Power is found from the product of voltage and _____.

6-112
(*current*) Maximum collector power refers to the maximum permissible power that can be dissipated in the internal collector circuit of the transistor. Hence, to determine the actual instantaneous power being dissipated, we must know the instantaneous collector _____ and the instantaneous collector current.

6-113
(*voltage*) Refer to the graph in Fig. 6-113. This is a graph of collector voltage versus collector current. We have seen this family of curves before. They are the static _____ characteristics of a typical transistor.

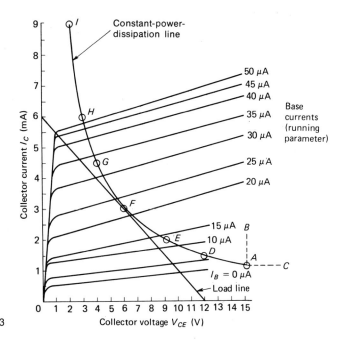

Fig. 6-113

6-114
(*output*) Assume that we are dealing with a transistor whose maximum collector power rating is 18 milliwatts (mW). This rating indicates that the transistor will be endangered if the instantaneous product of collector voltage and collector current should exceed _____ _____ .

6-115
(*18 mW*) Assume further that we might find the actual collector voltage V_{CE} to be 15 V at a given instant. We know that the maximum power rating is 18 mW. To find the maximum current that should be permitted in the collector circuit for these conditions we can apply the power formula: $P = V \times$ _____ .

6-116
(*I*) Since I is the unknown, the power formula should be rearranged to read:
$I = P/$_____ .

6-117
(*V*) Thus, if the collector voltage is 15 V and the maximum power permitted is 18 mW, then the maximum permissible current is $I = P/V = 0.018 \text{ W}/15 \text{ V} =$ _____ A or 1.2 mA.

USING TRANSISTOR CHARACTERISTIC CURVES AND CHARTS

6-118
(*0.0012*) Now, referring to Fig. 6-113, locate 15 V on the V_{CE} axis and 1.2 mA on the I_C axis. Using these as coordinates, mark a point on the graph to correspond to these coordinates. This is the point identified as _____ in Fig. 6-113.

6-119
(*A*) This point locates graphically the maximum collector current that may be permitted to flow when the collector voltage is _____ V to ensure that the maximum power rating of 18 mW will not be exceeded.

6-120
(*15*) Any point above point *A* in a vertical direction would locate a *forbidden* current at 15 V on the collector because such a current would be higher than 1.2 mA and the collector power would then be _____ than 18 mW.

6-121
(*greater* or *more, etc.*) Such a forbidden current point is identified by the letter _____ in the graph.

6-122
(*B*) Also, any point beyond *A* to the right in a horizontal direction would represent a *forbidden* voltage (point C) since this voltage would be greater than 15 V at a current of 1.2 mA, thus causing the power to exceed _____.

6-123
(*18 mW*) All points to the left of and below *A* would be permissible regions because they would represent currents lower than 1.2 mA at 15 V, or voltages lower than _____ V at 1.2 mA.

6-124
(*15*) Now assume that the applied collector voltage V_{CE} is 12 V. Again we can locate a maximum power point by using: $I = 0.018 \text{ W}/12 \text{ V} =$ _____ mA.

6-125
(*1.5*) We can plot this point on the curve as before. The point would be located above the 12-V point on the horizontal axis and opposite the _____-_____ point on the vertical axis.

6-126
(*1.5-mA*) Thus, this point is identified as _____ in Fig. 6-113.

166 A PROGRAMMED COURSE IN BASIC ELECTRONICS

6-127
(D) Point E is obtained in a similar manner. Again we use 18 mW as the power and this time assume the collector voltage to be 9 V. In this case, the maximum permissible current is $I = 0.018/9 = $ _____ mA.

6-128
(2.0) In the same way we can plot the remaining points in the series. That is, point F is obtained by assuming $V_{CE} = $ _____ V, finding I_C to be 3.0 mA by using the maximum collector power dissipation rating of 18 mW.

6-129
(6) Point G is obtained by assuming $V_{CE} = $ _____ V, obtaining a current of 4.5 mA for I_C, again by using the maximum collector power rating of 18 mW.

6-130
(4) Similarly, point H is found by assuming the collector voltage to be 3 V, finding the current in the collector circuit to be _____ mA, while taking 18 mW as the maximum power rating.

6-131
(6) Finally, point I is located by finding the current flowing when V_{CE} is equal to _____ V.

6-132
(2) When these points are all connected by a smooth curve, we call the curve thus obtained the _____-_____-_____ line.

6-133
(*constant-power-dissipation*) As we saw in discussing point A, all regions above and to the right of any point on this curve are forbidden regions because they indicate excessive _____ dissipation in the collector circuit.

6-134
(*power*) However, all points to the left and below the constant-power-dissipation line are permissible regions because the power dissipation for these points will be _____ than the maximum power rating of the transistor.

6-135
(*smaller* or *less*, etc.) As we have seen, the load line enables us to predict the instantaneous values of V_{CE} and I_C for any instantaneous value of base _____.

6-136
(*current*) Thus, if the load resistor is selected to give a load line that is at all points below or to the left of the constant-_____-dissipation line, the maximum collector power rating will never be exceeded.

6-137
(*power*) Furthermore, if the load line is made tangent to the constant-power-dissipation line at a desirable point, maximum _____ from the transistor will be realized, yet the maximum rating will not be exceeded.

6-138
(*power*) The load line in Fig. 6-113 is shown tangent to the constant-power-dissipation line at point _____.

6-139
(*F*) Since the bottom of the load line intersects the horizontal axis at the 12-V mark, the output or terminal voltage of the collector battery must be _____ V, in the example of Fig. 6-113.

6-140
(*12*) To have the load line become the tangent at point F for these conditions, the load line must intersect the vertical axis at the _____-mA level as shown.

6-141
(*6*) The load resistor needed to produce this load line may now be determined from $R = V_{CC}/I_C$ when $V_{CE} = 0$. Since $V_{CC} = 12$ V and the current $I_C = 0.006$ A when $V_{CE} = 0$, then $R =$ _____ Ω. (V_{CC} is collector battery potential.)

6-142
(*2,000*) Summarizing: In the example of Fig. 6-113, the maximum collector power dissipation rating has been assumed to be _____ mW.

6-143
(*18*) For this rating, we have found a whole series of points on the static output curves, each point representing a power dissipation of 18 mW for that particular collector voltage. When these points are connected by a smooth curve, we establish the constant-_____-dissipation line.

6-144
(*power*) If the load resistor is then selected so that the load line is _____ to the constant-power-dissipation line, we are sure that maximum power output will be realized without exceeding the transistor's maximum power rating at any time.

168 A PROGRAMMED COURSE IN BASIC ELECTRONICS

6-144
(*tangent*)

INTERELEMENT CAPACITANCES AND FEEDBACK
6-145
Figure 6-145 is a pictorial-schematic diagram of a transistor showing the junctions and the interelement capacitances that exist within it. As the diagram illustrates, there are a total of _____ interelement capacitances.

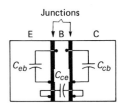

Fig. 6-145

6-146
(3) C_{eb} is the capacitance between the base and emitter and is called the emitter-base capacitance. C_{ce} is the collector-emitter capacitance. The third is _____, the collector-base capacitance.

6-147
(C_{cb}) Consider first the collector-base capacitance C_{cb}. The base semiconductor material is one plate of the capacitance, the PN junction is the dielectric, and the semiconductor material of the _____ is the other plate.

6-148
(*collector*) The capacitance of C_{cb} depends upon the *width of the PN junction*. This width is a variable factor and depends upon the voltage across the junction and the current through the junction. Thus, the value of C_{cb} is a function of both junction voltage and _____.

6-149
(*current*) You will recall that the PN junction between the collector and the base is *reverse-biased*. As the value of this reverse bias is increased, it is found that the width of the junction increases. This means that the "thickness" of the dielectric of C_{cb} also _____.

6-150
(*increases*) If the dielectric of a capacitor is thickened, the capacitance of this capacitor must _____.

6-151
(*decrease*) Thus, as the reverse bias between collector and base is increased, the interelement capacitance C_{cb} must _____.

USING TRANSISTOR CHARACTERISTIC CURVES AND CHARTS **169**

6-152

(*decrease*) The relationship between collector-base reverse bias and the value of C_{cb} is shown in Fig. 6-152 for a typical transistor. Note that the capacitance of C_{cb} decreases as the collector-to-base voltage V_{CB} is ——————.

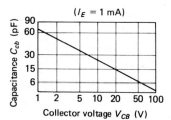

Fig. 6-152

6-153

(*increased*) If circuit conditions permit, therefore, the collector-to-base voltage should be made as ——————— as possible in order to keep C_{cb} as small as possible.

6-154

(*large*) Figure 6-154 shows how C_{cb} varies with emitter current I_E. Since most of the emitter current flows through the collector-base junction, an increase in emitter current causes an ——————— in the capacitance of C_{cb}. (Usually this effect is smaller than that shown in Fig. 6-154.)

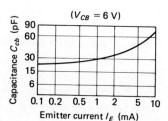

Fig. 6-154

6-155

(*increase*) The increase in emitter current causes C_{cb} to rise. This shows that an increase in emitter current has the effect of ——————— the width of the base-collector PN junction.

6-156

(*reducing*) When a transistor is used for high-frequency amplification, the action of C_{cb} has an important effect on its performance. Refer to Fig. 6-156. The signal collector current (or collector ac due to the signal) may be considered to originate at V_{CC}, proceed through R_L, through the transistor, and down to ground, as symbolized by arrows 1 and ———————.

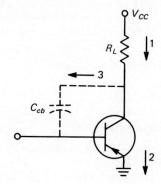

Fig. 6-156

170 A PROGRAMMED COURSE IN BASIC ELECTRONICS

6-157
(2) The collector current is the one we use for amplification. However, a second path for the collector ac exists through C_{cb}. This component of the collector current is shown by arrow _____.

6-158
(3) Thus, a fraction of output signal is fed back through C_{cb} to *reinforce* the input. This current is called *positive feedback* current. The magnitude of the feedback current depends on the size of C_{cb}. If C_{cb} is small, the feedback current will be _____.

6-159
(*small*) If the feedback current is small, its effect will be insignificant and may be ignored. Since C_{cb} for a standard transistor is small, its capacitive reactance will be _____ for low frequencies. ($X_C = 1/2\pi fC$)

6-160
(*large*) Thus, for low frequencies, when $X_{C_{cb}}$ is large, the feedback current is _____, so that the transistor performs just as its characteristics predict.

6-161
(*small*) As the frequency being amplified increases, however, the capacitive reactance of C_{cb} must _____.

6-162
(*decrease*) With decreasing capacitive reactance, C_{cb} provides increasing amounts of feedback current. When the feedback current is great, it causes instability and lowered sensitivity. Thus, at _____ frequencies, the performance of a transistor begins to deteriorate.

6-163
(*high*) Therefore, a transistor designed for high-frequency amplification must be fabricated and operated so that the amount of positive feedback is _____ to as low a value as possible.

6-164
(*reduced*) This is done in 2 ways. First, the value of C_{cb} in a transistor designed for high-frequency work must be kept as _____ as possible.

6-165
(*small* or *low*) Second, external circuits may be employed which reduce the amount of _____ feedback by neutralization methods.

6-166
(*positive*) Some low-frequency audio transistors have C_{cb} values as high as 50 picofarads (pF). High-frequency transistors, however, must have a much _____ value of C_{cb} than this. The average C_{cb} of high-frequency transistors is in the order of 2 or 3 pF.

6-167
(*smaller* or *lower*) The emitter-base interelement capacitance is symbolized in Fig. 6-145 by C_{eb}. The PN junction between emitter and base is forward-biased, causing the PN junction itself to be very narrow. Thus, C_{eb} must normally have a comparatively _____ value since it is a capacitor with its plates very close to each other.

6-168
(*large*) Normally, the input to a transistor is shunted by a low input resistance. Thus, even though C_{eb} is _____, it introduces little design difficulty since it is shunted by a low input resistance. Diffusion capacitance is important, however, in high-frequency design work, and there it is taken into account.

6-169
(*large*) This effect can be understood by considering C_{eb} as one *parallel-circuit* element and the input resistance as the other. In a parallel circuit consisting of a very low and a medium impedance, the net impedance is governed chiefly by the value of the _____-impedance element.

6-170
(*low*) The input resistance is very low compared with the reactance of C_{eb} for most of the audio range. Thus, the input resistance is the chief contributor to the final value of the _____ of the parallel circuit.

6-171
(*impedance*) This means that the reactance of _____ does not cause the net input impedance to differ very much from the value of the input resistance.

6-172
(C_{eb}) The input resistance, being a pure resistance, is not frequency-sensitive. That is, the voltage drops for various signal frequencies are _____ to each other if the voltages of these frequencies are the same initially.

6-173
(*equal*) Since the input resistance is the principal ingredient of the input impedance, the frequency response of the transistor will not be seriously affected by the presence of _____.

6-174
(C_{eb}) The collector-emitter interelement capacitance is symbolized by C_{ce}. This capacitance is from 5 to 10 times larger than C_{cb}. The effective capacitance C_{ce} is also changed by variations in emitter _____ and collector voltage.

6-175
(*current*) The high-frequency response of an amplifier deteriorates if the output is shunted by a high capacitance. In the common-emitter configurations, the output is taken from across the collector and _____ of the transistor.

6-176
(*emitter*) Since C_{ce} is large, and since it is the capacitor that shunts the output of a common-emitter amplifier, we would therefore expect the _____-frequency response of the CE amplifier to suffer.

6-177
(*high*) This is quite true. In the CB amplifier, however, the output is taken from across the collector and _____ of the transistor.

6-178
(*base*) The capacitance that shunts these 2 elements is C_{cb}. Since C_{cb} is considerably _____ in value than C_{ce}, the high-frequency shunting is not so pronounced.

6-179
(*smaller*) Thus, the _____ configuration is capable of better high-frequency response in amplifiers than the CE configuration.

6-179
(*CB* or *common-base*)

COMPARISON OF CONFIGURATIONS WITH RESPECT TO RESISTANCES
6-180
In Fig. 6-180, the block labeled *Transistor* represents a transistor in any configuration. That is, it may represent a transistor in the CE, CC, or _____ configuration.

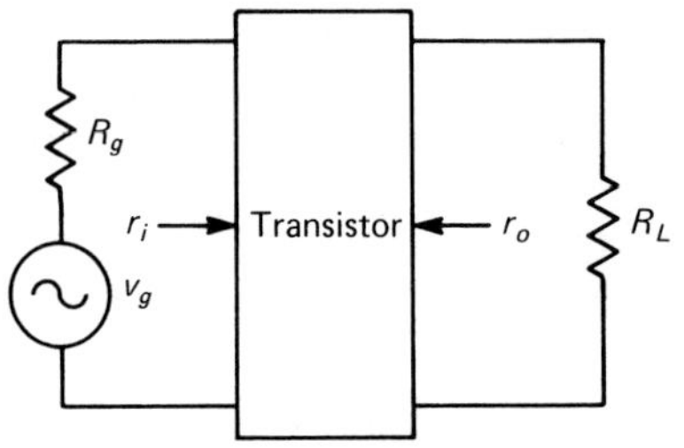

Fig. 6-180

USING TRANSISTOR CHARACTERISTIC CURVES AND CHARTS

6-181
(*CB*) The input signal is represented by v_g, while the input resistance of the source is identified by _____.

6-182
(R_g) Transistor output resistance is r_o; transistor _____ resistance is r_i; load resistance is R_L.

6-183
(*input*) In a vacuum tube, the input resistance to the tube is completely independent of the load resistance R_L. In a transistor amplifier, however, the input resistance is definitely affected by the _____ resistance R_L.

6-184
(*load*) Figure 6-184 shows how R_L affects input resistance r_i in all 3 configurations. For the common-collector and the common-_____ configurations, input resistance increases with increases of load resistance.

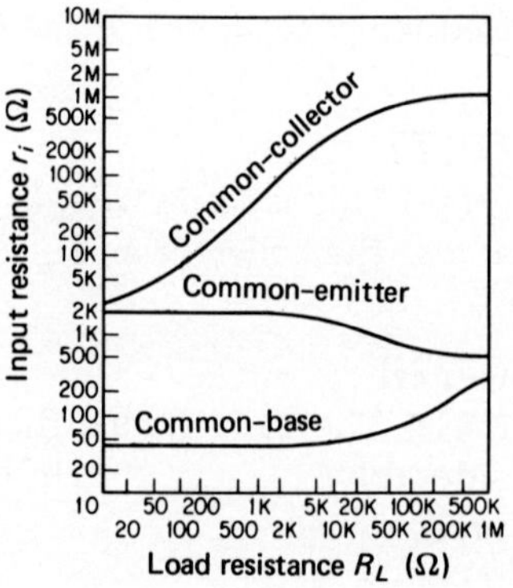

Fig. 6-184

6-185
(*base*) On the other hand, for the common-emitter configuration, the input resistance _____ as the load resistance is increased.

6-186
(*decreases*) For example, when the load resistance in the CC configuration is 2 kΩ, the input resistance is approximately 100 kΩ, but when the load resistance is raised to 500 kΩ, the input resistance becomes _____ megohm(s) (MΩ).

6-187
(*1*) Both the common-base and common-emitter configurations maintain relatively constant r_i for all values of R_L between 10 Ω and _____ kΩ.

6-188
(*2*) To match the impedance of a high-impedance source, the best configuration to use would be the _____ configuration, since this configuration has the highest input resistance of all 3, especially at high values of load resistance.

174 A PROGRAMMED COURSE IN BASIC ELECTRONICS

6-189
(*CC*) In most vacuum-tube amplifier arrangements, the resistance of the signal source has no effect upon the output resistance of the amplifier. In a transistor amplifier, however, R_g has a definite effect upon _____ as shown in Fig. 6-189.

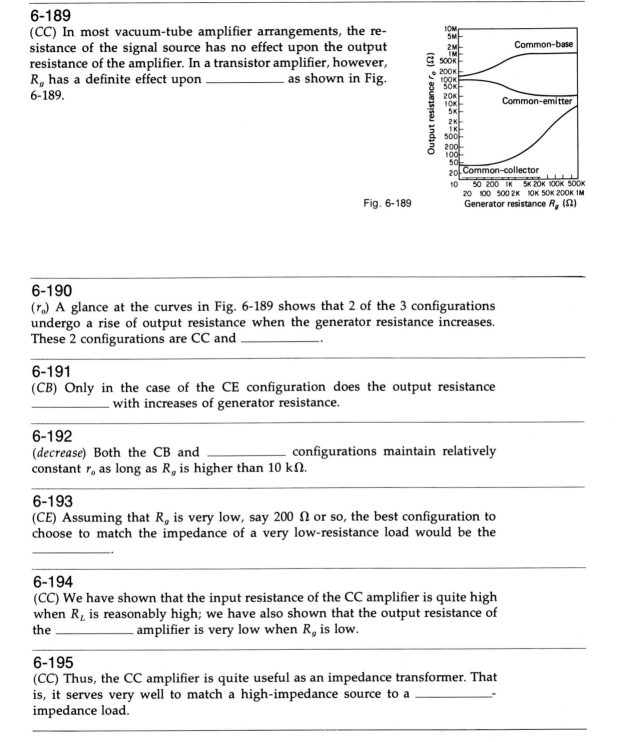

Fig. 6-189

6-190
(r_o) A glance at the curves in Fig. 6-189 shows that 2 of the 3 configurations undergo a rise of output resistance when the generator resistance increases. These 2 configurations are CC and _____.

6-191
(*CB*) Only in the case of the CE configuration does the output resistance _____ with increases of generator resistance.

6-192
(*decrease*) Both the CB and _____ configurations maintain relatively constant r_o as long as R_g is higher than 10 kΩ.

6-193
(*CE*) Assuming that R_g is very low, say 200 Ω or so, the best configuration to choose to match the impedance of a very low-resistance load would be the _____.

6-194
(*CC*) We have shown that the input resistance of the CC amplifier is quite high when R_L is reasonably high; we have also shown that the output resistance of the _____ amplifier is very low when R_g is low.

6-195
(*CC*) Thus, the CC amplifier is quite useful as an impedance transformer. That is, it serves very well to match a high-impedance source to a _____-impedance load.

6-196
(*low*) By the same reasoning, the best impedance match between a low-impedance source and a high-impedance load could be achieved by using the _____ configuration.

6-196
(*CB*)

COMPARISON OF CONFIGURATIONS WITH RESPECT TO GAIN
6-197
Figure 6-197 compares the voltage gain, power gain, and current gain available from a common-base amplifier for various values of _____ resistance.

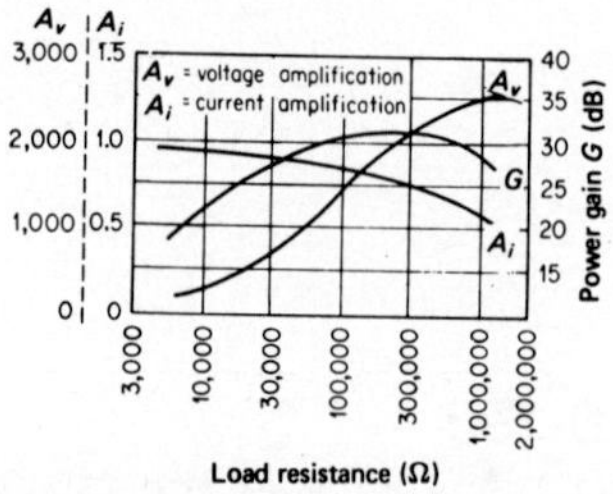

Fig. 6-197

6-198
(*load*) It will be recalled that current gain A_i is the ratio of signal output current to signal _____ current.

6-199
(*input*) Referring to the A_i curve, it is readily apparent that the current gain, even under the most favorable conditions, approaches but does not attain a value of _____.

6-200
(*1*) As the load resistance is increased, the current gain _____.

6-201
(*decreases*) The current gain has a value of 0.75 when the load resistance is _____ Ω.

6-202
(*300,000*) In contrast, the voltage gain increases as the load resistance _____.

6-203
(*increases*) In review, voltage gain is defined as the ratio of signal _____ voltage to signal input voltage.

176 A PROGRAMMED COURSE IN BASIC ELECTRONICS

6-204
(*output*) According to the curves in Fig. 6-197, a theoretical voltage gain of _____ is possible when the load resistance is 100,000 Ω.

6-205
(*1,500*) The curve labeled G in Fig. 6-197 shows how the _____ gain of the CB amplifier varies with load resistance.

6-206
(*power*) It is evident in this case that the power gain peaks at a specific value of load resistance. This occurs when the load resistance is approximately _____ Ω.

6-207
(*300,000*) For this value of load resistance, the power gain is approximately _____ dB.

6-208
(*32*) A similar series of curves for the common-emitter configuration is illustrated in Fig. 6-208. From the current-gain curve, it is at once apparent that the CE amplifier is capable of considerably more current gain than the _____ configuration.

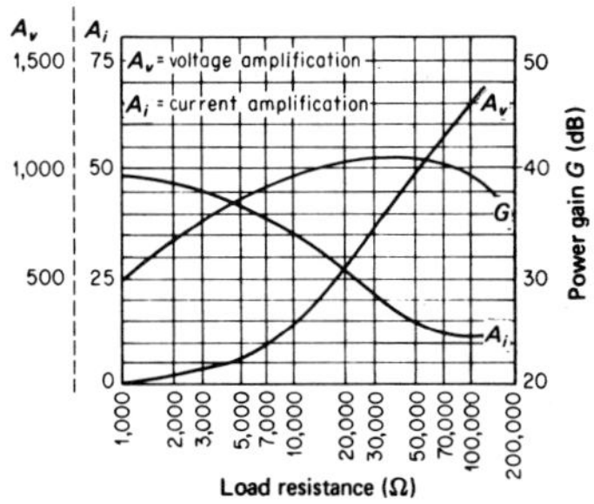

Fig. 6-208

6-209
(*CB*) For the particular transistor to which these curves apply, as a matter of fact, it is possible to obtain a current gain of nearly 50 when the load resistance is _____ Ω.

6-210
(*1,000*) In this configuration too, the current gain drops as the _____ is increased.

6-211
(*load resistance*) When the load resistance is 10,000 Ω in this CE amplifier, the current gain is _____.

6-212

(35) The voltage gain of the CE amplifier rises quite linearly for all values of load resistance above about 10,000 Ω. When the load resistance is approximately 100,000 Ω, the voltage gain is approximately _____.

6-213

(1,300) In this configuration, the power gain again shows a _____ at a specific value of load resistance.

6-214

(peak or maximum) The peak in power gain occurs when the load resistance is between 30,000 and _____ ohms.

6-215

(50,000) The peak value of power gain is approximately _____ dB.

6-216

(41) A similar set of curves for the common-collector configuration is given in Fig. 6-216. In this case, inspection of the curves reveals that it is the _____ gain which can approach unity but never exceed it.

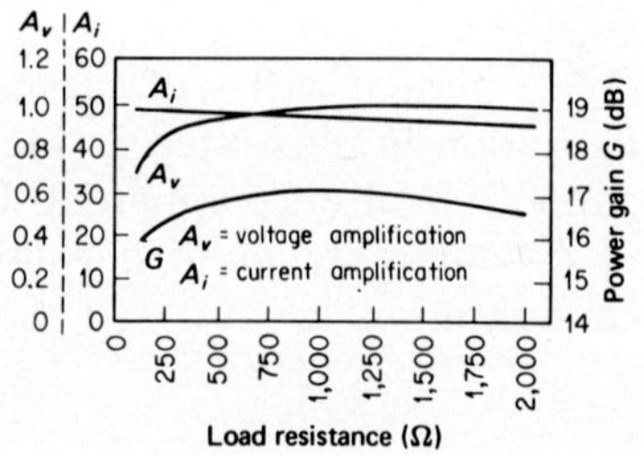

Fig. 6-216

6-217

(voltage) As in the other configurations, the _____ gain decreases as the load resistance is increased.

6-218

(current) Also, the power gain again peaks at a definite value of load resistance. In this particular CC amplifier, the power-gain peak occurs when the load resistance is in the vicinity of _____ Ω.

6-219

(1,000) For this value of load resistance, the power gain is very close to _____ dB.

178 A PROGRAMMED COURSE IN BASIC ELECTRONICS

6-220
(17) Using the chart of typical values in Fig. 6-220, it is possible to select the configuration that will provide optimum performance for specific applications. For instance, to match a low-impedance generator to a high-impedance load with reasonable voltage gain, you would do well to select a _____ configuration.

Item	CB Amplifier	CE Amplifier	CC Amplifier
Input resistance	30–150 Ω	500–1,500 Ω	2–500 Ω
Output resistance	300–500 Ω	30–50 Ω	50–1,000 Ω
Voltage gain	500–1,500	300–1,000	Less than 1
Current gain	Less than 1	25–50	25–50
Power gain	20–30 dB	25–40 dB	10–20 dB

Fig. 6-220

6-221
(CB) To obtain the highest voltage gain and highest power gain into a load of medium-high resistance, you would probably choose the _____ configuration.

6-222
(CE) To obtain excellent current gain into a low-impedance load, the best choice would be the _____ configuration.

6-223
(CC) The widest range of input impedances is available from the _____ configuration.

6-223
(CC)

EMITTER FOLLOWER
6-224
Let us briefly review the subject of impedance matching. In Fig. 6-224, Z1 is the impedance of the source, and Z2 is the impedance of the _____.

Fig. 6-224

6-225
(load) Power is to be transferred from the source to the load. It can be shown that maximum power is transferred from source to load when $Z1 = Z2$. If the source impedance is much greater than the load impedance, or if the load impedance is much greater than the source impedance, then very little _____ will be transferred from the source to the load.

USING TRANSISTOR CHARACTERISTIC CURVES AND CHARTS

6-226

(*power*) Consider the case illustrated in Fig. 6-226. Here the source impedance is shown to be _____ than the load impedance.

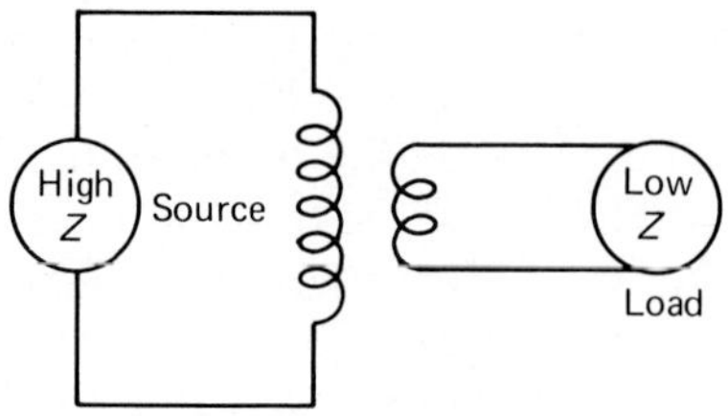

Fig. 6-226

6-227

(*greater*) To transfer power from source to load in the arrangement of Fig. 6-226, an *impedance-matching* device must be employed. As this figure shows, the impedance-matching device is a _____.

6-228

(*transformer*) The primary of this transformer is given a large number of turns. This makes its impedance relatively large. Thus, the high-impedance source can work into the primary, which is a _____-impedance load.

6-229

(*high*) The primary power is now converted into secondary power minus the losses of the transformer. The secondary may now be considered a new _____ which works into the low-impedance load.

6-230

(*source*) The secondary is given relatively few turns. Its impedance is, therefore, reasonably low. Now a low-impedance source (the secondary) works into a _____-impedance load.

6-231

(*low*) Transformers do not have good frequency response; that is, they cannot provide equal signal voltages for frequencies that vary over a wide range. This is largely due to the fact that transformers are inductive devices. Any inductance is frequency-sensitive. The voltage drops for different _____ may be quite different from each other.

6-232

(*frequencies*) As mentioned in Frame 6-195, the common-collector configuration can be used to match a high impedance to a low impedance and, moreover, gives *much* better frequency response than the transformer. In Fig. 6-232a, we have drawn the CC configuration exactly as was shown in Fig. 2-170. In Fig. 6-232b, the circuit has been cast in a form preferred by circuit designers. The circuits are the same, except that the _____-V battery has been left out.

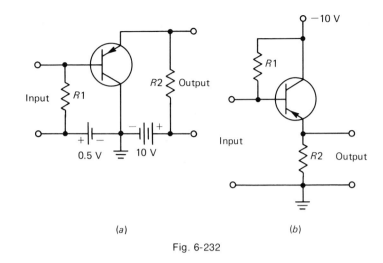

Fig. 6-232

6-233

(*0.5*) In the form of Fig. 6-232b, the circuit is called an *emitter follower*. The reason for this is that the output is taken from the resistor in the _____ lead.

6-234

(*emitter*) Consider the problem of converting a milliammeter into a voltmeter. A current-reading device has a low resistance whereas a voltmeter should have as _____ a resistance as possible.

6-235

(*high*) Suppose we wish to construct a voltmeter to read 10 V full scale and have an input resistance of at least 50,000 Ω. To construct the voltmeter, a milliammeter which has a full scale reading of 1 mA and an internal resistance of _____ Ω is chosen.

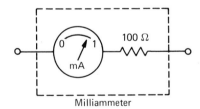

Fig. 6-235

6-236

(*100*) If the full scale voltage is to be 10 V, and the full scale current of the ammeter is 1 mA, then we must put a resistor in series with the ammeter. The value of this resistor is found from Ohm's law,

$$R = \frac{V}{I} = \frac{10 \text{ V}}{1 \text{ mA}} = \underline{\qquad} \text{ k}\Omega$$

USING TRANSISTOR CHARACTERISTIC CURVES AND CHARTS **181**

6-237

(10) The voltmeter shown in Fig. 6-237 would meet the requirement of reading full scale at 10 V. However, the resistance of the circuit is only _____ Ω.

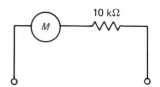

Fig. 6-237

6-238

(10,000) To raise the impedance level of the voltmeter, we can insert the voltmeter into an _____ _____ as shown in Fig. 6-238.

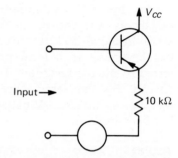

Fig. 6-238

6-239

(emitter follower) The meter would still read full scale at _____ V, but the impedance at the input would now be much higher, on the order of 1 MΩ or more.

6-240

(10) Note that in the above application, the emitter follower changed impedance levels for a dc signal. Recall that a transformer can only work with _____ signals.

6-241

(ac) In conclusion, an emitter follower is an _____-_____ device which has a better frequency response than the transformer. Further, since only a single transistor is required, the emitter follower is much more compact, much simpler, and much less expensive than the transformer.

6-241

(impedance-matching)

CRITERION CHECK TEST

___6-1 The equation for the dc load line of a transistor is $V_{CC} =$ (a) $I_C R_L + V_{BE}$, (b) $I_C/R_L + V_{CE}$, (c) $I_C R_L + V_{CE}$, (d) $I_C R_L + V_{BC}$.

___6-2 When a transistor is cut off, (a) maximum current flows, (b) maximum voltage appears across the load resistor, (c) maximum voltage appears across the transistor, (d) the base is heavily forward-biased.

____6-3 For the circuit of Fig. 6-21, if the base current is 100 μA, the collector-emitter voltage is about (a) 4.1 V, (b) 7.6 V, (c) 9.4 V, (d) 12.2 V.

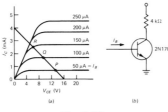

Fig. 6-21

Fig. 6-77

The following 5 questions refer to the figure below.

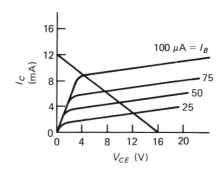

____6-4 If the quiescent base current is 25 μA, the quiescent collector current is about (a) 2 mA, (b) 4 mA, (c) 6 mA, (d) 8 mA.

____6-5 If the base dc goes from 25 μA to 75 μA, the collector current goes from (a) 2 mA to 3 mA, (b) 4 mA to 7 mA, (c) 5 mA to 9 mA, (d) 7 mA to 11 mA.

____6-6 The current amplification of this device is about (a) 20, (b) 40, (c) 60, (d) 80.

____6-7 The collector supply of this transistor is (a) 8 V, (b) 12 V, (c) 16 V, (d) 20 V.

____6-8 The load resistance is (a) 1,333 Ω, (b) 2,667 Ω, (c) 3,333 Ω, (d) 6,667 Ω.

____6-9 The output curve in Fig. 6-77 is an example of a (a) clipped waveform, (b) nonlinear waveform, (c) linear waveform, (d) unsegmented waveform.

____6-10 When a transistor is operated so that output current flows for 200° of the input signal, that is what class of operation? (a) A, (b) AB, (c) B, (d) C.

____6-11 A 2-transistor class B amplifier is commonly called a (a) push-pull amplifier, (b) symmetrical amplifier, (c) dual amplifier, (d) reciprocating amplifier.

____6-12 For each point on the constant-power-dissipation line (a) there must be a tangent load line for proper operation, (b) collector voltage and collector current are at their maximum values, (c) collector voltage times collector current equals the maximum collector power rating, (d) collector voltage and collector current are at their minimum values.

_____6-13 If the maximum power rating of a transistor is 1 W, and the collector current is 100 mA, then the maximum allowable collector voltage is (*a*) 1 V, (*b*) 3 V, (*c*) 10 V, (*d*) 100 V.

_____6-14 The symbol C_{cb} denotes, for a junction transistor, (*a*) the capacitance due to the point-contact junction at the collector, (*b*) the capacitance due to the PN junction of the collector base region, (*c*) a capacitor used to neutralize the collector capacitance, (*d*) stray capacitance caused by surface effects at the collector.

_____6-15 An increase in the width of the base-collector PN junction (*a*) increases C_{cb}, (*b*) decreases C_{cb}, (*c*) has no effect on C_{cb}, (*d*) affects all interelement capacitances.

_____6-16 To reduce the effect of C_{cb}, one usually (*a*) chooses an audio-frequency transistor, (*b*) uses external circuit elements to neutralize C_{cb}, (*c*) selects a transistor whose C_{cb} is zero, (*d*) puts a capacitor in parallel with C_{cb}.

_____6-17 The common-base amplifier offers better frequency response than the common-emitter amplifier because (*a*) C_{ce} is much larger than C_{cb}, (*b*) C_{cb} is much larger than C_{ce}, (*c*) the input resistance is larger, (*d*) the output resistance is larger.

_____6-18 For which configuration(s) does the input resistance depend strongly on the load resistance? (*a*) CE, (*b*) CC, (*c*) CE, (*d*) CE and CB.

_____6-19 The output resistance of the CC configuration (*a*) is higher than that of the other 2 configurations, (*b*) depends strongly on the generator resistance, (*c*) is independent of the generator resistance, (*d*) goes down as the generator resistance increases.

_____6-20 For the CB configuration in Fig. 6-197, maximum power gain occurs at a load resistance of (*a*) 3 kΩ, (*b*) 30 kΩ, (*c*) 300 kΩ, (*d*) 1,000 kΩ.

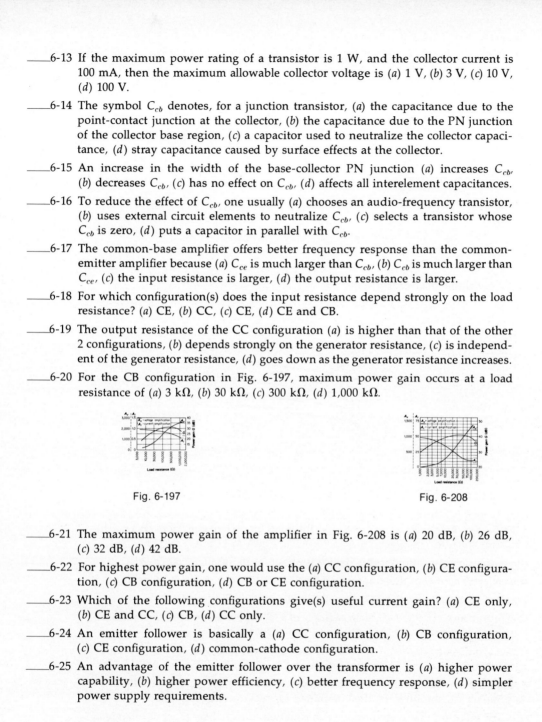

Fig. 6-197   Fig. 6-208

_____6-21 The maximum power gain of the amplifier in Fig. 6-208 is (*a*) 20 dB, (*b*) 26 dB, (*c*) 32 dB, (*d*) 42 dB.

_____6-22 For highest power gain, one would use the (*a*) CC configuration, (*b*) CE configuration, (*c*) CB configuration, (*d*) CB or CE configuration.

_____6-23 Which of the following configurations give(s) useful current gain? (*a*) CE only, (*b*) CE and CC, (*c*) CB, (*d*) CC only.

_____6-24 An emitter follower is basically a (*a*) CC configuration, (*b*) CB configuration, (*c*) CE configuration, (*d*) common-cathode configuration.

_____6-25 An advantage of the emitter follower over the transformer is (*a*) higher power capability, (*b*) higher power efficiency, (*c*) better frequency response, (*d*) simpler power supply requirements.

7 reading transistor specifications

OBJECTIVES

(1) From a given transistor specification sheet, state the intended functions of the transistor, the significance of variations in case size, and the symbols used to indicate maximum dc values. (2) Define the derating factor. (3) Compute the power rating using the derating factor. (4) Locate the maximum, minimum, and nominal values of hybrid parameters on a transistor specification sheet. (5) State a method for converting hybrid parameters in one configuration to parameters in another configuration. (6) Locate the power gain and noise figure on a transistor specification chart. (7) Locate the alpha cutoff frequency and output capacitance on a transistor specification sheet. (8) State the definition of pico and nano as units. (9) Contrast the high- and low-frequency gain in a transistor specification sheet. (10) Define the breakdown voltage of a transistor and locate it in a transistor specification sheet. (11) Locate values of the collector saturation current on a transistor specification sheet. (12) Define the collector saturation resistance and collector saturation voltage.

INTRODUCTION

Every major transistor manufacturer publishes *specifications* for products, generally in 2 distinct forms, each directed at a specific group of consumers. The simpler form, usually no more than a line of figures on a chart describing many different transistors, is intended for the technician or engineer who wants information relative to the "highlights" of the transistor. These include factors such as short-circuit forward-current-transfer ratio for the CB or CE configuration, or both (α_{fb} and α_{fe} respectively, or h_{fb} and h_{fe} respectively), breakdown voltages, maximum permissible collector current, alpha cutoff frequency, and power dissipation.

The data contained in the short-form specification is often sufficient for interpreting circuit diagrams in which the transistor appears. The project engineer may also use it for quick trials of new or modified circuits that call for definite parameters. The short form is also satisfactory for experimenting with replacement transistors manufactured by companies other than that called for on the original diagram.

The General Electric Types 2N332, 2N333, 2N334, 2N335, and 2N336 are silicon NPN transistors intended for amplifier applications in the audio and radio frequency range and for general purpose switching circuits. They are grown junction devices with a diffused base and are manufactured in the Fixed-Bed Mounting design for extremely high mechanical reliability under severe conditions of shock, vibration, centrifugal force, and temperature. These transistors are hermetically sealed in welded cases. The case dimensions and lead configuration conform to JETEC standards and are suitable for insertion in printed boards by automatic assembly equipment.

SILICON TYPES
2N332
2N333
2N335

DIMENSIONS WITHIN JETEC OUTLINE TO-5 JETEC BASE E5-44

NOTE 1: This zone is controlled for automatic handling. The variation in actual diameter within this zone shall not exceed .010.

NOTE 2: Measured from max. diameter of the actual device.

NOTE 3: The specified lead diameter applies in the zone between .050 and .250 from the base seat. Between .250 and 1.5 maximum of .021 diameter is held. Outside of these zones the lead diameter is not controlled.

absolute maximum ratings (25°C)

Voltages
Collector to Base (Emitter Open)	V_{CBO}	45	V
Emitter to Base (Collector Open)	V_{EBO}	1	V

Collector Current
	I_C	25	mA

Power
Collector Dissipation (25°C)	P_C	150	mW
Collector Dissipation (100°C)	P_C	100	mW
Collector Dissipation (150°C)	P_C	50	mW

Temperature Range
Storage	T_{STG}	−65°C to 200°C	
Operating	T_A	−65°C to 175°C	

		2N332			2N333			2N335			
		Min.	Nom.	Max.	Min.	Nom.	Max.	Min.	Nom.	Max.	
SMALL-SIGNAL CHARACTERISTICS											
Current Transfer Ratio	h_{fe}	9	15	22	18	30	44	37	60	90	
Input Impedance	h_{ib}	30	43	80	30	43	80	30	43	80	Ω
Reverse Voltage Transfer Ratio	h_{rb}	.25	1.5	5.0	.25	2.0	10.0	.5	3.0	10.0	$\times 10^{-6}$
Output Admittance	h_{ob}	0.0	.25	1.2	0.0	.2	1.2	0.0	.15	1.2	μmhos
Power Gain ($V_{CE}=20$ V; $I_E=-2$ mA; $f=1$ kHz; $R_G=1$ kΩ; $R_L=20$ kΩ) G_e			35			39			42		dB
Noise Figure	NF		20			15			12		dB
HIGH-FREQUENCY CHARACTERISTICS											
Alpha Cutoff Frequency ($V_{CB}=5$ V; $I_E=-1$ mA)	f_{ab}		10			12			14		mHz
Collector Capacity ($V_{CB}=5$ V; $I_E=-1$ mA; $f=1$ MHz)	C_{ob}		7			7			7		pF
Power Gain (Common Emitter) ($V_{CB}=20$ V; $I_E=-2$ mA; $f=5$ MHz)	G_e		14			14			13		dB
DC CHARACTERISTICS											
Common Emitter Current Gain ($V_{CE}=5$ V; $I_C=1$ mA)	h_{FE}		14			31			56		
Collector Breakdown Voltage ($I_{CBO}=50$ μA; $I_E=0$)	BV_{CBO}	45			45			45			volts
Collector Saturation Resistance ($I_B=2.2$ mA; $I_C=5$ mA)	R_{SC}		90	200		80	200		70	200	Ω
CUTOFF CHARACTERISTICS											
Collector Current ($V_{CB}=30$ V; $I_E=0$; $T_A=25$°C)	I_{CBO}		.002	2		.002	2		.002	2	μA
($V_{CB}=5$ V; $I_E=0$; $T_A=150$°C)	I_{CBO}			50			50			50	μA
SWITCHING CHARACTERISTICS ($I_{B1}=0.5$ mA; $I_{B2}=-0.5$ mA; $I_C=5.0$ mA)											
Delay Time	t_d		.7			.65			.6		μs
Rise Time	t_r		.65			.55			.5		μs
Storage Time	t_s		.4			.75			.9		μs
Fall Time	t_f		.13			.14			.15		μs

A design engineer, however, requires far more detailed specifications before he or she can select a particular type for a new application, or a new approach to an existing application. To satisfy these needs, transistor specifications are also available in *specification sheet* form, a sample of which is given in the figure accompanying this chapter.

Unfortunately, there are almost as many points of view concerning the information and its presentation as there are manufacturers. No two companies present the specifications in exactly the same way. Not only do they choose different parameters for listing, but they also utilize different symbols for the same parameter. The emphasis varies from one manufacturer to another, as well; an item of information that appears of prime importance to one seems to assume a much less important role to another.

Complete standardization lies in the future. The engineer and technician of the present, however, must learn to live with this difficult situation. Clearly, before beginning the study of a specification sheet, it is necessary to provide yourself with the equivalent of a vocabulary list with which you will be able to determine the exact meanings of the symbols used in the particular sheet at hand.

We shall discuss one particular specification sheet as a sample. It is the sheet for the family 2N332 to 2N335 manufactured by the General Electric Co. It was selected because of its ready availability, its completeness, and its generally excellent arrangement.

ABSOLUTE MAXIMUM RATINGS

7-1
As indicated in the upper right-hand block, the 2N332, 2N333, and 2N335 are NPN transistors fabricated from _____ as the basic semiconductor.

Refer to the specification sheet for this and the following frames in this chapter.

7-2
(*silicon*) Study the broad device capabilities and intended service as given in the upper portion of the sheet. These transistors are intended for service in both the audio- and _____-frequency ranges.

7-3
(*radio*) The transistors described may also be used for general-purpose _____ where they are connected in circuits that utilize their high-resistance *Off* and low-resistance *On* characteristics.

7-4
(*switching*) The transistors are hermetically sealed in welded cases. The outline drawing gives maximum and nominal dimensions. For example, the absolute maximum external diameter is given as _____ in.

7-5
(*0.370*) The nominal value of any measurement is the so-called design "target." The manufacturer sets up to achieve this measurement but allows for assembly variations that could make this measurement larger or _____ than the nominal value.

7-6
(*smaller*) Equipment manufacturers who will use the transistors are interested in maximum dimensions to ensure adequate space in assembly. For this reason, specification sheets show the nominal and maximum dimensions, but never the _____ dimensions.

7-7
(*minimum*) The broad description also states that these are grown-_____ junction transistors since the collector and emitter regions are grown while the base is diffused.

7-8
(*diffused*) We turn now to the *absolute maximum* ratings. All the ratings are based upon an ambient temperature of _____ °C.

7-9
(*25*) When the subscript of a symbol is given in capital letters (upper-case type), the characteristic or parameter is a dc value. Thus V_{CBO}, V_{CEO}, I_C, P_C, etc., are _____ voltages, currents, and wattages.

7-10
(*dc*) All the voltages given in the maximum ratings apply to the transistor electrodes named, but in each case the third electrode is left disconnected. This fact is indicated by the parenthetical phrases "(Emitter _____)" and "(Collector _____)."

7-11
(*Open, Open*) V_{CBO} is given as 45 V. The portion V_{CB} tells us that the voltage is being measured between collector and _____.

7-12
(*base*) The O portion of the subscript tells us that the remaining element, the emitter, is disconnected or _____.

7-13
(*open*) Thus, the dc collector voltage that must never be exceeded in the common-base configuration is _____ V.

7-14
(45) With the collector open, the maximum dc voltage that should ever be allowed to appear between the emitter and _____ is 1 V.

7-15
(base) The maximum permissible direct current flowing out of the collector terminal as measured by a milliammeter inserted between the collector and the battery is _____ mA.

7-16
(25) Dissipation ratings, such as those indicated by P_C, are *thermal* ratings. These ratings are stated in terms of watts or milliwatts and are intended to limit junction _____ to a safe value.

7-17
(*temperature*) Note that 3 different values for P_C are stipulated. One applies to an ambient temperature of 25°C, while the others are for temperatures of _____ and _____ °C.

7-18
(100, 150) The ratings show that the power-dissipation capability of the transistor _____ when the ambient temperature increases.

7-19
(*decreases*) In some specification sheets this decrease in power-dissipation ability is given in terms of "decrease in power handling per °C of temperature _____."

7-20
(*increase*) This is called the *derating factor* and is expressed in mW/°C. In this case an increase from 25 to 100°C causes a decrease in power handling from 150 to 100 mW. To find the derating factor the *change* in power is divided by the change in temperature, or 50 mW/75°C = _____ mW/°C.

7-21
(0.67 or $^2/_3$) Thus, the derating factor provides information as to the rate at which the permissible dissipation should be decreased for each degree Celsius _____ in temperature.

7-22
(*rise*) Using the derating factor calculated above, it will be found that at a temperature of 31°C power dissipation will decrease by _____ mW, and the total collector power dissipation should not exceed _____ mW.

READING TRANSISTOR SPECIFICATIONS

7-23

(*4, 146*) As another example, consider the 2N2193, a planar epitaxial transistor. The dissipation rating for a case temperature of 25°C is 2.8 W. Obviously, this transistor can dissipate considerably _____ power than the 2N332–2N335 family.

7-24

(*more*) The derating factor for the 2N2193 is 16.0 mW/°C increase in case temperature above 25°C. This means that, for every degree rise above 25°C, the dissipation rating must be reduced by _____ W.

7-25

(*0.016*) If this transistor, under actual operating conditions, were to run at 100°C, its temperature would then be _____ °C above its basic rated temperature.

7-26

(*75*) At a derating of 0.016 W for each degree, it would have to be derated $0.016 \times 75 =$ _____ W for this condition of operation.

7-27

(*1.2*) Thus, its maximum dissipation rating at 100°C is _____ W.

7-28

(*1.6*) Junctions may be damaged by extremes of temperatures. As the *temperature range ratings* show, this danger is present not only when the transistor is operating in a circuit, but also while the transistor is being _____.

7-29

(*stored*) The safe range of temperatures for storage, from −65 to 200°C, is somewhat _____ in extent than the safe range of operating temperatures.

7-30

(*larger* or *greater*, etc.) The maximum temperature to which any junction in the 2N335 should be allowed to rise during operation is _____ _____.

7-30

(*175°C*)

SMALL-SIGNAL CHARACTERISTICS

7-31
In turning to the electrical characteristics, it is important to note the *general test conditions* specified. In this case, these conditions are given on the same line as each test. The temperature at which all tests were run to obtain the ratings is _____ _____.

7-32
(25°C) For *power gain*, the collector-to-emitter dc voltage has been held at 20 V, the emitter current at −2 mA, and the frequency at _____ _____.

7-33
(1 kHz) The current amplification factor (current transfer ratio) is symbolized in this particular sheet as _____.

7-34
(h_{fe}) As for other items below, 3 different values are given for each transistor relative to h_{fe}. For example, in the case of the 2N335, the target value is 60, but some transistors may be expected to have h_{fe} values as low as _____ and as high as 90.

7-35
(37) In this family of transistors, the one having the smallest nominal current transfer ratio is the _____.

7-36
(2N332) Note that input impedance is given as h_{ib}. This parameter is given for the common-_____ configuration, as the symbol indicates.

7-37
(*base*) In an early discussion we used the figure 39 as a typical value for h_{ib}. Note that this typical value fits between the minimum and maximum values of h_{ib} specified for this family of transistors. The nominal input impedance for each member of this family is _____ Ω.

7-38
(43) In our discussions we have used the *reverse-voltage-amplification factor*. In the equations developed earlier, we have symbolized this factor with h_{re} for the CE configuration. For the CB configuration, this parameter would be symbolized by _____.

7-39
(h_{rb}) In this transistor specification sheet, the symbol h_{rb} is used and the parameter is called the _____-_____-_____ _____.

7-40

(*reverse-voltage-transfer ratio*) In a previous discussion, we gave 380×10^{-6} as a typical value of h_{rb}. Let us write the coefficient as 3.8 and change the number to read $3.8 \times$ _____ .

7-41

(10^{-4}) Thus, the typical value for h_{rb} is 3.8×10^{-4}. Note that this value falls well within the range of h_{rb} values given for this family of transistors. The widest range shown in the sheet for h_{rb} is from 0.25×10^{-4} to _____ $\times 10^{-4}$.

7-42

(*10*) This specification sheet uses the term *admittance* rather than *conductance* as we have done in previous discussions. Conductance is the reciprocal of *resistance*; admittance is the _____ of *impedance*.

7-43

(*reciprocal*) In our discussions we have employed the symbol h_{oe} for output conductance (or admittance) as applied to the common-emitter configuration. Here, we see the symbol h_{ob}. Evidently, this company states the output admittance for the _____ configuration.

7-44

(*CB* or *common-base*) In the equivalent circuit we developed earlier, the hybrid parameters for the CE configuration were worked out. These were h_{ie}, h_{re}, h_{fe}, and h_{oe}. We have just seen that this manufacturer states the hybrid parameters for the _____ configuration except for h_{fe}, which applies to the CE configuration.

7-45

(*CB*) By using conversion formulas presented in earlier discussions, the parameters for any configuration may be converted into the parameters of the other 2 configurations easily. For example, if we wanted to convert h_{fe} to match the other 3 parameters (that is, for the CB configuration), it would be converted to _____ .

7-46

(h_{fb}) The equation for this conversion is

$$h_{fb} = \frac{-h_{fe}}{1 + h_{fe}}$$

Let us find h_{fb} for the 2N335 using its nominal rating. The nominal value of h_{fe} for the 2N335 is _____ .

7-47
(60) Substituting in the equation we have: $h_{fb} = -60/(1+60)$. Thus, the value of h_{fb} is quickly found to be ——————.

7-48
(−0.98) This is a typical value for h_{fb}. In an actual circuit, the values of R_G (generator resistance) and R_L (—————— ——————) would be known.

7-49
(*load resistance*) Knowing these resistances, we could then calculate the voltage gain A_v, current gain A_i, power —————— G, input resistance r_i, and output resistance r_o. (Student should review these formulas.)

7-50
(*gain*) This manufacturer states the power gain in the common-—————— circuit by giving the value for G_e.

7-51
(*emitter*) The values of G_e given are correct, however, only for the conditions stated. That is, the collector-emitter dc voltage is 20 V, the emitter current is −2 mA, the test frequency is —————— ——————, R_G is 1 kΩ, and the load resistance is 20 kΩ.

7-52
(*1 kHz*) For small-signal conditions, at a frequency of about 1 kilohertz (kHz) used as the input signal, we could therefore expect the output signal from a 2N335 to be —————— dB above the input signal in power content.

7-53
(42) As a last item in this section, consider the noise figure, NF. (This rating is frequently called the *noise factor* and may also be symbolized by ——————.)

7-54
(F_o) A lower noise figure means better noise quality in amplifiers. Of the 3 transistors in this family, it is evident that the —————— has the best noise quality.

7-54
(2N335)

HIGH-FREQUENCY AND DC CHARACTERISTICS

7-55
The first item under *high-frequency characteristics* is the —————— cutoff frequency.

7-56

(*alpha*) In addition, the symbol $f_{\alpha b}$ informs us that the manufacturer is stating the cutoff frequency of the _____-_____ configuration.

7-57

(*common-base*) The design and fabrication of the 2N335 is such that its forward-current-amplification factor (transfer ratio) will drop to 0.707 times its value at 1 kHz when the frequency rises to _____ MHz.

7-58

(*14*) Of the 3 transistors in this family, the one that is most limited with regard to the highest frequency it will amplify without excessive dropoff is the _____.

7-59

(*2N332*) It should also be noted that the specification sheet informs us that the general test conditions are $V_{CB} = 5$ V and the emitter current is _____ mA.

7-60

(*−1*) The next factor is C_{ob}. This is the capacitance from collector to base, measured across the output terminals in the _____ configuration.

7-61

(*CB*) This capacitance is given as _____ pF for all 3 transistors in the family.

7-62

(*7*) Many manufacturers have begun to use a new series of abbreviations for decimal multipliers. For example, a picofarad = 10^{-12} farad. Thus, a picofarad (or pF) is identical with a _____ farad.

7-63

(*micromicro* or $\mu\mu$) Another relatively new abbreviation you will encounter is the *nano* prefix. This prefix represents a multiplier of 10^{-9}. Thus, a capacitance of 7 pF may be written as a capacitance of _____ nF.

7-64

(*0.007*) Similarly, a capacitance of 25 nF may be written as a capacitance of _____ pF.

7-65

(*25,000*) The values given for the next item disclose that at high frequencies (in this case _____ _____), there is a decline in the performance characteristics of the transistor.

7-66
(*5 MHz*) Note that at the higher frequency it has a _____ value, as compared with its value at 1 kHz.

7-67
(*smaller*) Going one step further, the manufacturer specifies a power gain of _____ rather than 42 for the 2N335.

7-68
(*13*) At this comparatively high frequency, the power gain G_e is still significant but is much _____ than it is at the lower frequencies.

7-69
(*smaller*) Looking further into the family, we note a departure from the former trends. With regard to high-frequency power gain, the *poorest* transistor of the family is the _____.

7-70
(*2N335*) Your attention is called to the fact, however, that the difference between 14 and 13 dB is not particularly significant and that these transistors may be considered to perform equally well in this regard at _____ frequencies.

7-71
(*high*) Turning next to the *dc characteristics*, we encounter the rating for the *collector breakdown voltage*. For this specification, the collector-base current is 50 µA, the emitter current is _____, and the ambient temperature is 25°C.

7-72
(*zero*) The breakdown voltage test is made, in this case, by applying a gradually increasing reverse bias to the _____-base junction.

7-73
(*collector*) While this voltage is being applied, the emitter is floating or disconnected. This fact is made clear in two ways. First, the symbol I_{CBO} tells us that the current is being measured between collector and base with the emitter _____.

7-74
(*open*) Second, the fact that the emitter is open is pointed out by the value of the emitter current. Since this current is _____, the emitter must be open.

7-75

(*zero*) The breakdown voltage is the voltage where the junction which previously drew a leakage or saturation current of 50 μA suddenly begins to conduct heavily. That is, at the breakdown voltage point, the collector-base current will suddenly _____ to a high value.

7-76

(*rise* or *increase, etc.*) This high reverse current is known as the *avalanche current* and in most instances should be avoided if transistor damage is not to occur. This means that the collector _____ voltage rating should never be exceeded.

7-77

(*breakdown*) The item under *cutoff characteristics* is the collector cutoff current I_{CBO}. As shown by the symbol, this is the collector-base current with emitter _____.

7-78

(*open*) The nomenclature we have employed for I_{CBO} in previous discussions was *leakage* or _____ current.

7-79

(*saturation*) The values given for I_{CBO} in the 2 lines illustrate dramatically the importance of ambient temperature. At a temperature of 25°C, the maximum saturation current to be expected is only 2 μA. However, at a temperature of _____ _____, the value of I_{CBO} can rise as high as 50 μA.

7-80

(*150°C*) Many specification sheets, under the heading of dc characteristics, provide a value for either *collector saturation voltage* or *collector saturation resistance*. For the 2N335, the latter value is given and is symbolized by _____.

7-81

(R_{SC}) Collector saturation voltage $V_{CE(SAT)}$ (not given on this sheet) is the collector-to-emitter voltage drop which exists when a certain large specified base current, I_B, flows. The collector _____, I_C, will also be specified.

7-82

(*current*) Collector saturation resistance R_{SC} is defined as the ratio between the collector saturation voltage and the specified current. That is,

$$R_{SC} = \frac{?}{I_C}$$

7-83

($V_{CE(SAT)}$) This characteristic is important in switching circuits where transistors are either Off (or nonconducting) or On (_____ under saturation conditions).

7-84

(*conducting*) That is, it is important for the engineer who is designing a transistor-switching arrangement to know either the voltage drop across the emitter-collector electrodes, or the _____ between these electrodes at the time that the transistor is On, or in saturation.

7-85

(*resistance*) Thus, the value of R_{SC} is actually the internal resistance of the transistor "switch." A perfect switch, when closed, has a resistance of 0 Ω. The nominal resistance of the 2N335 "switch" is _____ Ω.

7-86

(*70*) The absolute limit allowed for R_{SC} in production is _____ Ω of saturation resistance.

7-86

(*200*) Note to student: Switching characteristics are related to pulse techniques and will not be discussed at this time.

CRITERION CHECK TEST

All the questions are based on the specifications for the 2N332 transistor.

____7-1 The ratings given with symbols using capital letters and capital subscripts denote (*a*) ac parameters, (*b*) dc parameters, (*c*) RMS values, (*d*) effective values.

____7-2 If a transistor is used for amplifying very weak signals, it is important to know its (*a*) power rating, (*b*) rise time, (*c*) noise figure, (*d*) delay time.

____7-3 The noise figure given for the 2N332 transistor is (*a*) 20 dB, (*b*) 20 V, (*c*) 20 mA, (*d*) 20 Ω.

____7-4 The maximum allowable collector-to-base voltage at 25°C is (*a*) 25 V, (*b*) 35 V, (*c*) 40 V, (*d*) 45 V.

____7-5 The value given for the base-collector capacitance is (*a*) 7 nF, (*b*) 7 pF, (*c*) 0.07 μF, (*d*) 7 μF.

____7-6 The nominal value of the dc saturation resistance is (*a*) 50 mhos, (*b*) 50 Ω, (*c*) 90 Ω, (*d*) 200 Ω.

____7-7 Above 100°C, the power dissipation derating factor is (*a*) 1 mW/°C, (*b*) 2 mW/°C, (*c*) 3 mW/°C, (*d*) 4 mW/°C.

____7-8 Nominal h_{fe} is _____, maximum h_{fe} is _____: (*a*) 12, 17, (*b*) 14, 27, (*c*) 15, 22, (*d*) 18, 36.

first principles of electron tubes

OBJECTIVES

(1) Describe the origin of thermionic electrons in an incandescent lamp. (2) Sketch a two-element vacuum tube diode, labeling filament, filament battery, plate, and B battery, and trace the flow of electrons through the plate circuit. (3) Depict a curve of plate current versus plate voltage for a typical vacuum tube diode. (4) Discuss the operation of a vacuum tube diode as a rectifier. (5) Draw a schematic of a diode in series with an ac generator and load resistor and determine the load resistor voltage waveform. (6) Relate the work function of a metal to thermionic emission. (7) List several desirable properties of material to be used as a vacuum tube filament. (8) List at least two advantages of using separate elements for the filament and cathode. (9) Give the reason for evacuating a vacuum tube. (10) Distinguish between a hard and soft tube, and give the circuit symbol for each. (11) Trace the development of space charge in a vacuum tube with no plate voltage. (12) Depict a series of curves which plot vacuum tube plate current versus filament voltage for several values of plate voltage. (13) Define and discuss the phenomenon of plate current *temperature saturation* and why a vacuum tube is operated in this region. (14) Depict a graph which plots vacuum tube plate current versus plate voltage for a fixed value of filament voltage. (15) Define and discuss the phenomenon of *plate saturation*. (16) Obtain the dc plate resistance of a vacuum tube. (17) Sketch a test circuit to obtain plate characteristic curves, labeling and indicating the functions of meters and variable voltage supply. (18) Discuss the peak-inverse-voltage (PIV) rating of a diode. (19) Relate the input voltage waveform to the peak inverse voltage.

VACUUM TUBE DIODES

Our study of electron tubes starts with *thermionic emission,* a phenomenon that is fundamental to the operation of all types of standard electron tubes found in radio, radar, sonar, television, etc. As you will discover, thermionic emission as it occurs in an electron tube is, most often, a one-way process where electron motion can occur only from a heated body to some other body which is not heated. Such a unidirectional characteristic enables us to utilize thermionic emission for converting alternating current to pulsating direct current, a very important process in all aspects of electronics.

8-1
When an electric current of sufficient intensity is passed through a thin filament of tungsten, the filament can be made to glow as in an incandescent lamp. The light produced results from the conversion of electric energy into _____ energy, which in turn causes emission of light.

8-2
(*heat*) When a thin wire of any material is heated to a sufficiently high temperature, electrons are forced out of its surface into the surrounding space. Electrons are charges of electricity carrying a _____ sign.

8-3
(*negative*) In an ordinary incandescent lamp, the electrons that "boil off" the white hot surface move inside the glass envelope freely since the air has been evacuated from the space. Such a space where gases have been removed is called a _____.

8-4
(*vacuum*) An ion is a charged particle. Electrons emitted from hot filaments are called thermionic electrons. This word comes from the roots *therm*, meaning heat, and *ionic*, meaning having the nature of a _____ particle.

8-5
(*charged*) The heated filament from which the electrons are emitted is initially a neutral body. Upon losing electrons by thermionic emission, the filament must take on a _____ charge.

8-6
(*positive*) Thus, if the electrons are slow-moving when they come off the filament, they will tend to circle back and return to the filament since a positive body will _____ negative electrons.

8-7
(*attract*) Some electrons will escape from the positive field of the filament to be absorbed in the glass envelope. To do this, the electrons must have a relatively high _____.

8-8
(*speed*) If a metal plate is inserted in the glass envelope, some of the fast-moving electrons that escape from the field of the filament will be absorbed by the plate. Thus, the plate, if isolated, will take on a _____ charge.

8-9

(*negative*) If this plate is connected outside the tube through a milliammeter back to the filament (Fig. 8-9), the meter will show a small reading. This means that _____ are being drawn from the plate back to the filament via path A–B–C–D–E.

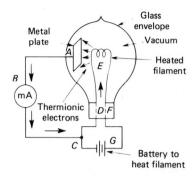

Fig. 8-9

8-10

(*electrons*) The battery shown in Fig. 8-9 is there *merely to heat the filament*. It has no effect on the electron flow from the metal plate, through the meter, and back to the f_____.

8-11

(*filament*) The heating current flowing from the battery through the filament flows in a circuit called the *filament circuit,* represented by the path C–D–E–F–G. The circuit with the meter in it is a separate one called the *plate circuit*. The electrons flowing in this circuit are the ones resulting from th_____ e_____ from the hot filament.

8-12

(*thermionic emission*) All meters bear signs that indicate their terminal polarities. Electrons should always enter a meter at the "−" terminal. Since electrons in the plate circuit enter the meter at the top terminal (point B), the meter should be connected in the circuit so that its _____ terminal is at B.

8-13

(*negative*) Twenty years after thermionic emission was discovered, the first practical *diode* or *2-element vacuum tube* was invented by Fleming in England (Fig. 8-13). This tube contains a heated filament and a _____ inside an evacuated glass envelope.

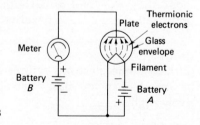

Fig. 8-13

8-14

(*plate*) The filament of the tube in this circuit is heated by battery _____.

8-15
(A) The heating circuit of Fig. 8-13 starts at the "−" end of battery A. Electrons flow from this point, through the _____ and back to the "+" end of battery A.

8-16
(*filament*) The *plate circuit* of Fig. 8-13 is considered to start with the thermionically emitted electrons from the filament which move through the vacuum to the plate, through the meter, through battery _____, and finally back to the filament.

8-17
(B) Note that the "+" terminal of battery B is connected (through the meter) to the plate of the tube. This makes the plate positive with respect to the filament. The filament is negative with respect to the plate because the filament is connected to the _____ terminal of battery B.

8-18
(*negative*) When no battery is used in the plate circuit (as in Fig. 8-9), electrons trapped by the plate get there purely by chance. In the new circuit (Fig. 8-13), electrons are drawn to the plate because electrons are negative and the plate is _____.

8-19
(*positive*) Thus, a far greater number of electrons reach the plate and travel back to the filament via the plate circuit. In this circuit, therefore, the meter will show a substantially _____ plate current than in the circuit of Fig. 8-9.

8-20
(*larger*) Refer to Fig. 8-20. By connecting the battery in the circuit, an increase of plate current is observed. We may now view the path of the electrons from filament to plate by thermionic emission as a kind of _____ that is equivalent to the *space opposition* offered to the flow of electron current.

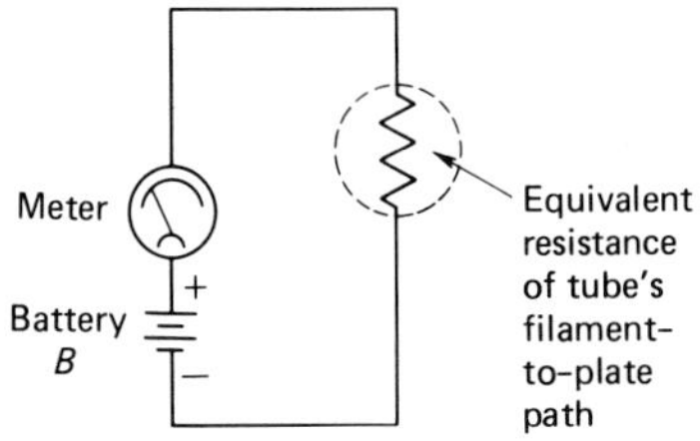

Fig. 8-20

8-21
(*resistance*) This equivalent resistance is very low because electrons can pass very freely through a vacuum and few of them will escape from the plate circuit as long as the plate is kept positive with respect to the _____.

FIRST PRINCIPLES OF ELECTRON TUBES

8-22

(*filament*) In the circuit of Fig. 8-22, a significant change has been made. This change consists in a reversal of the polarity of _____ _____.

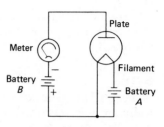

Fig. 8-22

8-23

(*battery B*) Since the positive terminal of battery *B* is now connected to the filament, and the negative terminal to the plate, the plate is _____ with respect to the filament.

8-24

(*negative*) Electrons carry a negative charge. As electrons are emitted from the hot filament, they will be _____ (attracted, repelled) by the plate since both the plate and the electrons have the same sign of charge.

8-25

(*repelled*) If electrons are not attracted to the plate, no current can flow from filament to plate. Thus, the space inside the tube now resembles an *open circuit*. Contrasted with a *complete circuit* in which current flows easily, the resistance of an open circuit is very _____.

8-26

(*high*) Thus, when the plate of a diode is positive with respect to the filament, the equivalent resistance of the tube is very low. When the plate is negative with respect to the filament, the equivalent resistance is very _____. (This is summarized in Fig. 8-26.)

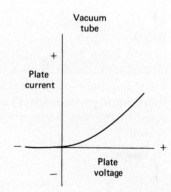

Fig. 8-26

8-26
(*high*)

202 A PROGRAMMED COURSE IN BASIC ELECTRONICS

SIMPLE DIODE RECTIFIER

8-27
A diode can be used as a *rectifier*. A rectifier changes alternating current to direct current. To start the rectifier circuit, we draw the diagram in Fig. 8-27. In this diagram, battery B (henceforth called the B battery) of the previous circuits has been replaced by a(n) _____ generator.

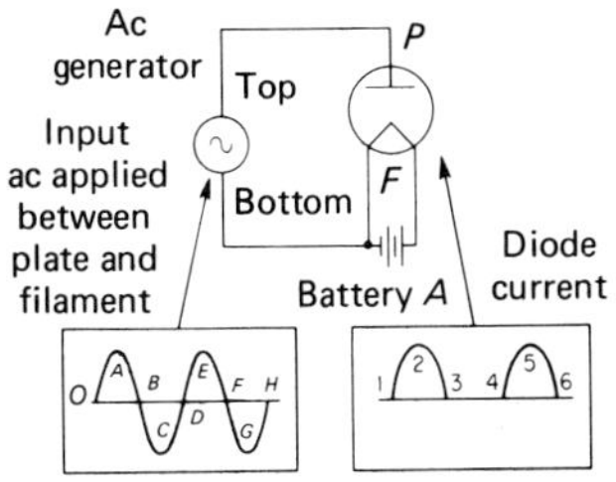

Fig. 8-27

8-28
(*ac*) The output of the generator is ac and is applied between the filament and plate. If we assume that for input wave condition A (see insert box), the top of the generator is "+" and the bottom is "−," then for this condition the plate will be _____ with respect to the filament.

8-29
(*positive*) Thus, for most of the wave interval OAB, the plate is positive with respect to the filament, and the diode will display a _____ equivalent resistance.

8-30
(*low*) This means that the tube will conduct just as though its filament were connected to its plate by a resistor. Current will therefore flow through the tube and the circuit. As the input voltage rises and falls from O to A to B, the tube current will similarly rise and fall as shown in the *diode current* insert as the interval from _____ to _____ to _____ (Fig. 8-27).

8-31
(*1, 2, 3*) When the input ac voltage reverses in polarity, the top of the generator becomes negative and the bottom positive. This occurs when the input wave is covering the interval from B to _____ to _____ .

8-32
(*C, D*) For the interval BCD, the plate is therefore _____ with respect to the filament.

8-33
(*negative*) When the plate is negative with respect to the filament, the tube cannot conduct and displays a very _____ equivalent resistance.

FIRST PRINCIPLES OF ELECTRON TUBES

8-34
(*high*) Thus, for the input interval BCD, the output current through the tube is _____.

8-35
(*zero*) This interval is shown in Fig. 8-27 in the *diode current* insert as the period from _____ to _____.

8-36
(*3, 4*) On the third half-cycle, the input voltage covers the interval from D to _____ to _____.

8-37
(*E, F*) Since this half-cycle is an exact duplicate of the first one, the plate again goes positive with respect to the filament. The tube conducts once more, current flows, and the output current in the tube takes the form of the interval from _____ to _____ to _____ in the insert in Fig. 8-27.

8-38
(*4, 5, 6*) The circuit has been drawn somewhat differently in Fig. 8-38. Note first that this circuit has a new part. This is labeled _____ and stands for load resistance.

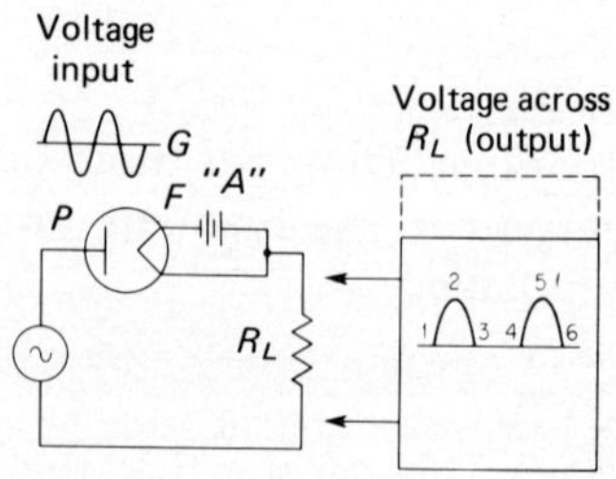

Fig. 8-38

8-39
(R_L) Each time the diode conducts (plate positive with respect to filament), a very large current tends to flow in the tube. Since this would endanger the tube and generator, _____ is inserted in series to hold the current down to a safe value.

8-40
(R_L) The diode current shown in Fig. 8-38 flows only in one direction, although it rises and falls from some maximum value to zero once in each cycle. Since a unidirectional current is dc regardless of its fluctuations, the current in the tube in Fig. 8-38 is pulsating _____.

8-41
(*dc*) As the pulsating dc flows through the load resistor R_L in Fig. 8-38, it produces a pulsating voltage drop having the same waveform. Thus, the voltage drop across R_L may be used as a new source of pulsating _____.

204 A PROGRAMMED COURSE IN BASIC ELECTRONICS

8-42
(*dc*) Thus, in Fig. 8-38, the voltage drops shown in the insert are pulsating dc. These pulses of voltage appear for every interval of the input ac when the plate is _____ with respect to the filament.

8-43
(*positive*) The gap labeled 3–4 in the insert in Fig. 8-38 represents zero voltage. This gap appears in the output voltage waveform during each interval in which the plate is _____ with respect to the filament.

8-43
(*negative*)

VACUUM TUBE HEATERS
8-44
Some metals emit electrons freely at relatively low temperatures. Such metals are said to have a low *work function*. For example, tungsten emits electrons at a relatively low temperature; hence tungsten has a low _____ _____ .

8-45
(*work function*) The metal tantalum also has a low work function. Thus, tantalum emits electrons freely at _____ temperatures.

8-46
(*low*) A good thermionic emission material for filaments must have not only a low work function but also a high melting and vaporization point. Tantalum has a low vaporization point, while that of tungsten is high. Hence, _____ is more suitable as a filament material for vacuum tubes.

8-47
(*tungsten*) *Pure* tungsten, although possessing a lower work function than most other pure metals, still must be heated to white heat for copious emission. When thorium oxide is added to tungsten, emission occurs at lower temperatures. Thus, thoriated tungsten has a _____ work function than pure tungsten.

8-48
(*lower*) Modern small receiving tubes use an emitter composed of oxides of barium and strontium. These substances have very low work functions compared even with thoriated tungsten, hence require very low _____ for operation.

FIRST PRINCIPLES OF ELECTRON TUBES

8-49

(*temperatures*) Pure tungsten requires a temperature of about 2300°C for sufficient electron emission; thoriated tungsten requires about 1600°C; filaments of the _____ _____ _____ _____ type need temperatures of 900°C or less.

8-50

(*barium and strontium oxide*) Figure 8-50 shows a vacuum tube of the *directly heated* type. In this tube type, the heating current flows through the filament and heats it to a high enough temperature to cause thermionic _____.

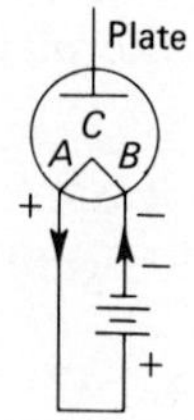

Fig. 8-50

8-51

(*emission*) In this tube, the filament is the emitter. As current flows through the filament, however, a voltage drop equal to the heater battery voltage must occur between the points labeled _____ and _____.

8-52

(*A, B*) Thus, some parts of the filament are more positive than other parts. Point *A* is more positive than point *C*, and point *C* is more _____ than point *B*.

8-53

(*positive*) Point *A* being positive tends to hold back emission by attraction; similarly, electrons will not be emitted as copiously from point _____ as they are from point *B*.

8-54

(*C*) If the filament is heated by ac, the emission is more uniform but another defect appears; since the ac passes through zero twice in each cycle, the filament tends to cool and heat at a rate equal to twice the power-line frequency. This causes the electron emission to vary at the same rate. Thus, the magnitude of the current flowing from filament to plate must also _____.

8-55
(*change* or *vary*) A much improved vacuum tube can be made by enclosing the heater in a sleeve of metal (Fig. 8-55). The filament is now only a heater, not an emitter. The sleeve, or cathode, is oxide-coated and is made red hot by the heater. Thus, in this type of tube, thermionic electrons come from the _____.

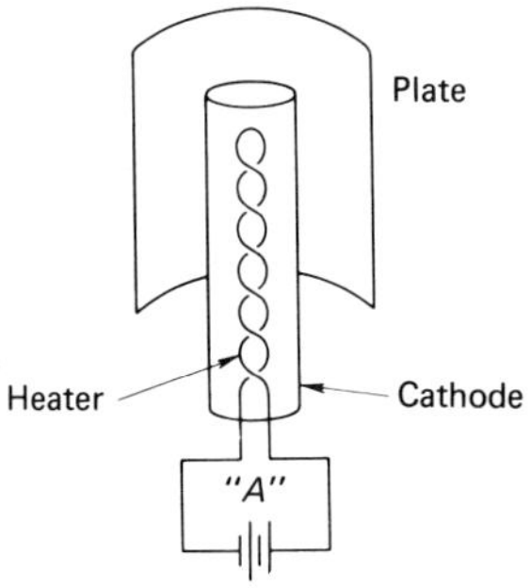

Fig. 8-55

8-56
(*cathode*) The cathode is an *equipotential* surface. Since it is not connected to the A battery, there is no voltage drop along its surface. Because its potential is the same all over, the electron emission is quite _____.

8-57
(*uniform*) The mass of the cathode is considerably higher than the mass of the original filament. Such an emitter has *thermal inertia*. Because of its large mass it tends to retain its heat even when the ac dips through zero. Such inertia makes for _____ rather than intermittent emission with ac heating.

8-58
(*steady, constant,* or *unvarying*) Figure 8-58 shows the schematic representation of an *indirectly heated diode*. As may be plainly seen, the _____ battery does not form part of the circuit from cathode to plate.

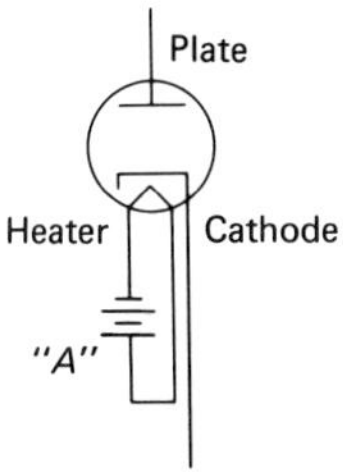

Fig. 8-58

8-59
(*A*) The only function of the filament in this type of tube is that of heating the _____.

8-60
(*cathode*) Heater current does not flow through the cathode; hence there is no voltage drop due to A battery current in the cathode. This makes the cathode an _____ surface.

FIRST PRINCIPLES OF ELECTRON TUBES

8-61
(*equipotential*) When ac is used to heat the filament or heater of an indirectly heated tube, the cathode tends to maintain steady temperature, and hence steady emission, even though the heater current is periodic. The characteristic of constant temperature due to large mass is called thermal _____.

8-62
(*inertia*) A rectifier circuit using an indirectly heated tube is shown in Fig. 8-62. Electron current flows in the direction indicated by the arrow, producing a voltage drop as shown across R_L. In this circuit, the cathode is symbolized by the letter _____.

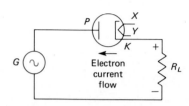

Fig. 8-62

8-63
(*K*) In Fig. 8-62, the heater symbol is shown inside the glass envelope. Customarily, however, the heater circuit is drawn separately. Thus, heater power would be supplied to the points denoted by _____ and _____ in Fig. 8-63.

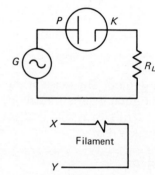

Fig. 8-63

8-63
(*X, Y*)

SPACE CHARGE IN THE VACUUM TUBE
8-64
If oxygen were present in a tube, the heater would quickly oxidize and deteriorate until it burned out. To prevent this, as much of the air (containing oxygen) is removed as is practical. A space that is free of all gases is called a _____.

8-65
(*vacuum*) Tubes that have no gas in their envelopes are called vacuum tubes, or *hard* tubes, and are symbolized as in Fig. 8-65. The type of tube shown belongs to the class of _____ heated types of tubes.

Fig. 8-65

208 A PROGRAMMED COURSE IN BASIC ELECTRONICS

8-66

(*indirectly*) In some diodes, an inert gas (not oxygen) is admitted into the envelope at low pressure after all the air is removed. These are *gas* tubes, or *soft* tubes, symbolized as in Fig. 8-66. From this drawing, it is clear that the presence of gas is denoted by the _____ inside the envelope.

Fig. 8-66

8-67

(*dot, disk, or circle*) Consider a vacuum tube with only its heater operating, and with plate and cathode unconnected. As electrons are emitted from the neutral cathode, the latter is left with a _____ charge, having lost negative bodies (Fig. 8-67).

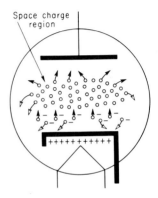

Fig. 8-67

8-68

(*positive*) The positive cathode exerts an attracting force on the electrons moving away from it, causing some of them to stop moving and others to return. The dark arrows in Fig. 8-67 show the emitted electrons; the _____ arrows near the cathode represent the electrons returning to the cathode.

8-69

(*light*) Because of the large number of electrons that are slowed down or stopped by the slightly positive cathode, a cloud of electrons forms in the space between the _____ and plate.

8-70

(*cathode*) This cloud of electrons is called the *space charge*. As more electrons enter the cloud, the magnitude of the total charge on the cloud becomes ever more _____ since it consists entirely of electrons.

8-71

(*negative*) The highly negative space charge repels other electrons. When the space charge becomes large enough, the force it exerts becomes strong enough to prevent new electrons from being emitted from the _____.

FIRST PRINCIPLES OF ELECTRON TUBES

8-72

(*cathode*) Some of the electrons in the space-charge cloud start to drift away due to mutual repulsion of like charges. The electrons which drift away are absorbed by the glass. This reduction of the negative space charge then allows an equal number of new electrons to be emitted from the _____ and enter the electron cloud.

8-73

(*cathode*) Thus a balance is soon reached wherein just enough electrons leave the cathode to make up for those that are drifting out of the _____-_____ region.

8-74

(*space-charge*) The circuit of Fig. 8-74 helps us understand the effects of space charge. The resistor R is a potentiometer that can adjust the voltage applied to the _____ of the tube.

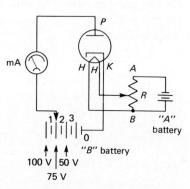

Fig. 8-74

8-75

(*heater*) As the wiper of the potentiometer is moved toward point A, the voltage applied to the heater _____.

8-76

(*increases*) Variation of the heater voltage enables the user to set the temperature of the heater at any figure between wide limits. By adjusting the heater voltage, however, he also changes the temperature of the _____ of the tube.

8-76
(*cathode*)

TEMPERATURE SATURATION

The complete circuit traversed when electrons flow from the cathode to the plate, through the meter and B battery, and back to the cathode, is called the *plate circuit*, and the potential difference between the plate and cathode is called the *plate voltage*.

8-77
Meter *mA* (Fig. 8-74) measures the current flowing in the _____ circuit of the diode.

8-78
(*plate*) By moving the tap on the *B* battery (Fig. 8-74), it is possible to change the _____ voltage from 100 to 75 V, or to 50 V.

8-79
(*plate*) Refer to Fig. 8-79. Let us start with the filament voltage set for 2 V and the plate voltage tap on the 50-V point. Assume that the milliammeter shows a reading of 2 mA for this condition. Now, if the heater voltage is changed to 4 V, the plate current rises to 8 mA, showing that the *hotter* cathode is emitting _____ electrons.

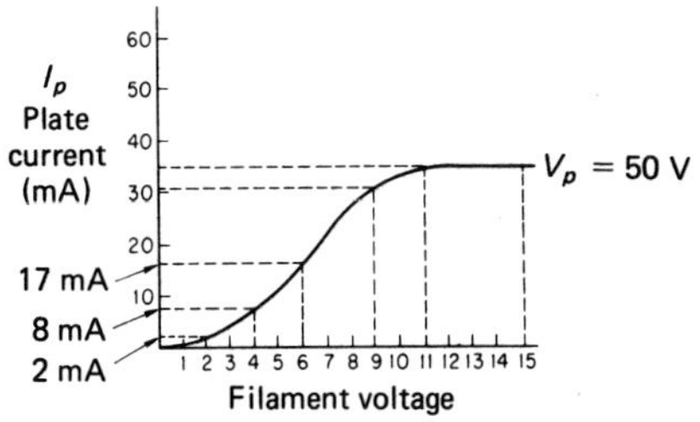

Fig. 8-79

8-80
(*more*) At a heater voltage of 6 V, the plate current has risen to 17 mA as this process continues. When the heater voltage is up to 9 V, the cathode temperature is high enough to provide a plate current of _____ mA.

8-81
(*30*) At a heater voltage of 11 V, the plate current is up to 35 mA. But, at 12, 13, 14, and 15 V, the plate current refuses to rise above _____ mA.

8-82
(*35*) This leveling off is the result of space charge. When the filament voltage is less than 11 V, the plate of the tube—with its positive voltage of 50 V—is able to attract electrons out of the space charge as fast as the _____ can supply them.

8-83
(*cathode*) As the filament voltage reaches 11 V, the cathode has become so hot that it is supplying _____ to the space charge as fast as the plate can pull them away.

8-84
(*electrons*) At 12 V, the cathode supplies even more electrons to the space charge. Since the plate voltage is unchanged, however, it cannot draw any more electrons than before. Thus, the plate current does not rise beyond _____ mA.

8-85
(35) The same condition is true at filament voltages of 13, 14, and 15 V. The tube is now in a condition of *temperature saturation*. Thus, even if the heater voltage were increased to 16 volts, the plate current would be _____ mA.

8-86
(35) Temperature saturation is defined as a condition in which, for a given plate voltage, no further increase of plate current can occur regardless of the temperature to which the _____ is raised.

8-87
(*cathode*) In Fig. 8-79 plate current is symbolized by I_P, and the voltage between plate and cathode (or simply plate voltage) is symbolized by _____.

8-88
(V_P) The filament voltage at which temperature saturation occurs for the tube whose characteristic is given in Fig. 8-79 is _____ V.

8-89
(11) If the plate voltage is now raised from 50 to 75 V, the plate becomes considerably more positive. In this condition, it exerts a _____ force on the electrons in the space charge than it did before.

8-90
(*greater*) Refer to Fig. 8-90. Since the force exerted by the plate is greater than before, the plate can pull electrons out of the space charge faster. Thus, for the new voltage of 75 V for V_P, the cathode temperature must be _____ beyond what it was before, in order to reach temperature saturation.

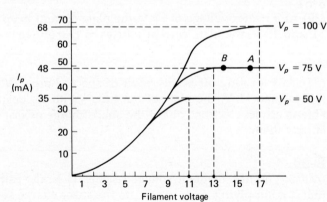

Fig. 8-90

8-91
(*increased*) In Fig. 8-90, it is evident that temperature saturation for a plate voltage of 75 V does not occur until the filament voltage reaches _____ V.

8-92
(13) The temperature saturation current in this case is _____ mA.

8-93
(48) The same trend continues when the plate voltage is raised to 100 V. Here, the cathode temperature must be raised still further before saturation occurs. In this case, the filament voltage must be raised to _____ V to cause saturation.

8-94
(17) The saturation plate current for a V_P of 100 V is evidently _____ mA.

8-95
(68) In summary, when the cathode temperature is raised with plate voltage held constant, a point is reached where no further increase of temperature causes an increase in plate current. In this condition, the tube is said to be temperature-_____.

8-96
(*saturated*) One should note that an important advantage can be gained by operating in the temperature saturation region. Consider operating the vacuum tube at point A, in Fig. 8-90, where the plate voltage V_P is 75 V and the plate current I_P is _____ mA.

8-97
(48) Suppose the filament ages after many hours of use so that the effective temperature of the filament moves to point B. But, the operating point is still at $V_P =$ _____ V and $I_P = 48$ mA.

8-98
(75) Thus, operating the tube in the temperature-saturated region has made the operating point fairly _____ (dependent, independent) of the filament output.

8-98
(*independent*)

FIRST PRINCIPLES OF ELECTRON TUBES **213**

PLATE SATURATION

8-99
Keeping the heater temperature constant, let us now observe the effect of changing plate voltage upon plate current. In Fig. 8-99, the plate current, symbolized by I_P, is shown on the Y axis because plate current is the _____ variable in the relation between I_P and V_P.

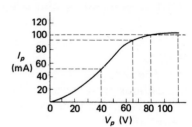

Fig. 8-99

8-100
(*dependent*) From Fig. 8-99, it can be seen that, if the filament or heater temperature is maintained constant, the plate current _____ as the plate voltage is increased.

8-101
(*increases*) From Fig. 8-99, it is clear that the plate current is _____ when the plate voltage is zero.

8-102
(*zero*) As the plate current rises, the rise is reasonably linear over a part of the curve. The center of the linear part of the curve occurs where the plate voltage is about 40 V. At this point the plate current is approximately _____ mA.

8-103
(*50*) When the plate current reaches about 92 mA, the curve shows a definite tendency to flatten out. It is noted that, at this point, the plate voltage is about _____ V.

8-104
(*65*) After the plate voltage reaches about 80 V and the plate current attains a value of about 100 mA, further increase in plate voltage causes very _____ increase in plate current.

8-105
(*little*) At a plate voltage of 65 V, the curve marks the beginning of a region known as the region of *plate saturation*. In the plate saturation region, increases of plate voltage have very little effect upon _____ _____.

214 A PROGRAMMED COURSE IN BASIC ELECTRONICS

8-106
(*plate current*) When the diode reaches the plate saturation region, the plate is taking all the _____ that the cathode can emit at that particular temperature.

8-107
(*electrons*) The tiny increase in plate current occurring between the 80- and 110-V points is due to the strong attractive field now produced by the high plate voltage. Under such a strong electrostatic field, the cathode is encouraged to emit additional _____, which results in a slightly increased plate current.

8-108
(*electrons*) The emission of extra electrons from the cathode due to the high intensity electric field produced by the plate is known as *field emission* and sometimes as the *Schottky effect*. Unlike *thermionic emission*, field emission does not require that the cathode be at a _____ temperature.

8-109
(*high*) The curve shown in Fig. 8-109 is the plate voltage–plate current characteristic curve of a diode called a 5Y3GT. This curve shows the dependence of plate _____ upon plate voltage over a wide range of values.

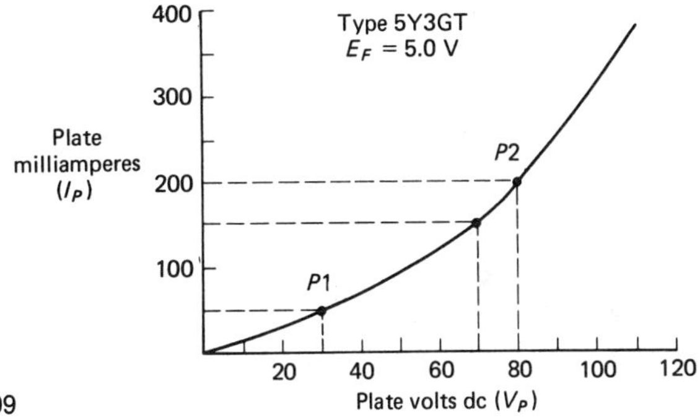

Fig. 8-109

8-110
(*current*) The curve is correct only for a 5Y3GT and only when the filament voltage is _____ V.

8-111
(*5.0*) The curve shows that the plate current of a 5Y3GT, like most diodes, is _____ when the plate voltage is zero.

8-112
(*zero*) Since no tendency to flatten out is observed on this curve, it is evident that the plate voltage has not been raised sufficiently to approach the condition of _____ _____.

FIRST PRINCIPLES OF ELECTRON TUBES

8-113
(*plate saturation*) Curves like this are valuable because they show how much plate current flows for any given value of plate voltage within the curve range. For example, the plate current flowing when the plate voltage is 69 V is _____ mA.

8-114
(*150*) At P1, the plate current is 50 mA. This is the result of the application of a plate voltage of _____ V.

8-115
(*30*) At P2, the plate voltage is 80 V. This results in a plate current of _____ mA.

8-115
(*200*)

PLATE RESISTANCE

As previously mentioned, the space between the cathode and plate of a diode offers a definite resistance to the flow of electrons between these 2 elements. When such a resistance is measured under dc conditions, i.e., when the plate voltage is dc and when the observed plate current is also dc, the resistance is then known as the *dc plate resistance* of the tube. The symbol for dc plate resistance is R_p.

8-116
The dc plate resistance of a diode is the opposition in ohms offered by the tube to the flow of plate current. It is found from Ohm's law: $R_p = V_P/I_P$. Thus, at point P1, the plate resistance is $R_p = 30/0.05 =$ _____ Ω.

8-117
(*600*) The dc plate resistance of the 5Y3GT at point P2 is $R_p = V_P/I_P =$ _____ Ω.

8-118
(*400*) This shows that the dc plate resistance of a diode is not *constant* over the entire range of its V_P versus I_P curve. This curve shows that the _____ resistance varies from point to point over the range covered.

8-119
(*plate*) Figure 8-119 shows the circuit used to obtain the values needed for plotting the V_P versus I_P curve of the 5Y3GT discussed in the previous sections. In this circuit, meter M1 is connected across the diode from plate to cathode, hence is used to measure the _____ _____ of the tube.

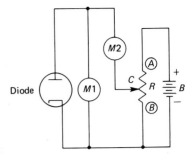

Fig. 8-119

8-120
(*plate voltage*) In the same figure, M2 is in series with the tube and voltage divider R. Thus, M2 measures the _____ _____ of the tube.

8-121
(*plate current*) The plate voltage is at a maximum when the wiper of R is moved all the way to position _____.

8-122
(*A*) To reduce the plate voltage, the wiper of R should be moved toward point _____ on the voltage divider.

8-123
(*B*) When the wiper of R is at position C, the plate current of the tube must flow through that portion of R included between points _____ and C.

8-123
(*A*)

RECTIFIER PEAK VOLTAGE

8-124
Figure 8-124 is a review diagram of a simple diode rectifier system. In this diagram, V is the diode and G is the ac _____.

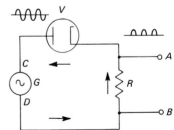

Fig. 8-124

8-125
(*source* or *generator*) In Fig. 8-124, R is the load resistor for the diode and the arrows show the direction of electron _____ flow.

8-126
(*current*) The useful voltage appears as a pulsating drop across R. Since the electron current flows upward in R, the lower end of R must become _____ in polarity.

8-127
(*negative*) The output terminals are labeled A and B. Terminal _____ is the "+" terminal.

8-128
(*A*) To produce the electron current direction shown, terminal C of the generator must be positive in polarity and terminal D _____ in polarity.

8-129
(*negative*) In Fig. 8-129, the generator is shown with polarity such that the plate of the diode is negative and the cathode positive. For this condition, the diode cannot conduct and the electron current is equal to _____.

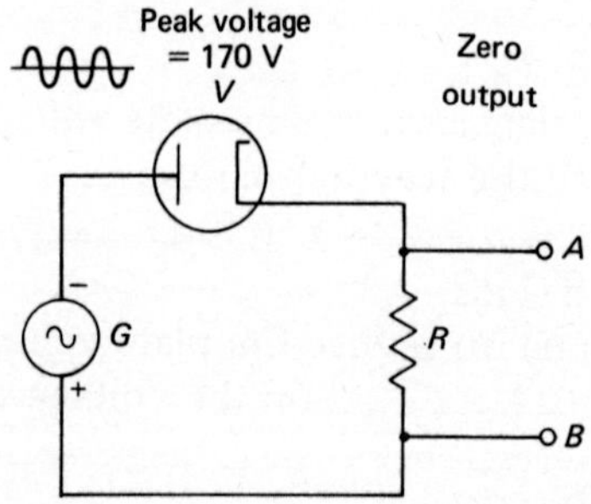

Fig. 8-129

8-130
(*zero*) When there is no current flowing through R, the voltage drop across this load resistor is _____.

8-131
(*zero*) If the voltage drop across R is zero, then during the nonconducting half-cycle the full output voltage of the generator must appear across the diode elements. If the peak generator voltage is 170 V, then during the nonconducting half-cycle the voltage that appears across the diode must reach a value of _____ V.

8-132
(*170*) The voltage that appears across a diode rectifier during the nonconducting half-cycle is called the *peak inverse voltage* (abbrev. PIV). For the situation described in Fig. 8-129, the peak inverse voltage is _____ V.

8-133
(170) The elements of a diode must be spaced far enough apart so that application of a given PIV does not cause arc-over from plate to cathode. Hence, all rectifiers are rated by their manufacturer as to the maximum _____ they will stand without arc-over.

8-134
(PIV) If a tube is rated at a maximum PIV of 400 V, this means that the tube may arc over and destroy the cathode if the peak inverse voltage applied to it exceeds _____ V.

8-135
(400) A rectifier is to be used to rectify the output of a transformer that has a rated secondary voltage of 320 V peak. The rectifier used must therefore have a peak-inverse-voltage rating of at least _____ V.

8-136
(320) Peak voltage = 1.41 × effective voltage. If the transformer above had an output rating of 320 V *effective*, then the peak-inverse-voltage rating of the rectifier would have to be at least _____ V.

8-137
(451.2) Considering the nominal output voltage of the ac lines in the United States to be 120 V effective, a rectifier used on the ac lines should have a peak inverse voltage of at least _____ V.

8-137
(170)

CRITERION CHECK TEST

____8-1 A heated filament will emit electrons. These electrons are called (a) free electrons, (b) thermionic electrons, (c) loose electrons, (d) heated electrons.

____8-2 In Fig. 8-9 the function of the battery is to (a) provide a voltage from plate to glass envelope, (b) provide a voltage between filament and plate, (c) attract the electrons to the plate, (d) heat the filament.

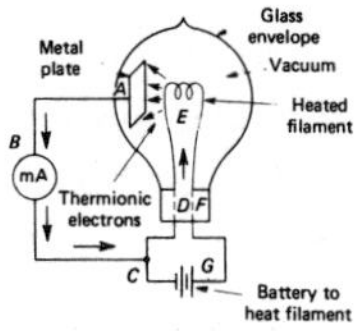

Fig. 8-9

_____8-3 In Fig. 8-20 the polarity of the B battery (a) causes a large current to flow in the plate circuit, (b) causes a large effective resistance, (c) causes the filament temperature to increase greatly, (d) repels electrons back to the filament.

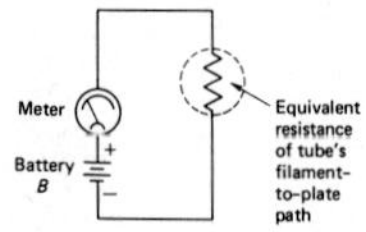

Fig. 8-20

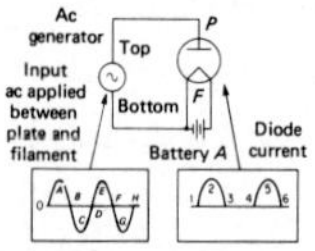

Fig. 8-27

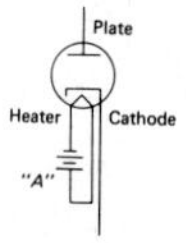

Fig. 8-58

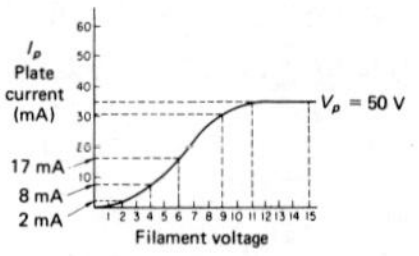

Fig. 8-79

_____8-4 When the plate of a diode is made negative with respect to the filament, (a) electrons are repelled back to the filament, (b) a large plate current flows, (c) the filament cools off, (d) the glass envelope assumes a large negative voltage.

The next 3 questions refer to Fig. 8-27.

_____8-5 The letter P stands for (a) plate, (b) potential, (c) product, (d) polarity.

_____8-6 The voltage O–A–B–C–D is a plot of the (a) plate voltage, (b) filament voltage, (c) plate current, (d) electron current.

_____8-7 The diode current is zero in the region (a) 1–2–3, (b) 3–4, (c) 4–5, (d) 5–6.

_____8-8 In Fig. 8-58 the plate receives thermal electrons from the (a) B battery, (b) heater, (c) cathode, (d) A battery.

_____8-9 A hard tube is defined as a tube with (a) a tungsten filament, (b) five electrodes, (c) no gas in the envelope, (d) a metal envelope.

The following 3 questions refer to Fig. 8-79.

_____8-10 Filament voltage is a direct measure of (a) plate current, (b) plate voltage, (c) filament temperature, (d) plate temperature.

_____8-11 What is the plate current when the filament voltage is 9 V? (a) 2 mA, (b) 8 mA, (c) 17 mA, (d) 31 mA.

_____8-12 At what filament voltage does the plate current level off? (a) 2, (b) 5, (c) 9, (d) 11.

_____8-13 A vacuum tube is normally operated in the temperature saturation region to (a) protect against filament aging, (b) keep the tube envelope hot, (c) disperse the space charge, (d) prevent plate distortion.

_____8-14 Plate saturation results when the (a) filament voltage is too high, (b) plate voltage is too low, (c) space-charge region is depleted, (d) plate temperature is too low.

_____8-15 The dc plate resistance of a vacuum tube is defined as (a) V_P/I_P, (b) $I_P \times V_P$, (c) I_P/V_P, (d) $I_P^2 \times V_P$.

_____8-16 Saturation is a condition where an increase in plate voltage will produce (a) a rise in electron emission, (b) a decrease in electron emission, (c) no appreciable change in plate current, (d) tube cutoff.

_____8-17 The peak-inverse-voltage rating of a diode is defined as the maximum allowable (a) filament voltage, (b) negative voltage across the load resistor, (c) positive voltage applied to plate, (d) negative voltage applied to plate with respect to the cathode.

_____8-18 If a transformer used in a rectifier circuit has an effective rating of 200 V, then the diode must have a PIV rating of at least (a) 200 V, (b) 280 V, (c) 360 V, (d) 440 V.

operating characteristics of electron tubes

OBJECTIVES

(1) Sketch a typical vacuum tube triode, labeling grid, plate, cathode, and heater. (2) Draw and discuss the electrostatic field lines in a triode with a negative grid, a positive grid, and a grid which is at zero volts with respect to the cathode. (3) Discuss a method of preventing large grid current flow. (4) Relate the placement of a grid in a triode to the triode's ability to amplify voltages. (5) Draw a family of plate characteristic curves and locate a point on a curve for a given set of parameters. (6) Define and compute from characteristic curves the plate resistance and amplification factor. (7) Draw the ac equivalent circuit of a triode. (8) Determine the operating point of a triode. (9) Compute the voltage gain of a triode. (10) Draw and discuss a triode circuit which uses a cathode resistor to provide grid bias. (11) Compute the value of the cathode resistor and the polarity of the voltage drop across it. (12) State the function of the cathode bypass capacitor. (13) State the names of at least three different multielement electron tubes. (14) List two advantages of the pentode over the triode. (15) Draw a family of plate characteristics for a pentode and describe the behavior of the plate current with changing plate voltage. (16) Define and compute the transconductance of a pentode. (17) Derive a relation amongst the three vacuum tube parameters g_m, r_p, and μ. (18) Compare transistors and triodes with respect to electrode functions, class of carriers, and input resistance. (19) Compare the transistor and pentode with respect to collector (plate) characteristics.

TRIODE FUNDAMENTALS

From the very start, when radio communication was shown to be practicable by Marconi and others of his time, it became evident that a great need would be felt for a device that could *amplify* signals of various frequencies. There can be little doubt that if Lee DeForest had not given his invention of the triode to the world when he did, the progress of electronics would have been seriously set back. Although a triode can do many other things besides amplifying signals, all its other functions are inextricably bound to its ability to amplify and can be traced back to it without difficulty. Our first concern, of course, is to learn something about the structure of the triode; then we can follow this comprehension with fundamental experiments that illustrate the characteristics that make amplification possible.

9-1
A triode vacuum tube contains a heater, a cathode, a plate, and a fourth element constructed of spiral wound wire between the cathode and plate. From Fig. 9-1a, this fourth element is evidently called a _____.

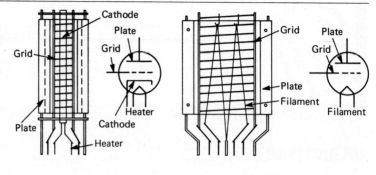

Fig. 9-1   (a)   (b)

9-2
(*grid*) The triode in Fig. 9-1 shows an *indirectly* heated triode in which a separately heated cathode is the electron emitter. Some triodes, as in Fig. 9-1b, do not have a separate cathode and are called *directly* heated types. In these tubes, the _____ is the emitter of electrons.

9-3
(*filament*) In Fig. 9-3, a series of drawings shows the electrostatic field effects of the grid (wire cross sections between plate and grid). In *a*, the grid is very negative compared with the cathode. The upper arrows, between grid and plate, show the field between these electrodes; the lower arrows show the field between the _____ and _____.

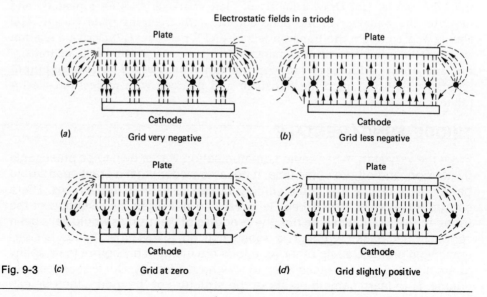

Fig. 9-3

222 A PROGRAMMED COURSE IN BASIC ELECTRONICS

9-4
(*grid, cathode*) The arrows show the direction an electron would move if placed anywhere along the line on which the arrow rests. For instance, an electron placed on any line in *a* between the grid and plate would move toward the plate; an electron on any line between grid and cathode would move toward the _____.

9-5
(*cathode*) This is so because the grid is the most negative object in the tube. Electrons would therefore be _____ from the grid and _____ toward either the plate or cathode, depending on where they were originally located.

9-6
(*repelled, attracted*) In *a*, the grid is so negative that we say that all the field lines *terminate* on the grid. This means that there are no field lines going directly from the cathode to the plate or, conversely, from the plate to the _____.

9-7
(*cathode*) This means that no electron can travel from the cathode directly to the _____ since no field lines exist to carry it over this space.

9-8
(*plate*) Thus, any electrons emitted from the cathode must return to the cathode. Hence, for the conditions shown in *a*, the plate current of the tube must be _____.

9-9
(*zero*) Now consider *b* in Fig. 9-3. The grid is slightly negative but not negative enough to cause *all* the lines to terminate on it. Hence, there are some lines that go directly from the cathode to the _____.

9-10
(*plate*) An electron that manages to emerge from the cathode on one of the lines that go directly from the cathode to the plate will, therefore, follow the line and proceed to the plate. Thus, for the conditions shown in *b*, a small amount of _____ _____ must flow.

9-11
(*plate current*) In *c* of Fig. 9-3, the grid has the same potential as the cathode. With the potential difference between the grid and cathode equal to zero, the only other force that still exists that can move electrons is that produced by the positive _____ voltage.

9-12
(*plate*) Most of the field lines in condition *c* now originate at the cathode and terminate at the _____.

9-13
(*plate*) Thus, most of the electrons emitted by the cathode will now fall on field lines that go directly from the cathode to the plate. Hence, in this case the plate current can become much _____ than it was in case *b*.

9-14
(*larger*) A few of the field lines terminate on the grid merely because the grid is an electrical conductor. Electrons that move along these lines may flow into the grid circuit causing a small grid _____ to flow.

9-15
(*current*) If the grid is made slightly positive as in *d* of Fig. 9-3, both the plate and grid set up fields that move electrons away from the _____.

9-16
(*cathode*) Since both grid and plate are positive with respect to the cathode in *d*, electrons will flow into both grid and plate circuits. Thus, for this condition both grid and _____ currents may be quite large.

9-17
(*plate*) The term *space current* applies to the total current flowing from the cathode. Thus, if grid current and plate current are both flowing, space current must be the _____ of these two.

9-18
(*sum*) In summary, if the grid is very negative with respect to the cathode, the plate current and grid current are both equal to _____.

9-19
(*zero*) As the grid becomes less and less negative, the plate current must _____ in magnitude.

9-20
(*increase*) When the grid is zero volts with respect to the cathode, the plate current may be quite large. At the same time, a _____ current appears.

9-21
(*grid*) If the grid is made positive with respect to the cathode, not only is the plate current large, but the _____ current also may be quite large.

9-22
(*grid*) Because grid current is undesirable, the grid circuit is usually adjusted (or biased) so that the grid is negative. Thus, biasing a grid refers to maintaining a _____ (positive, negative) voltage on the grid.

9-23
(*negative*) The potentials of the tube elements may be measured in terms of the voltage between each of the nonemitting elements and the _____ as a reference.

9-24
(*cathode*) The average grid voltage is symbolized by V_G and, here, is the voltage between grid and cathode. The average plate voltage is symbolized as V_P and, here, is the voltage between plate and _____.

9-25
(*cathode*) As in the case of the diode, making the plate more positive usually causes the plate current to _____.

9-26
(*increase*) And, as we have just seen, making the grid less negative causes the plate current to _____.

9-27
(*increase*) Thus, plate current is essentially under the control of 2 voltages: V_P and _____.

9-28
(V_G) In the design of a triode, the grid is located much closer to the cathode than the plate. Thus, a given change of grid voltage will cause a much _____ change of plate current than would the same change in plate voltage. This is the basic reason that a triode can amplify.

9-28
(*greater*)

TRIODE CHARACTERISTIC CURVES

9-29

A family of curves for the plate circuit of a triode is shown in Fig. 9-29. Note that each member of the family is obtained by using a certain value of grid voltage held constant while V_P is varied. The curve for the case that $V_G = 0$ is labeled by the letter _____.

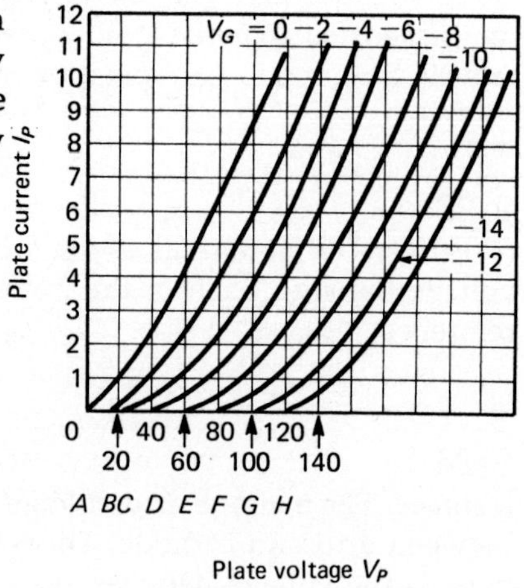

Fig. 9-29

A family of $V_P - I_P$ curves

9-30

(*A*) The various plate currents given by curve *A* are obtained by varying the plate voltage from 0 V where the curve starts to 120 plate volts where the curve ends. This curve, then, provides information on the plate currents for $V_G = 0$ for any plate voltage between _____ and 120 V.

9-31

(*0*) For example, using curve *A* we find that the plate current is 1 mA when the plate voltage is 20 V; also, the plate current is 3 mA when the plate voltage is about 48 V. Thus, the plate current is 5 mA when the plate voltage is _____ V. (Estimate to the nearest volt.)

9-32

(*68*) Similarly, the plate current is 7 mA when the plate voltage is 87 V and 9 mA when the plate voltage is _____ V. (Estimate to the nearest volt.)

9-33

(*105*) The remainder of the family is generated by repeating the V_P changes and observing the I_P variations resulting from these changes while V_G is held at a fixed value that differs from curve to curve. The total range of V_G steps in this family is from $V_G = 0$ to $V_G =$ _____ V.

226 A PROGRAMMED COURSE IN BASIC ELECTRONICS

9-34
(−14) A family of plate characteristic curves is very useful. For example, we can answer questions like this: If the plate voltage is 120 V and the bias is −6 V, what is the plate current? The answer is 4 mA. Or, if the plate voltage is 140 V and the grid bias is −8 V, what is I_P? The answer is _____ _____.

9-35
(4 mA) Similarly, if the plate voltage is 80 V and $V_G = 0$ V, then $I_P = 6.5$ mA. Thus, if $V_P = 100$ V and $V_G = -2$ V, then $I_P =$ _____ _____.

9-36
(6 mA) If the plate voltage is 140 V and the grid bias is −6 V, then $I_P =$ _____ _____.

9-37
(6 mA) Another type of question is this: What grid bias is necessary to hold the plate current down to 5 mA when the plate voltage is 90 V? The answer is _____ V.

9-38
(−2) Thus, the grid bias required to hold I_P to 8 mA when the plate voltage is 110 V is _____ _____.

9-39
(−1 V) Here is still another type of question: The grid bias on the tube is −4 V, and the plate current is 10 mA. What is the plate voltage? The answer is 152 V. Thus, if the bias is −8 V and the plate current is 1 mA, the plate voltage must be _____ V.

9-40
(90) Finally, if the bias is −10 V and the plate current is 5.8 mA, the plate voltage must be _____ V.

9-40
(180)

TRIODE EQUIVALENT CIRCUIT

9-41

In many applications, it is useful to replace an electron tube by an equivalent circuit. As in the case of the transistor, we shall use static curves to obtain the equivalent circuit. Figure 9-41 shows a family of plate curves for a type _____ triode.

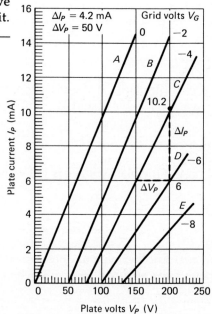

Fig. 9-41

9-42

(6C5) We see that each curve, A, B, C, and D, is a fairly straight, diagonal line. Recall that the v-i characteristic of a _____ is a straight, diagonal line.

9-43

(*resistor*) Thus, the first element of the triode equivalent circuit is a simple resistor, called the *plate resistance* r_p. Using the fact that the slope of the v-i characteristic yields the value of _____, we can compute the plate resistance.

9-44

(*resistance*) On curve C, Fig. 9-41, dotted lines have been drawn to compute the slope. The ordinate is denoted by ΔI_P, and the abscissa by _____.

9-45

(ΔV_P) From the graph, Fig. 9-41, ΔI_P is about 4.2 mA. The abscissa, ΔV_P, is about _____ V.

228 A PROGRAMMED COURSE IN BASIC ELECTRONICS

9-46
(50) The ratio of $\Delta V_P/\Delta I_P$ is the plate resistance, r_p. Thus,

$$r_p = \frac{\Delta V_P}{\Delta I_P} = \frac{50 \text{ V}}{4.2 \times 10^{-3} \text{ A}} = 12{,}000 \underline{\qquad}$$

9-47
(Ω) The plate resistance of this triode is then approximately 12,000 Ω. For a simple resistor, when the current is reduced to zero, the voltage is reduced to \underline{\qquad}.

9-48
(zero) For curve C ($V_G = -4$ V), when the current is reduced to zero, the voltage is *not* zero, but \underline{\qquad} V.

9-49
(75) For curve D ($V_G = -6$ V), when the current is reduced to zero, the voltage is \underline{\qquad} V.

9-50
(100) For curve E ($V_G = -8$ V), when the current is reduced to zero, the voltage is about \underline{\qquad} V.

9-51
(130) For each curve, when the current is reduced to zero, the voltage is not \underline{\qquad}. Thus, besides the plate resistor, a voltage source is needed to account for the zero-current voltage.

9-52
(zero) The value of the voltage source can be obtained by using the identity

$$\Delta V_P = \left(\frac{\Delta V_P}{\Delta V_G}\right)\Delta V_G$$

The ratio $\Delta V_P/\Delta V_G$ is called the *amplification factor* and is given the symbol μ. The expression for ΔV_P may also be written as

$$\Delta V_P = \underline{\qquad} \times \Delta V_G$$

9-53
(μ) We may compute μ from the graph of Fig. 9-41. In going from curve C to curve D, the grid voltage goes from -4 V to \underline{\qquad} V.

9-54
(−6) Hence, the change in grid voltage is $\Delta V_G = -6 - (-4) = -2$ V. For curve C, the plate voltage at zero current is about 75 V, while for curve D the corresponding plate voltage is about _____ V.

9-55
(100) The change in plate voltage is then

$$\Delta V_P = \underline{} - \underline{} = 25 \text{ V}$$

9-56
(100, 75) Using the definition of μ, one finds

$$\mu = \frac{\Delta V_P}{\Delta V_G} = \frac{25}{-2} = \underline{} \text{ V}$$

9-57
(−12.5) The amplification factor for the 6C5 tube is about 12.5. The minus sign indicates that there is a _____° phase shift between plate voltage and grid voltage.

9-58
(180) Figure 9-58 shows the complete ac equivalent circuit for a triode. A voltage source of value μv_g is connected in _____ with a plate resistance r_p.

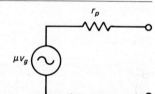

Fig. 9-58

9-59
(series) Note that instead of using the "Δ" symbol, lowercase letters and subscripts have been used for voltages and currents. Lowercase letters are conventionally used to denote instantaneous changing ac quantities; therefore the "_____" symbol is not needed.

9-60
(Δ) Summarizing: There are 2 important parameters for a triode: (1) the amplification factor, μ, defined as

$$\mu = \frac{\Delta V_P}{\Delta V_G} = \frac{v_p}{v_g}$$

(2) the plate resistance, r_p, defined as

$$r_p = \frac{\Delta V_P}{\Delta I_P} = \frac{v_p}{?}$$

230 A PROGRAMMED COURSE IN BASIC ELECTRONICS

9-60
(i_p)

TRIODE GAIN AND PLATE DISSIPATION
9-61
Consider the circuit shown in Fig. 9-61 containing a type _____ triode. The load resistance is denoted by R_L and the grid voltage is an ac voltage, v_g, in series with a fixed grid-bias battery, V_{GG}.

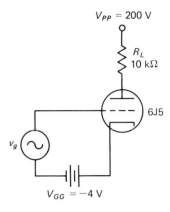

Fig. 9-61

9-62
(6J5) Just as with transistor circuits, the first step in analyzing this triode is to draw the *load line*. Recall that _____ points are needed to construct the load line.

9-63
(2) The first point is found by setting the plate current equal to zero. All the voltage drop then appears across the triode; that is, $V_P =$ _____ V. This corresponds to point A on Fig. 9-63.

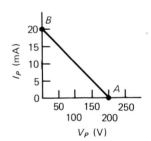

Fig. 9-63

9-64
(200) The second point is found by setting the plate voltage equal to zero. Then all the voltage drop appears across _____ .

9-65
(R_L) Using Ohm's law for R_L, one has $200 = I_P \times R_L = I_P \times 10{,}000$ so that

$$I_P = \frac{200 \text{ V}}{10{,}000 \text{ }\Omega} = 0.020 \text{ A} = 20 \text{ _____}$$

OPERATING CHARACTERISTICS OF ELECTRON TUBES

9-66
(*mA*) This point is denoted by *B* on Fig. 9-63. The solid diagonal line which connects points *A* and *B* is called the _____ _____.

9-67
(*load line*) In Fig. 9-67, the load line has been drawn on the characteristic curves of the 6J5 triode. The first point to locate on the load line is the dc operating point established by the − _____ -V grid bias battery.

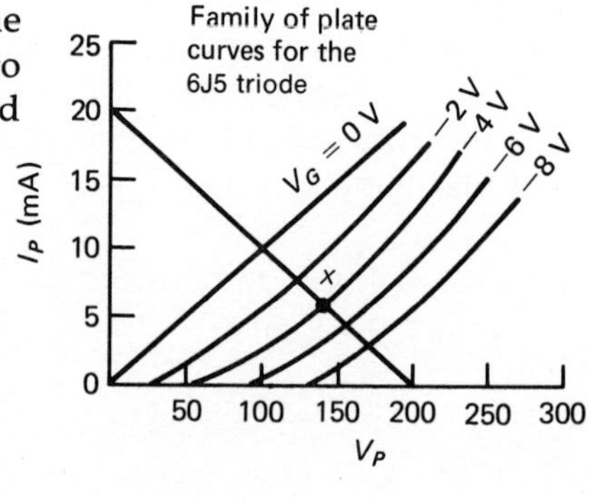

Fig. 9-67

9-68
(4) The intersection of the load line with the $V_G = -4$ curve is denoted by point _____ on Fig. 9-67.

9-69
(*X*) The plate voltage at point *X* is about 140 V and the plate current is about _____ mA.

9-70
(6) Further, using the formula, power = voltage × current, the *plate dissipation* at the dc operating point is 140 V × 6 mA = _____ W.

9-71
(*0.84*) Figure 9-71 shows a typical set of values found in a tube manual for the 6J5 triode. The maximum plate dissipation allowable is _____ W.

Typical operating values for 6J5

Plate voltage	90 V
Grid voltage	0
Amplification factor, μ	20
Plate resistance	6,700 Ω
Plate current	10 mA
Maximum allowable plate dissipation	4 W

Fig. 9-71

9-72
(4) Since the plate dissipation for the circuit of Fig. 9-61 is only _____ W, the tube is operating well within safe limits. If the computed plate dissipation had been greater than 4 W, a circuit redesign would have been necessary.

9-73
(0.84) Next, we can compute the ac swing of the plate circuit due to the ac source voltage, v_g. In Fig. 9-73a, an ac signal is shown applied to the _____ circuit of the triode.

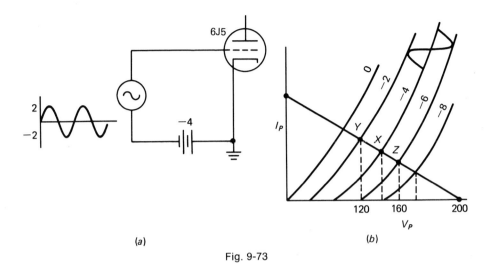

Fig. 9-73

9-74
(grid) The peak-to-peak amplitude of this ac signal is _____ V. In Fig. 9-73b, the same signal is drawn on the characteristic curves.

9-75
(4) In Fig. 9-73b, the ac signal swings around the −4-V dc bias. Since the peak-to-peak signal is 4 V, the *net* grid voltage swings from −2 to −_____ V.

9-76
(6) Because of the ac grid swing, the operating point moves along the load line. When the grid swing reaches −2 V, the operating point has moved along the load line to point _____.

9-77
(Y) At point Y, the plate voltage is 120 V. When the grid swing reaches −6 V, the operating point has moved to point Z on the load line and the plate voltage is _____ V.

9-78
(160) Since the plate voltage experiences a swing from 120 V to 160 V, the peak-to-peak ac plate swing is _____ V.

9-79
(*40*) The voltage gain of this triode is now computed. The peak-to-peak swing of the plate (output) is 40 V. The peak-to-peak swing of the input is 4 V. The voltage gain is then 40 V/4 V or _____ .

9-80
(*10*) Reviewing: The load line was drawn on the characteristic curves. The operating point was determined by the _____ bias voltage.

9-81
(*grid*) Reviewing: The plate dissipation was calculated and checked for safe operation. The voltage gain was computed by finding the ratio of the _____ swing to grid swing.

9-82
(*plate*) The reader will notice that in Fig. 9-71 the amplification factor μ is given as 20, yet the voltage gain was found to be only _____ .

9-83
(*10*) The reason for this *apparent* discrepancy is the plate resistance. The plate resistance is _____ (internal, external) to the triode.

9-84
(*internal*) The plate resistance causes a voltage drop internally in the triode. This reduces the voltage available to the _____ resistor, which in turn reduces the voltage gain to the computed value of 10.

9-84
(*load*)

TRIODE BIASING

9-85
One very important point about triode amplifiers. The reader probably has noted that in all our discussions the grid voltage was always kept _____ (positive, negative).

9-86
(*negative*) The reasons for this are several. First, in the positive grid region the variation of plate resistance is much greater than if one stays in the _____ grid region.

9-87
(*negative*) Secondly, if the grid is positive, electrons will be _____ _____ (attracted to, repelled from) the grid, thus creating a grid current.

9-88
(*attracted to*) If one recalls from Ohm's law that the resistance is V/I, as the current increases for a given voltage, the resistance _____.

9-89
(*decreases*) Hence, when the grid is negative, electrons are repelled, no current flows, and the input grid resistance is extremely _____.

9-90
(*high*) But if the grid is positive, current flows, and the resistance _____ (increases, decreases).

9-91
(*decreases*) In many circuit applications, it is very desirable that the input (grid) resistance of the amplifier be very high. Thus, one almost always operates the grid in the _____ (positive, negative) region.

9-92
(*negative*) But one must amplify positive voltages as often as negative voltages. For example, one might have to amplify a signal such as that in Fig. 9-92. Note that the maximum positive voltage is +4 and the maximum negative voltage is _____.

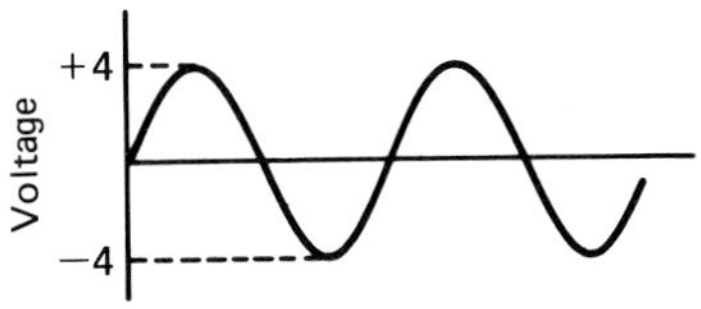

Fig. 9-92

9-93
(−4) To ensure that the grid voltage is *always* negative, one could put in series with v_g a fixed battery (called a *bias battery*) of _____ V.

9-94
(−4) In fact, this is exactly the purpose of the −4-V battery in Frame 9-61. In conclusion, a _____ battery can be used to put the operating point of an amplifier in the most useful region of operation.

OPERATING CHARACTERISTICS OF ELECTRON TUBES

9-95

(*bias*) A second method of providing negative bias is the cathode resistor. Figure 9-95 depicts a triode circuit with a supply battery V_{PP}, a load resistor R_L, a signal source v_g, and a new element, a cathode resistor denoted by _____.
The direction of I_P shown is that for conventional current. The electron flow direction is opposite to this.

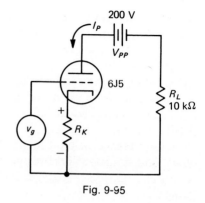

Fig. 9-95

9-96

(R_K) Because of the plate current I_P, the voltage drop in the resistor R_K will have a polarity as indicated in Fig. 9-95. The cathode is made positive by the drop in R_K, or the grid is made _____ with respect to the cathode.

9-97

(*negative*) Let us determine the value of R_K for the circuit of Fig. 9-95. Assume that the input signal, v_g, is given by Fig. 9-92. In order that the grid never swing positive, the bias voltage must be at least _____ V.

9-98

(−4) In Frame 9-92, we determined that with a supply voltage of 200 V, a load resistance of 10 Ω, and a −4-V grid bias, the plate current is 6 mA. Then, to provide the proper negative bias, the voltage drop across R_K must be _____ V.

9-99

(4) Using Ohm's law, the voltage drop across R_K is $I_P \times R_K$. Hence, we have $4 = I_P \times R_K$. Then $4 = 6 \times 10^{-3} \times R_K$, or

$$R_K = \frac{4 \text{ V}}{6 \times 10^{-3} \text{ A}} = \underline{\qquad} \; \Omega$$

236 A PROGRAMMED COURSE IN BASIC ELECTRONICS

9-100
(667) In Fig. 9-100, the final form of the triode amplifier circuit is drawn. Note that a cathode bypass capacitor, denoted by C_K, has been added to *bypass* _____ for ac signals. If C_K is not present, the amplification will be reduced. Finally, a coupling capacitor, C_c, is used to block any dc interaction between this amplifier stage and any succeeding stages.

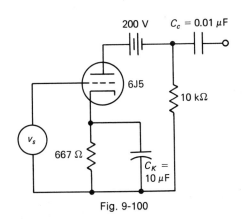

Fig. 9-100

9-101
(R_K) Summarizing: In Fig. 9-100, the 200-V battery is a supply battery which provides plate _____ .

9-102
(*current*) The source v_s is the signal to be amplified by the type _____ triode.

9-103
(6J5) The 667-Ω resistor provides the _____ (positive, negative) bias for the triode.

9-104
(*negative*) The cathode bypass capacitor, _____ , prevents a reduction in gain for ac signals.

9-105
(C_K) The capacitor, C_c, couples the ac signal to the next stage, but blocks the flow of _____ .

9-106
(*dc*) The 10-kΩ resistor is the load resistor for the _____ circuit.

9-106
(*plate*)

OPERATING CHARACTERISTICS OF ELECTRON TUBES **237**

PENTODE FUNDAMENTALS

9-107
Several vacuum tube devices have been developed with more than 3 elements. There is the tetrode with 4 elements, the pentode with 5 elements, the hexode with _____ elements, etc.

9-108
(6) Each of these devices has a characteristic curve _____ _____ (different from, the same as) the triode.

9-109
(*different from*) However, the method of circuit analysis is the same as for the triode. Specifically, one looks in the tube manual for the characteristic curves, draws the load line, determines the operating point, then determines the gain by reading in the tube manual typical values of plate resistance r_p and amplification factor _____.

9-110
(μ) We shall finish this chapter with a brief description and analysis of the pentode, a vacuum tube device with _____ elements.

9-111
(5) Figure 9-111 depicts a pentode. Note that the 2 extra elements which the pentode has are called the screen grid and _____ grid.

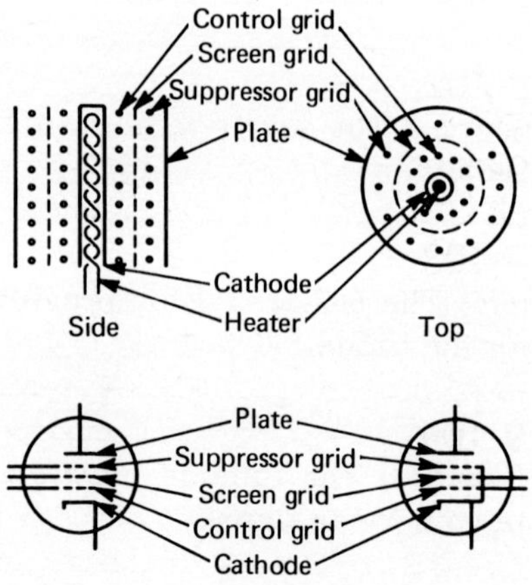

Fig. 9-111

9-112
(*suppressor*) Each one of these extra grids provides some advantage over the simpler triode. We shall briefly mention these advantages without going into details. The screen grid enables the pentode to give better performance than the triode when the input signal is varying at a very high rate or, said another way, for signals of _____ (high, low) frequency.

9-113
(*high*) The presence of the suppressor grid enables the pentode to deliver more current per unit of applied voltage. Hence, in the output (power) stage of vacuum tube amplifiers it is very common to find a _____ .

9-114
(*pentode*) Thus the most common applications of the pentode are in the output stages of an amplifier or to improve on the _____-frequency response of triodes.

9-115
(*high*) Next, consider the plate characteristics of a typical pentode as given in Fig. 9-115. As indicated on the chart, the tube type is a _____ .

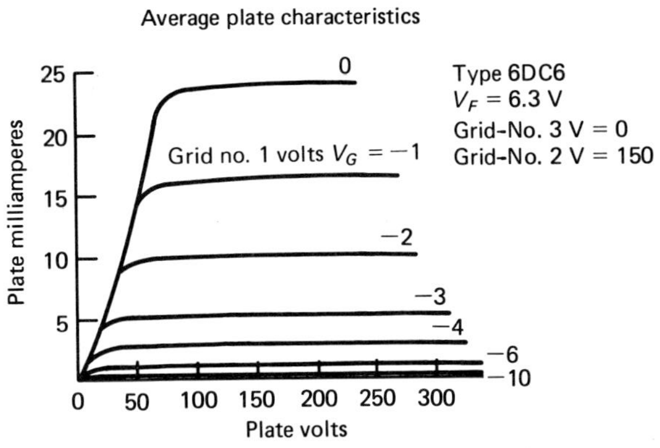

Average plate characteristics

Type 6DC6
$V_F = 6.3$ V
Grid-No. 3 V = 0
Grid-No. 2 V = 150

Fig. 9-115

9-116
(*6DC6*) This tube type operates with _____ V applied to its filament as shown on the chart.

9-117
(*6.3*) Grids of a pentode are numbered outward from the cathode, starting with the control grid as No. 1. On this basis, the screen grid is No. _____ .

9-118
(*2*) Grid No. 3 must, therefore, be the _____ grid (Fig. 9-111).

9-119
(*suppressor*) The 6DC6 pentode is designed to operate with _____ volts on its suppressor grid, the reference point being the cathode of the tube.

9-120
(*zero*) The curves in Fig. 9-115 have been drawn with the screen voltage constant throughout. The screen voltage used was _____ V with respect to cathode.

OPERATING CHARACTERISTICS OF ELECTRON TUBES **239**

9-121
(*150*) In a typical amplifier circuit, the 6DC6 is operated with a control grid bias of −2 V, and a plate voltage of about 200 V. Under these conditions, the static plate current would be _____ mA.

9-122
(*10*) A unique effect is observed in these curves when they are compared with those of a triode. With the bias at −2 V, the plate voltage may be changed from about 100 V all the way up to over 250 V without causing any change in the _____.

9-123
(*plate current*) The same effect is observed for virtually any bias voltage. That is, for V_G of any value from 0 to −10 V, the particular curve for that bias shows that the plate current is not much affected by changes of _____ voltage.

9-124
(*plate*) All pentodes behave this way. Thus, we can state that the plate current of a pentode is relatively independent of the _____ _____.

9-124
(*plate voltage*)

TRANSCONDUCTANCE

9-125
In Fig. 9-125, the characteristic curves of the type 6DC6 pentode are again presented. As noted in the previous section, the *v-i* characteristics are straight, _____ (vertical, horizontal) lines.

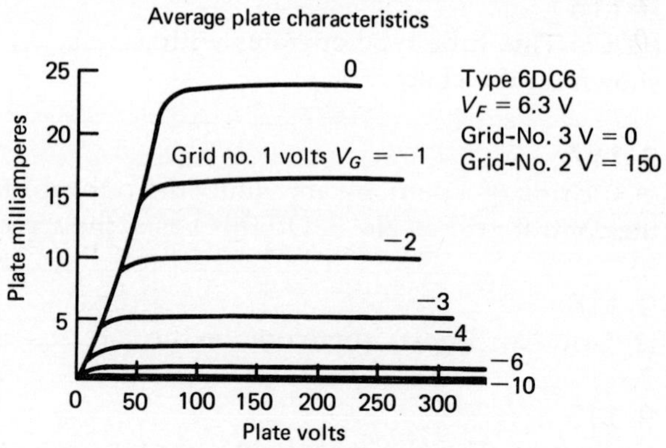

Fig. 9-125

9-126
(*horizontal*) Recall that the simple circuit element whose *v-i* characteristic is a straight, horizontal line is a current source. The plate circuit of a pentode therefore behaves like a _____ _____.

9-127
(*current source*) Since the pentode is an amplifying device, it is useful to express this current source in terms of the grid (input) voltage. For example, when the grid voltage is −1 V, the plate current is about _____ mA (Fig. 9-125).

9-128
(*17*) When the grid voltage is −2 V, the plate current is _____ mA (Fig. 9-125).

9-129
(*10*) To express the plate current in terms of the grid voltage, one can use the mathematical identity $\Delta I_P = (\Delta I_P/\Delta V_G) \Delta V_G$. Conventionally, the ratio $\frac{\Delta I_P}{\Delta V_G}$ is defined as the *transconductance*, and assigned the symbol g_m. The plate current then can be expressed as $\Delta I_P = $ _____ $\times \Delta V_G$.

9-130
(g_m) The transconductance can be computed from Fig. 9-125:

$$g_m = \frac{\Delta I_P}{\Delta V_G} = \frac{(17-10)\text{ mA}}{(-1-(-2))\text{ V}} = \frac{7\text{ mA}}{1\text{ V}} = \frac{7 \times 10^{-3}\text{ A}}{1\text{ V}} = \underline{\hspace{1cm}} \frac{\text{A}}{\text{V}}$$

9-131
(*0.007*) The ratio of volt/ampere is equal to ohms. The ratio of ampere/volt is then reciprocal ohms or _____ .

9-132
(*mhos*) The transconductance of this pentode is then 0.007 mho. In tube manuals, the transconductance is always listed in micromhos. A micromho (μmho) is equal to one-millionth of a mho, or 1 μmho = _____ mho.

9-133
(*0.000001*) Expressing the transconductance of the 6DC6 in μmhos, one has 0.007 mho = _____ μmho.

9-134
(*7,000*) Summarizing: The transconductance of a tube is listed in units of μmhos and is a measure of the gain; the higher the transconductance, the _____ (lower, higher) the gain.

9-135
(*higher*) In Fig. 9-135 we have listed typical values as given in a tube manual. The transconductance is given as 7,000 μmho and the plate resistance as _____ kΩ.

6DC6 Pentode Typical Operation	
Plate voltage	300 V
Suppressor voltage	0 V
Screen voltage	200 V
Maximum plate dissipation	2 W
Transconductance	7,000 μmho
Plate resistance	500 kΩ
Grid voltage	−3 V

Fig. 9-135

9-136
(*500*) A simple relationship exists between the transconductance, plate resistance, and amplification factor. Consider the product of g_m and v_p:

$$g_m \times v_p = \frac{\Delta I_P}{\Delta V_G} \times \frac{\Delta V_P}{\Delta I_P} = \frac{\Delta V_P}{?}$$

$$\mu = \frac{\Delta V_P}{\Delta V_G}$$

$$g_m = \frac{\Delta I_P}{\Delta V_G}$$

$$v_p = \frac{\Delta V_P}{\Delta I_P}$$

Fig. 9-136

9-137
(ΔV_G) But $\Delta V_P/\Delta V_G$, by definition, is equal to the amplification factor μ. Hence, $\mu = g_m \times v_p$. In a tube manual, if any 2 of the 3 tube parameters are given, the third can be calculated by using the relation $\mu = g_m \times$ _____.

9-137
(v_p)

COMPARISON OF TRANSISTORS AND ELECTRON TUBES
9-138
In an electron tube (Fig. 9-138), the current carriers are always electrons. These electrons are boiled off the hot _____ of the tube.

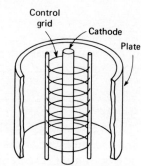

Fig. 9-138

9-139
(*cathode*) The destination of most of the electrons from the cathode is the plate of the tube. In traveling from the cathode to the plate, the electrons must pass through the spaces in the _____ of the tube.

9-140
(*grid*) In the transistor (Fig. 9-140), the current carriers may be either electrons or _____.

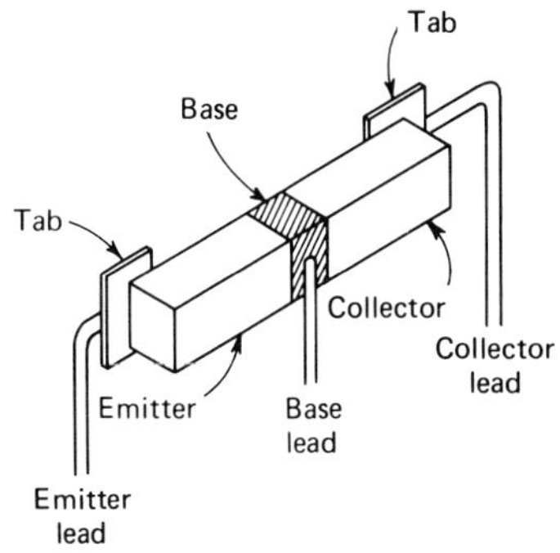

Fig. 9-140

9-141
(*holes*) In either case (NPN or PNP), the carriers originate in the emitter electrode and move toward the collector. All the carriers that arrive at the collector must pass through the _____ electrode.

9-142
(*base*) Thus, the emitter is comparable to the cathode, the base is comparable to the grid, and the _____ is comparable to the plate.

9-143
(*collector*) In the electron tube, plate current is determined mainly by the difference in voltage between the cathode and grid. That is, the _____ of the tube is considered the controlling element.

9-144
(*grid*) In the transistor, collector current is determined mainly by the voltage difference between the emitter and base. The emitter is like the cathode; hence the _____ is the controlling element.

9-145
(*base*) The electron tube requires a heater to boil electrons off the cathode. The transistor does not need a _____ element because its carriers are not thermionic in nature.

9-146

(*heater*) In a tube, the plate is always positive with respect to the cathode. In a transistor, the collector may be either positive or _____ with respect to the emitter, depending upon whether the transistor is NPN or PNP.

9-147

(*negative*) For most electron tube applications, grid current does not flow, because the grid is biased in a negative direction with respect to the _____ of the tube.

9-148

(*cathode*) For most transistor applications, emitter-to-base current does flow because of recombinations of holes and _____ in the base electrode.

9-149

(*electrons*) Since current does not flow in the grid circuit of a tube, its input resistance is high. Since current does flow in the emitter-base circuit of a transistor, its input resistance is _____.

9-150

(*low*) The input resistance of an electron tube is much higher than its output resistance, since grid current seldom flows while the _____ current of the tube may be quite high.

9-151

(*plate*) The input resistance of a transistor is lower than its output resistance in the common-base configuration, because the emitter is forward-biased while the collector is _____-biased with respect to the base.

9-152

(*reverse*) Figure 9-152 illustrates the $V_P - I_P$ curve for a pentode electron tube. This curve shows that the _____ current of the pentode rises quickly at first, then levels off.

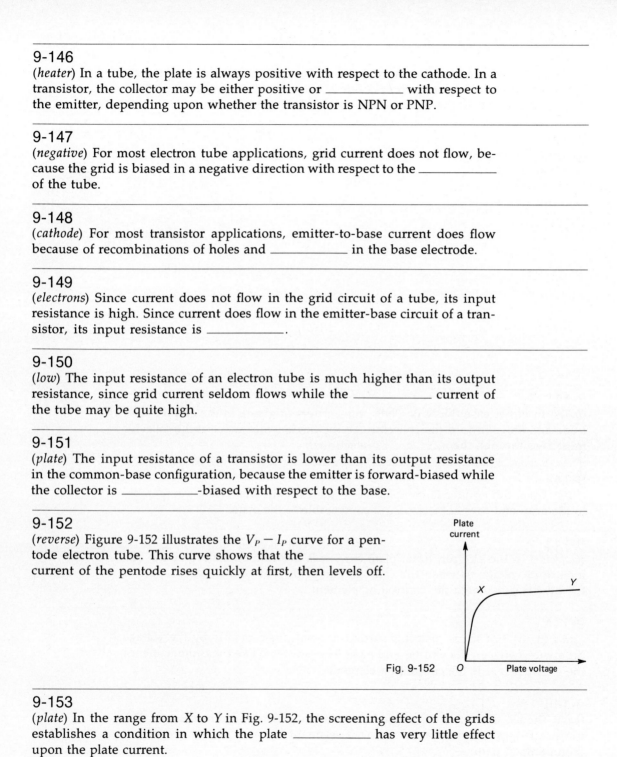

Fig. 9-152

9-153

(*plate*) In the range from X to Y in Fig. 9-152, the screening effect of the grids establishes a condition in which the plate _____ has very little effect upon the plate current.

9-154

(*voltage*) The collector voltage–collector current curve of a transistor is similar to the pentode curve. In the region from X to Y, the collector _____ has very little effect upon the collector current (Fig. 9-154).

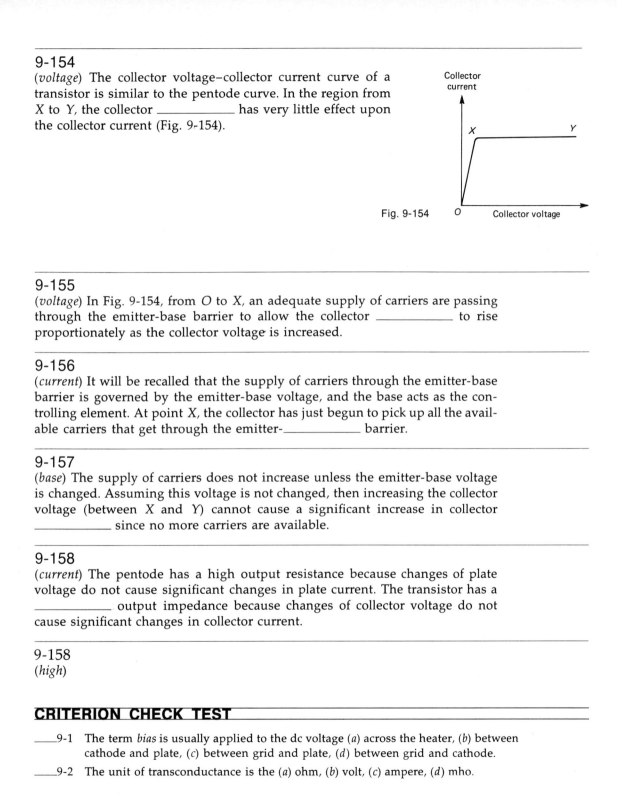

Fig. 9-154

9-155

(*voltage*) In Fig. 9-154, from O to X, an adequate supply of carriers are passing through the emitter-base barrier to allow the collector _____ to rise proportionately as the collector voltage is increased.

9-156

(*current*) It will be recalled that the supply of carriers through the emitter-base barrier is governed by the emitter-base voltage, and the base acts as the controlling element. At point X, the collector has just begun to pick up all the available carriers that get through the emitter-_____ barrier.

9-157

(*base*) The supply of carriers does not increase unless the emitter-base voltage is changed. Assuming this voltage is not changed, then increasing the collector voltage (between X and Y) cannot cause a significant increase in collector _____ since no more carriers are available.

9-158

(*current*) The pentode has a high output resistance because changes of plate voltage do not cause significant changes in plate current. The transistor has a _____ output impedance because changes of collector voltage do not cause significant changes in collector current.

9-158

(*high*)

CRITERION CHECK TEST

____9-1 The term *bias* is usually applied to the dc voltage (*a*) across the heater, (*b*) between cathode and plate, (*c*) between grid and plate, (*d*) between grid and cathode.

____9-2 The unit of transconductance is the (*a*) ohm, (*b*) volt, (*c*) ampere, (*d*) mho.

____9-3 To find the amplification factor of a tube, we use the following formula: (a) $\Delta V_P/\Delta V_G$, (b) $\Delta V_G/\Delta V_P$, (c) $\Delta I_P/\Delta V_G$, (d) $\Delta V_G/\Delta I_P$.

____9-4 The definition for r_p is (a) $\Delta V_P/\Delta I_P$, (b) $\Delta V_P/\Delta V_G$, (c) $\Delta V_P/\Delta V_G$, (d) $\Delta I_P/\Delta V_G$.

____9-5 A typical value of ac plate resistance for a triode might be in the neighborhood of (a) 1,000 Ω, (b) 10,000 Ω, (c) 100,000 Ω, (d) 1,000,000 Ω.

____9-6 The input voltage v_g and the output voltage v_o of a triode are (a) in phase with each other, (b) 90° out of phase with each other, (c) 180° out of phase with each other, (d) 270° out of phase with each other.

____9-7 The function of the capacitor in parallel with a cathode resistor is to (a) increase degeneration, (b) decrease degeneration, (c) increase regeneration, (d) decrease amplification.

____9-8 A triode is normally operated with a (a) positive grid, negative plate, (b) positive grid, positive plate, (c) negative grid, negative plate, (d) negative grid, positive plate.

____9-9 In a pentode circuit, the voltage on the suppressor grid is usually (a) negative, (b) zero, (c) equal to the plate voltage, (d) greater than the plate voltage.

____9-10 If the maximum plate dissipation is 2 W, what is the maximum plate current when the plate voltage is 200 V? (a) 1 mA, (b) 10 mA, (c) 100 mA, (d) 1 A.

____9-11 If the grid of a triode is operated in the positive region, (a) the grid may overheat, (b) the grid resistance will decrease, (c) the plate current will decrease sharply, (d) electrons will be repelled back to the cathode.

____9-12 The common-collector configuration is analogous to the (a) common-grid configuration, (b) common-plate configuration, (c) common-cathode configuration, (d) common-heater configuration.

____9-13 In Fig. 9-67, if the grid voltage is 0 V, then the plate voltage and current are (a) 100 V, 10 mA; (b) 120 V, 7 mA; (c) 150 V, 5 mA; (d) 140 V, 6 mA.

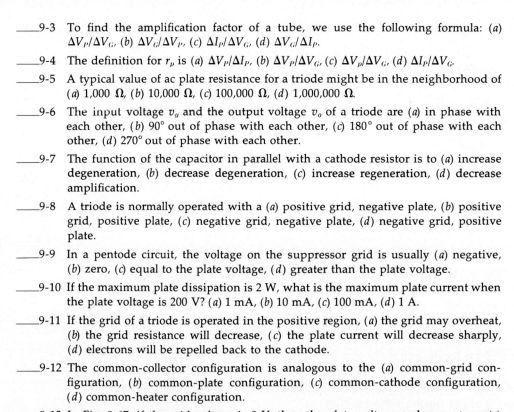

Fig. 9-67

10 audio amplifiers I

OBJECTIVES

(1) Contrast preamplifiers with power amplifiers in respect to operating power levels. **(2)** Define class A, AB, and B amplifiers, signal-to-noise ratio, and noise factor. **(3)** Draw graphs showing the dependence of the noise factor on collector current and voltage, signal source resistance, and frequency. **(4)** Diagram an emitter follower, a CE configuration with an unbypassed emitter resistor, a circuit with a series input resistor, a FET input circuit, and a cathode follower circuit. **(5)** Sketch a typical audio amplifier frequency response curve indicating reference frequency and frequencies where the response amplitude becomes unacceptable. **(6)** Sketch and describe a low-frequency compensating circuit. **(7)** Indicate those features which produce good low-frequency response, high-frequency degeneration, bias stability, and low noise. **(8)** Sketch and describe a high-frequency compensating circuit. **(9)** Describe the function of the high-pass filter and indicate which components provide base bias. **(10)** Sketch and discuss a typical *RC* coupling network. **(11)** Sketch and discuss the transformer-coupled configuration. **(12)** Sketch and list advantages of the impedance-coupled network. **(13)** Discuss a direct-coupled two-stage transistor dc amplifier. **(14)** Sketch and discuss a vacuum tube circuit with a voltage-divider gain control and a transistor circuit with a current-divider gain control. **(15)** State the 3 requirements to be met by a satisfactory volume control and sketch and describe several *RC*-coupled circuits which do not meet one or more of these requirements. **(16)** Modify an unsatisfactory volume control network so as to meet the three requirements. **(17)** Demonstrate that reflected impedance is the important consideration in designing volume controls for transformer-coupled stages. **(18)** Sketch and describe a volume control for a transformer-coupled amplifier circuit. **(19)** Give several examples of tone controls and modify a frequency-compensating network to produce a tone control. **(20)** Sketch a single-stage *RC* phase splitter circuit and a push-pull amplifier. **(21)** Sketch and describe a two-stage CE, CB phase-splitter circuit and a two-stage phase-inverter circuit using only the CE configuration.

PREAMPLIFIER PRINCIPLES

10-1
An amplifier that operates in the range of frequencies from 10 to approximately 20,000 cycles per second (hertz, abbrev. Hz) may be classed as an _____ amplifier.

10-2
(*audio*) As shown in Fig. 10-2, an audio amplifier may usually be classed as a preamplifier, a driver, or a _____ amplifier.

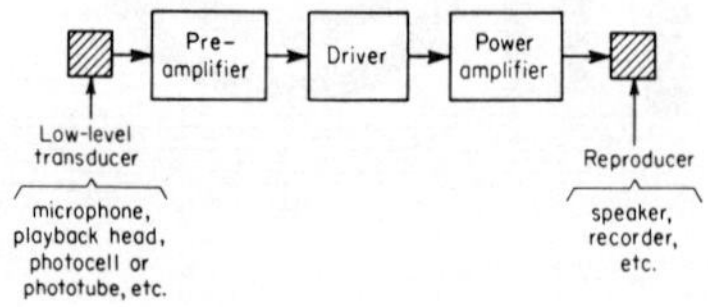

Fig. 10-2

10-3
(*power*) An audio amplifier is classified as a preamplifier if it is preceded by a low-level _____ such as a microphone, a playback head, or a photoelectric device.

10-4
(*transducer*) The input circuit of a transistor audio amplifier has an input resistance which is generally lower than that of an equivalent electron tube. Hence, current generally flows in the input circuit. When a signal causes current to flow, _____ must be consumed.

10-5
(*power*) For this reason, transistors are usually considered as current or power amplifiers in contrast to electron tubes, which are considered as either _____ amplifiers or power amplifiers.

10-6
(*voltage*) Preamplifiers operate at power levels measured in picowatts (pW) (micromicrowatts) or microwatts (μW); drivers operate at levels measured in milliwatts; _____ amplifiers operate at levels measured in hundreds of milliwatts or in watts.

10-7
(*power*) Transistor or tube amplifiers may be operated in class A. The class A transistor amplifier is always biased so that it operates on the _____ portion of its collector characteristics.

10-8
(*linear*) In class A, _____ current (as well as base and emitter current) flows during the entire input cycle and also during periods when no signal at all is present.

248 A PROGRAMMED COURSE IN BASIC ELECTRONICS

10-9
(*collector*) A single transistor used in class A is generally satisfactory; when a single transistor is used, the amplifier is termed *single-ended*. Two transistors may also work side-by-side in class A. In this application, the transistors (or tubes) are usually connected in a _____-_____ circuit.

10-10
(*push-pull*) Class B transistor amplifiers are normally biased so that collector current is cut off during intervals when the input signal is zero. Collector current flows only during the half-cycle of the input signal which aids the forward _____ applied to the transistor.

10-11
(*bias*) A class B amplifier cannot be used in single-ended audio amplifier circuits except in tuned power amplifier circuits. Class B amplifiers are generally used in _____-_____ circuits as audio amplifiers.

10-12
(*push-pull*) Class AB amplifiers are biased so that collector current flows for more than _____ the signal input cycle. Class AB audio amplifiers must be used in push-pull circuits just like class B amplifiers.

10-13
(*half*) An extremely important characteristic of a preamplifier is the amount of *noise* it introduces into the system. The best preamplifier in this respect will, of course, produce the _____ possible amount of noise.

10-14
(*least* or *smallest, etc.*) The *input* signal to any amplifier contains a certain amount of *noise power* N_i as well as the signal power S_i. First, let us set up a ratio of input signal power to input noise power and call this the input signal-to-noise ratio. In the given symbols this ratio is _____.

10-15
(S_i/N_i) Since a large amount of signal and a small amount of noise are desirable, this ratio should be as _____ as possible for best results.

10-16
(*large*) That is, the input transducer should have the characteristics necessary to give it a large _____-to-noise ratio.

10-17

(*signal*) Let us assume that the input transducer is satisfactory in this regard. As the signal and noise are amplified in the transistor, both will be increased in amplitude. Designating the output signal power by S_o and the output noise power by N_o, we can set up an output signal-to-noise ratio using these symbols thus: _____.

10-18

(S_o/N_o) We now define a new quantity: *noise factor*, F_o. The noise factor is equal to the input signal-to-noise ratio divided by the output signal-to-noise ratio. Symbolically,

$$F_o = \frac{S_i/N_i}{?}$$

10-19

(S_o/N_o) Let us first see what F_o would be for an ideal or perfect transistor. Assume that the input signal power is 0.05 μW and the input noise power is 0.002 μW. This would give an input signal-to-noise ratio of 0.05 μW/0.002 μW = _____.

10-20

(25) If the transistor is perfect, it will introduce *no noise of its own*, but it will amplify the signal and noise equally. Suppose the power gain is 10,000 for both. Thus, the value of S_o would be $0.05 \times 10^{-6} \times 10^4$ W = 0.05×10^{-2} W = 0.0005 W = _____ mW.

10-21

(0.5) Using the same power gain (10,000), the output noise power N_o would be $0.002 \times 10^{-6} \times 10^4$ W = 0.002×10^{-2} W = 0.00002 W = _____ mW.

10-22

(0.02) Since the output signal power is 0.5 mW and the output noise power is 0.02 mW, then the output signal-to-noise ratio is 0.5/0.02 = _____.

10-23

(25) Thus, the output signal-to-noise ratio is equal to the input signal-to-noise ratio if the transistor does not in itself generate any _____.

10-24

(*noise*) Hence, when we divide the input S to N ratio by the output S to N ratio to obtain the noise factor F_o, we get an answer of _____.

10-25
(*unity* or *one*) This tells us that an ideal, noise-free transistor amplifier has a noise factor of unity. Now let us see what happens to the size of F_o when the transistor introduces noise of its own making. Since this is *output* noise, the factor _____ will grow larger in the equation

$$F_o = \frac{S_i/N_i}{S_o/N_o}$$

10-26
(N_o) As N_o becomes larger because of internally generated noise, the value of the fraction S_o/N_o must become _____.

10-27
(*smaller*) But if S_o/N_o becomes smaller, then the value of F_o must become _____.

10-28
(*larger*) Hence, as the internally generated noise in a transistor amplifier becomes more intense, the noise factor for the system _____ in size.

10-29
(*increases*) This leads to the conclusion that a large noise factor F_o indicates a noisy system and a _____ noise factor promises quieter operation.

10-30
(*small*) The noise factor of a transistor amplifier is generally expressed in decibels. The noise factor in decibels may be found by taking ten times the logarithm of the ratio for the noise factor. As we have seen, the value of the F_o ratio for an ideal transistor is unity. The log of 1 is zero. Thus, the noise factor of an ideal transistor is _____ dB.

10-31
(*0*) The noise factor of an actual transistor amplifier is affected by the *operating point*. You will remember that the operating point is established by the zero-signal base or emitter current and the collector _____.

10-32
(*voltage*) Figure 10-32 shows how noise factor varies with collector voltage. In this set of curves, the running parameter is _____.

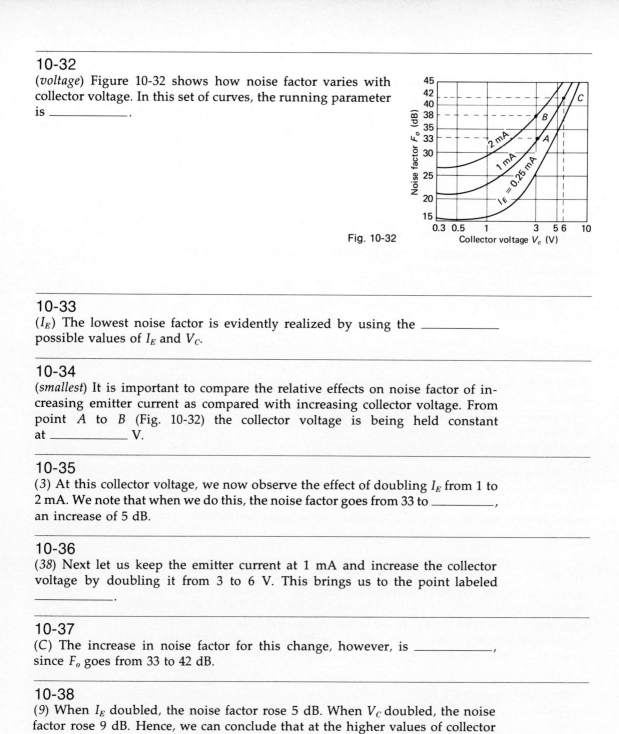

Fig. 10-32

10-33
(I_E) The lowest noise factor is evidently realized by using the _____ possible values of I_E and V_C.

10-34
(*smallest*) It is important to compare the relative effects on noise factor of increasing emitter current as compared with increasing collector voltage. From point A to B (Fig. 10-32) the collector voltage is being held constant at _____ V.

10-35
(3) At this collector voltage, we now observe the effect of doubling I_E from 1 to 2 mA. We note that when we do this, the noise factor goes from 33 to _____, an increase of 5 dB.

10-36
(38) Next let us keep the emitter current at 1 mA and increase the collector voltage by doubling it from 3 to 6 V. This brings us to the point labeled _____.

10-37
(C) The increase in noise factor for this change, however, is _____, since F_o goes from 33 to 42 dB.

10-38
(9) When I_E doubled, the noise factor rose 5 dB. When V_C doubled, the noise factor rose 9 dB. Hence, we can conclude that at the higher values of collector voltage, increasing V_C increases the noise factor _____ rapidly than increasing I_E.

252 A PROGRAMMED COURSE IN BASIC ELECTRONICS

10-39
(*more*) Figure 10-39 shows how the noise factor depends upon the internal _____ of the source of signal.

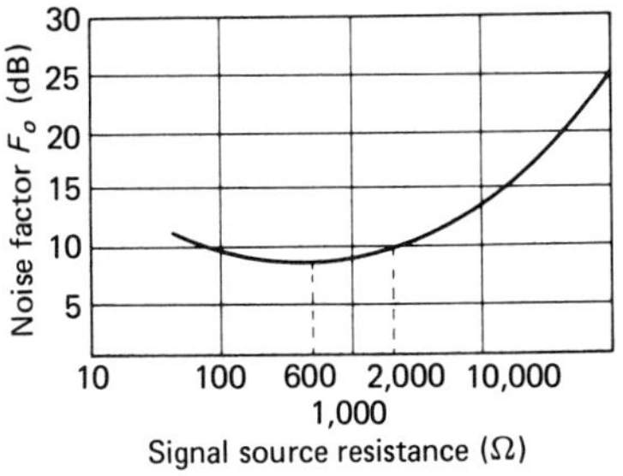

Fig. 10-39

10-40
(*resistance*) There is clearly a specific value of source resistance which produces the minimum amount of noise. For this particular transistor, least noise is generated when the source resistance is _____ Ω.

10-41
(*600*) The curve shows, furthermore, that the noise factor can be kept below 10 dB if the internal resistance of the source is made any value between 100 Ω and _____ Ω.

10-42
(*2,000*) The noise factor of a transistor amplifier also depends upon the _____ at which the amplifier operates as shown in Fig. 10-42.

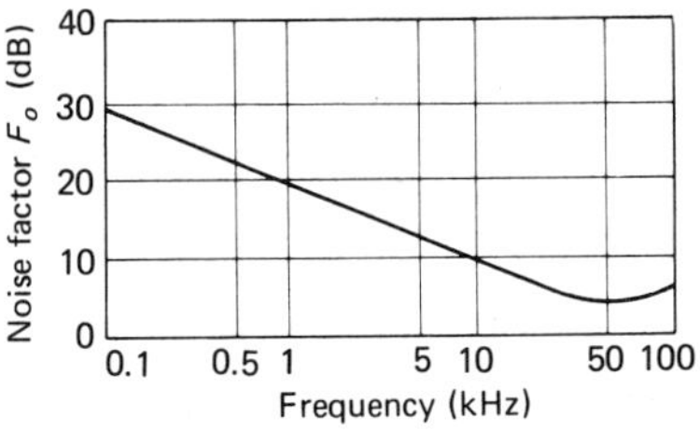

Fig. 10-42

10-43
(*frequency*) At very low frequencies, the noise factor appears to be quite _____.

10-44
(*high* or *large*) The noise factor gradually decreases as the frequency increases up to about _____ kHz.

10-45
(*50*) Above this frequency the noise factor begins to _____ once again.

AUDIO AMPLIFIERS | 253

10-46

(*increase*) Thus, in comparing audio amplifiers with respect to noise factor, it must be remembered that more noise is to be expected at the _____ audio frequencies.

10-47

(*low*) If the curve of Fig. 10-42 is extrapolated to 0 Hz (or dc), it shows that it is very difficult to design a dc amplifier with low _____.

10-48

(*noise*) Another important characteristic that determines the choice of preamplifier is *input resistance*. If the signal source is of the low-resistance type, then the input resistance of the preamplifier should be _____.

10-49

(*low*) The CB configuration has an input impedance that is normally between 30 and 150 Ω; this impedance is sufficiently low to match most very low _____ sources.

10-50

(*impedance*) If the impedance of the source ranges around 1,000 Ω, the best choice of configuration would be the _____ type, since the input impedance for this configuration is between 500 and 1,500 Ω.

10-51

(*CE*) Many sources, such as microphones and pickup heads of the crystal variety, are high-impedance sources. The preamplifier selected for use with this type of source should have a high _____ impedance.

10-52

(*input*) Several circuits can provide a high input impedance. One we have already discussed is the emitter follower. You will recall that an emitter follower has a high input resistance and a _____ output resistance.

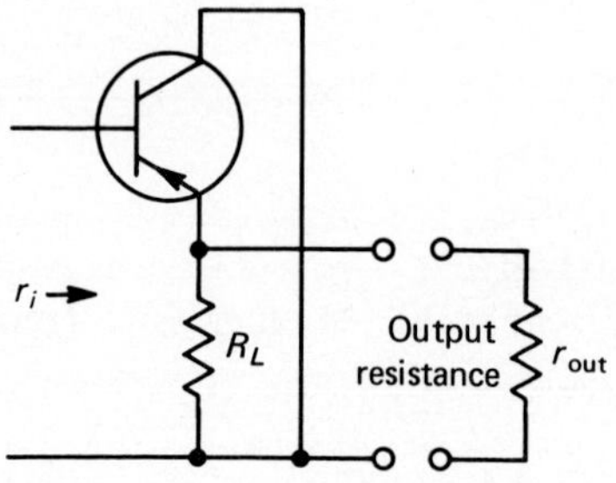

Fig. 10-52

10-53

(*low*) One precaution must be heeded, though, when the emitter follower is used. The input resistance, r_i, depends on the *net* emitter resistance. The net emitter resistance is R_L in parallel with the output _____.

10-54
(*resistance*) If the output resistance, r_{out}, varies considerably, then the performance of the emitter follower can be downgraded. Figure 10-54 depicts a circuit which circumvents this difficulty. Besides the emitter resistor R_E, there is now a collector resistor denoted by _____.

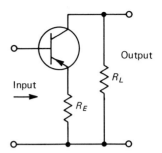

Fig. 10-54

10-55
(R_L) The output is taken from across resistor R_L, so that the circuitry following this preamplifier stage will *not* load down R_E. Thus the input resistance r_i will be determined solely by the resistor _____.

10-56
(R_E) Further, note that this circuit will exhibit good bias stability because of the emitter resistor R_E. Finally, this circuit can produce some voltage gain, whereas the emitter follower (a CC configuration) produces a voltage gain of less than _____ (Fig. 9-194).

10-57
(*1*) Another circuit used to obtain a high input impedance is shown in Fig. 10-57. If R_L is 500 Ω, then for a typical transistor the base-emitter resistance in the CE configuration is about _____ Ω (Fig. 9-194).

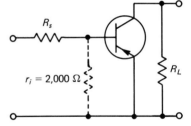

Fig. 10-57

10-58
(*2,000*) To achieve a total input resistance of, for example, 20,000 Ω, a series resistor R_s may be added. Since $r_i = 2,000$ Ω, R_s would have a value of _____ Ω.

10-59
(*18,000*) The advantage of this circuit over the emitter-follower arrangement is that the input resistance is governed chiefly by the fixed value of R_s. Thus, variations in transistor parameters or load variations from the input of the following stage would have very little effect on the _____ resistance.

10-60
(*input*) A disadvantage of this circuit is that there is a large value of base lead resistance present. We have found that large resistances in the base lead tend to cause bias instability. You will recall that good bias stability can be obtained by making the base lead resistance as _____ as possible.

10-61
(*small*) Yet another circuit used to obtain high input impedance employs the FET. Recall that the input resistance of a FET can be as high as hundreds of megohms. Suppose, as in the previous example, we want the input resistance r_i to be 20,000 Ω. Then, since the FET input resistance is so high, r_i will be determined by the fixed resistor, _____, which should have a value of 20,000 Ω.

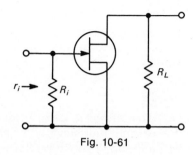

Fig. 10-61

10-62
(R_i) And finally, obtaining high input resistances with electron tube circuits is straightforward since the input resistance of a triode or pentode is very high. Figure 10-62 shows a tube circuit whose function is similar to that of the transistor emitter follower. It provides a high input resistance and a relatively _____ output resistance.

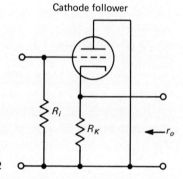

Cathode follower

Fig. 10-62

10-63
(*low*) In fact, because of the very high grid resistance, the input resistance would be determined by the fixed resistor, _____, and the output resistance, r_o, would be roughly equal to R_K. This circuit configuration is called a *cathode follower*.

10-63
(R_i)

FREQUENCY RESPONSE CURVES
10-64
The nominal range of human hearing extends from 20 to 20,000 Hz. This means that the lowest frequency a person with acute but normal hearing can hear is 20 Hz, and the highest frequency is _____ _____.

10-65
(*20,000 Hz*) A full orchestra is ordinarily capable of producing musical tones which extend over this range. For perfect fidelity of reproduction of this music, an audio amplifier would have to be able to amplify every frequency included between _____ and 20,000 Hz.

10-66
(*20*) In addition, for perfect fidelity of reproduction, the amplification over the audio range would have to be uniform. Thus, the amplifier must be designed so that it does not tend to amplify any one frequency or group of frequencies more or _____ than any other.

10-67
(*less*) To test an audio amplifier for this characteristic, one uses a constant-output variable signal generator from 20 to 20,000 Hz. As the name of the instrument implies, the frequency is capable of being set anywhere in the audio range while the amplitude of the signal is expected to be _____.

10-68
(*constant*) A typical test setup is shown in Fig. 10-68. An audio generator, labeled by the letter *A*, is connected to an amplifier, labeled by the letter _____. The output of the amplifier is fed to a meter.

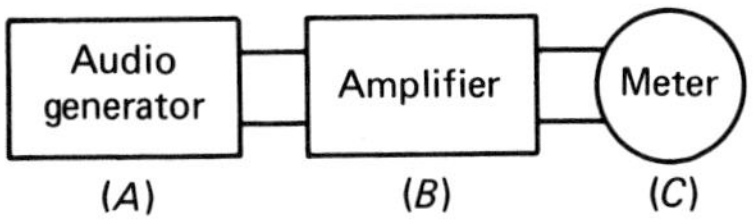

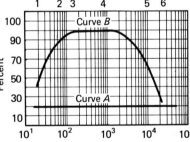

Fig. 10-68

10-69
(*B*) The frequency of the generator is swept through the audio range, and the output is read by the meter. A result of such a test is shown in Fig. 10-69. Curve *A* represents the constant output of the _____ _____. Curve *B* represents the output of the amplifier.

Fig. 10-69

10-70
(*audio generator*) Note that we have arbitrarily chosen the output of the generator to be _____ percent. By convention, we have set the output of the amplifier at 1,000 Hz (point 4) to be 100 percent.

10-71
(*20*) Hence the amplification of the amplifier at 1,000 Hz is 5. Now if we look at frequencies lower than 1,000 Hz, we see that the output of the amplifier remains uniform until we reach a frequency of 180 Hz, which is point _____ on curve *B*.

10-72
(*3*) At a frequency of 100 Hz, we find that the output is down to _____ percent of its value at 1,000 Hz, enabling us to locate point 2 on curve *B*.

10-73
(*95*) We finally find that at 20 Hz, the output of the amplifier has dropped to _____ percent of its value at 1,000 Hz.

10-74
(*44*) Starting once again at 1,000 Hz, but going up in frequency this time, we find that at 10,000 Hz (point 5), the output is down to _____ percent of its value at 1,000 Hz. And finally at 20,000 Hz, the amplifier output is down to a relative value of 30 percent.

10-75
(*65*) A natural question at this point is how does one use such a response curve to judge amplifier quality. The answer to that question lies in the fact that the ear cannot discern small amplitude variations in the sound it receives. By convention, the response of an amplifier is considered "flat" if the amplifier's output in the range of 20 to 20,000 Hz remains within 70 percent of the output at the reference frequency (_____ Hz).

10-76
(*1,000*) The amplifier depicted in Fig. 10-69 cannot be considered to be a "high fidelity" amplifier suitable for reproduction of recorded orchestral music because its frequency response is not flat (within 70 percent) over the range from 20 to _____ Hz.

10-77
(*20,000*) The amplifier shows a low-frequency dropoff, falling below 70 percent at about _____ Hz. Further, it shows a high-frequency drop, falling below 70 percent at 9,000 Hz.

10-78
(*35*) It should be emphasized that, while this amplifier could not be rated as high fidelity, it *would* produce perfectly intelligible speech. In the next section we shall see an example of improving the overall _____ response of an audio system.

10-78
(*frequency*)

DIRECT-COUPLED PREAMPLIFIERS WITH FREQUENCY COMPENSATION

10-79
Let us see how the two-stage direct-coupled preamplifier in Fig. 10-79 makes use of the facts presented in the previous discussion of preamplifier principles. The input signal source in this circuit is the microphone, identified as _____ in the drawing.

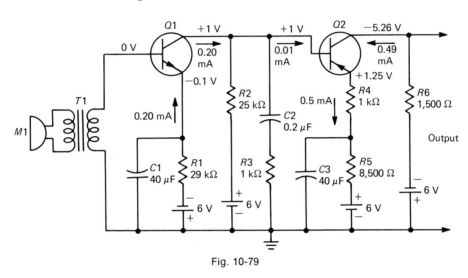

Fig. 10-79

10-80
(*M1*) This particular microphone is assumed to have poor low-frequency output for which compensation is to be made. First we note that the collector voltage (+1 V) and the emitter current (0.2 mA) are both very _____ as compared with their usual values in normal transistor circuits.

10-81
(*small*) We have found that small values for collector voltage and emitter current lead to a _____ noise factor, all other things being equal.

10-82
(*small*) Thus, a small noise factor is planned for. Next we note that the emitter resistor R1 is very heavily bypassed. This ensures that there will be little _____-frequency attenuation at this point in the circuit.

10-83
(*low*) There is no coupling capacitor between the output of Q1 and the input of Q2. This further ensures that the amount of low-frequency _____ will be negligible.

10-84
(*attenuation*) Transistor Q2 is a degenerated CE stage. Although R5 is heavily bypassed, it should be noted that the second emitter resistor _____ is not bypassed.

10-85
(*R4*) The negative feedback or degeneration produced by R4 causes the input impedance of Q2 to be relatively _____.

10-86
(*high*) The series combination consisting of C2 and R3 forms a high-frequency attenuation circuit through which the high frequencies will be partially shunted to _____.

10-87
(*ground*) Since the input impedance of Q2 is made high by the degenerative action of resistor R4, C2 may be made relatively small and still be effective for sufficient shunting action on the _____ frequencies.

10-88
(*high*) Thus, the high frequencies are partially shunted to ground, while the low frequencies are attenuated as little as possible. Such an amplifier provides greater gain for the low frequencies; hence it compensates for the poor _____-frequency output of the microphone.

10-89
(*low*) The functions of the parts may be summarized thus: The output of M1 is coupled to the base of Q1 through the component identified as _____.

10-90
(*T1*) Transistor Q1, arranged for low-noise-factor operation, amplifies the input signal. The emitter-swamping resistor for Q1 is resistor _____.

10-91
(*R1*) Degeneration due to the emitter-swamping resistor in the first stage is prevented by the part identified as _____.

10-92
(C1) The output signal from Q1 is developed across resistor R2. The 6-V battery connected to R1 provides the _____ bias needed in the base-emitter circuit.

10-93
(*forward*) Collector voltage for Q1 is provided by the battery connected to resistor _____.

10-94
(R2) C2 and R3 form a high-frequency attenuation circuit. R4 provides degeneration, while R5 is the emitter-_____ resistor for Q2.

10-95
(*swamping*) The signal is further amplified by Q2. The output signal is finally developed across _____.

10-96
(R6) A 2-stage, RC-coupled preamplifier is shown in Fig. 10-96. Judging from the transistor symbols and the polarity of the batteries, these transistors are _____ types.

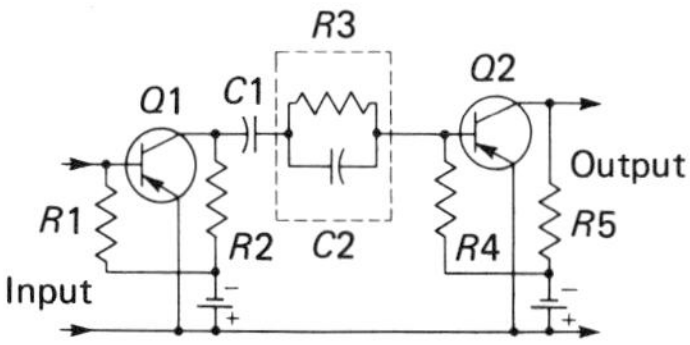

Fig. 10-96

10-97
(PNP) This circuit is intended to compensate for poor high-frequency output of the transducer connected to the input. It does this by means of an equalizer network (in dashed lines) containing the components R3 and _____.

10-98
(C2) Compensation is accomplished by the equalizer by attenuating the lower frequencies so that equalization of output is obtained. Hence, the equalizer network must be a _____-pass filter.

10-99
(*high*) The primary purpose of C1 is that of a blocking capacitor. That is, it prevents the _____ voltages applied to the various transistor electrodes in one stage from affecting the operation of the other stage.

10-100
(*dc*) The value of C2 is chosen so that its reactance will be low for the higher-frequency range and reasonably high for the lower frequencies. Thus, the _____ frequencies can pass through C2 with little attenuation.

10-101
(*higher*) Because the reactance of C2 at the low frequencies is so _____, it can be considered an open circuit for these frequencies.

10-102
(*high* or *large*) Thus, the _____ frequencies are forced to pass to the next stage through R3.

10-103
(*low*) R3 is given a value large enough to cause appreciable attenuation of the low frequencies. The amount of attenuation of the low frequencies realized depends upon the relative values of capacitor C2 and resistor _____.

10-104
(*R3*) Circuit analysis of the remainder of the preamplifier follows: Base-emitter bias for Q1 is determined by resistor _____.

10-105
(*R1*) The output signal of transistor Q1 is developed across resistor _____.

10-106
(*R2*) Capacitor C1 provides some low-frequency attenuation since it has a higher reactance for low frequencies than for high frequencies, but its main function is that of a _____ capacitor.

10-107
(*blocking*) Base-emitter bias for Q2 is provided by resistor R4, while the signal _____ is developed across resistor R5.

10-107
(*output*)

COUPLING NETWORKS

10-108
The coupling system shown in Fig. 10-108 will be known henceforth as an *RC coupling network*. The network itself consists of R1, R2, and _____.

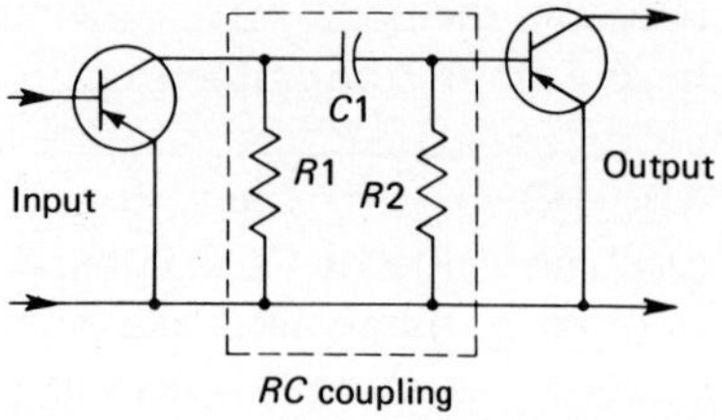

Fig. 10-108

10-109
(C1) The function of the coupling system is to transfer the signal from the output of one stage to the input of the succeeding stage without permitting the interaction of the _____ voltages between stages.

10-110
(dc) In the RC coupling network, the collector load resistor is R1, the coupling capacitor is C1, and the dc return resistor for the next stage is _____.

10-111
(R2) Since the collector dc must flow through R1, and since R1 must have a reasonably high resistance in order to develop the signal-voltage drop, there must be some _____ wasted by dissipation in R1.

10-112
(power) This makes for relatively low efficiency. Efficiency is defined in this connection as the ratio of ac output power to the _____ power delivered to the stage.

10-113
(dc) The dc blocking capacitor C1 prevents the dc voltage of the collector of the first stage from appearing on the _____ electrode of the second stage.

10-114
(base) Capacitor C1 should be considered as being in series with the input resistance of the following stage, while the signal voltage is delivered to the base of the second transistor as in a voltage divider. The reactance of C1 must be kept very _____, to avoid developing a large signal-voltage drop across it.

10-115
(small) Since the input resistance is normally less than 1,000 Ω, the reactance of C1 must be very small indeed. To obtain a very small reactance, it is necessary to make the *capacitance* of C1 very _____.

10-116
(large) In an electron-tube RC coupler, the value of C1 ranges from 0.001 to 0.01 microfarad (μF) normally. Because of the low input resistance R2, it is necessary to make C1 in a transistor coupler of the order of 10 μF or more. However, since the voltages involved are very low, even a 10-μF capacitor may be quite _____ in physical size.

AUDIO AMPLIFIERS | 263

10-117

(*small*) The resistance of R2 is usually made 7 to 15 times as large as the input resistance. This high value is required to prevent bypassing the signal around the base-emitter input of the second stage. However, R2 cannot be made too high because, if this is done, the bias temperature _____ is poor.

10-118

(*stability*) Low-frequency response is limited by the presence of C1. Since C1 has an increasing _____ for low frequencies, these frequencies will be somewhat attenuated.

10-119

(*reactance*) High-frequency response is limited by interelement capacitances. The collector-emitter capacitance of the first stage tends to shunt the high frequencies to ground; the base-emitter _____ of the second stage has a similar effect.

10-120

(*capacitance*) RC-coupled amplifiers are characterized by relatively high gain, small size, and economy of construction. They are generally confined to low-power applications because they have _____ power efficiency.

10-121

(*low*) The coupling system shown in Fig. 10-121 is known as *transformer coupling*. The primary winding of the transformer serves as the collector _____ impedance across which the output signal from the first stage develops the signal voltage.

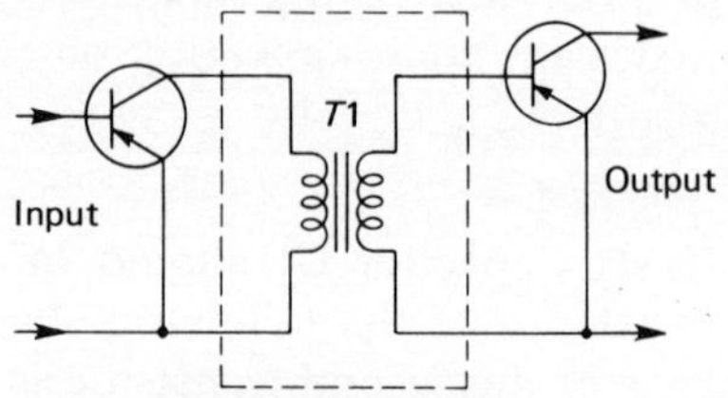

Fig. 10-121

10-122

(*load* or *output*) The secondary winding of T1 introduces the ac signal into the base-emitter circuit; at the same time, this winding serves as the dc _____ path for the base-emitter circuit.

10-123

(*return*) Temperature stabilization is excellent in a transformer-coupled stage, because the resistance in the base lead is _____, especially when the transformer secondary is properly designed.

10-124

(*low*) A further improvement in current stability factor can be achieved by inserting an emitter-_____ resistor in series with the emitter lead.

10-125
(*swamping*) Assuming that the primary of T1 achieves its impedance by means of inductance rather than resistance, the resistance in series with the collector lead will be _____ in size.

10-126
(*small*) With the resistance in the collector circuit small, the dc power dissipation will be _____.

10-127
(*small*) This, of course, results in _____ collector circuit efficiency.

10-128
(*high*) With a properly designed transformer, the collector circuit efficiency may approach the ideal value. This explains why _____ coupling is preferred in circuits that use battery power.

10-129
(*transformer*) A transformer is a natural impedance-matching device. Thus, transformer coupling facilitates a close match of the load to the output of the transformer and the source to the input of the transistor, thereby bringing about the _____ possible power gain for a given stage.

10-130
(*greatest* or *highest, etc.*) The reactance of the primary winding of the coupling transformer varies with frequency. At high frequencies, the reactance is relatively large; at low frequencies, the reactance may be appreciably _____ than it is at high frequencies.

10-131
(*smaller*) The signal-voltage drop is a function of reactance in the transformer winding. It will be remembered that the voltage developed across an inductive reactance _____ as the reactance increases for a given frequency.

10-132
(*increases*) Thus, at low frequencies, the reactance is smaller and the voltage developed across the primary winding by the signal is smaller. This tends to cause the _____-frequency response of the system to fall off.

10-133
(*low*) At high frequencies, the collector-to-emitter capacitance tends to shunt the reactance of the primary of the transformer. This causes the _____ frequency response of the transformer-coupled amplifier to fall off.

10-134

(*high*) In general, the _____ response of a transformer-coupled stage is not as good as that of the *RC*-coupled stage because of the dropoffs described in the previous frames.

10-135

(*frequency*) Beside the poorer frequency response, transformers are more costly and larger than resistors and capacitors used in coupling. However, because of the absence of large resistances, there is _____ dc power wasted in the transformer than in the corresponding load resistor used in *RC* coupling.

10-136

(*less*) This accounts for the fact that the use of transformer coupling is generally confined to applications requiring high power efficiency and high output _____.

10-137

(*power*) The high-frequency response of a transformer is further adversely affected by *leakage reactance* between primary and secondary windings. This effect is due to incomplete flux linkage. That is, leakage reactance is present when all the primary flux lines do not link with the turns of the _____ winding.

10-138

(*secondary*) Such leakage reactance causes high-frequency dropoff. The amount of leakage reactance present depends upon the design of the transformer. In well-designed transformers, _____ reactance is minimized by using suitable core iron and properly wound coils.

10-139

(*leakage*) Figure 10-139 illustrates an *impedance coupling network*. In this arrangement, the collector load resistor is replaced by a _____.

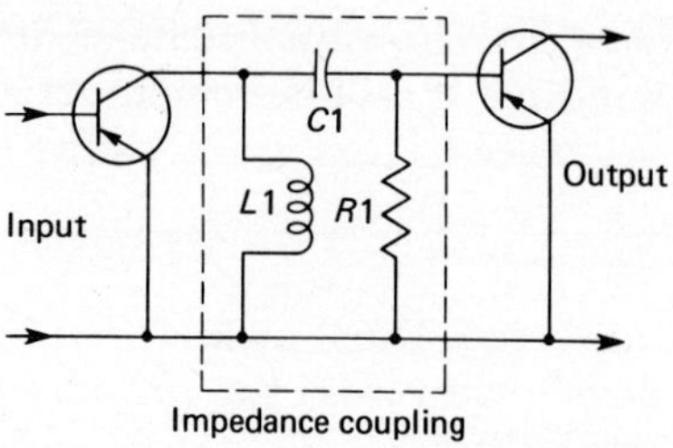

Fig. 10-139 Impedance coupling

10-140

(*coil* or *inductor, etc.*) Such an inductor has as little dc resistance as the primary of a transformer. Hence, the dc power loss in the collector circuit of the first transistor is just as _____ in this system as it is for transformer coupling.

10-141
(*small* or *low*, etc.) Like the transformer coupling system, however, the impedance coupler produces low-frequency dropoff because of changing coil reactance with frequency; also, _____-frequency dropoff is produced by the collector-to-emitter interelement capacitance.

10-142
(*high*) On the other hand, there is no secondary winding present. This means that there can be no _____ reactance as there is in a transformer.

10-143
(*leakage*) Thus, this particular cause of high-frequency dropoff is eliminated when an inductor is used instead of a _____.

10-144
(*transformer*) With this source of high-frequency dropoff eliminated, the frequency response of a(n) _____-coupled amplifier is somewhat better than that of a transformer-coupled amplifier.

10-145
(*impedance*) Direct coupling between two transistors is illustrated in Fig. 10-145. Note that the first transistor (left side) is an NPN type while the second is a _____ type.

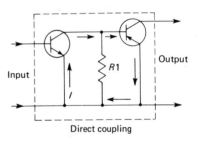

Fig. 10-145   Direct coupling

10-146
(*PNP*) Direct coupling is useful for dc amplification and amplification of very low frequencies. Such low frequencies cannot be coupled successfully from one stage to the next by *RC*, transformer, or impedance coupling networks. The arrows in Fig. 10-145 show the direction of _____ flow in the coupling system.

10-147
(*current*) Direct coupling is simpler with transistors than it is with vacuum tubes. There is only one kind of tube but two kinds of transistors, NPN and PNP. Note the battery connections in Fig. 10-147. The first transistor is an NPN type; hence it must be polarized so that its collector is _____ with respect to its emitter.

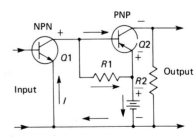

Fig. 10-147

AUDIO AMPLIFIERS | 267

10-148
(*positive*) As indicated in Fig. 10-147, this polarization is achieved. The PNP transistor, however, requires that its collector be _____ polarized with respect to its emitter.

10-149
(*negatively*) This is also achieved as shown. Some of the collector current of $Q1$ flows through the base-emitter circuit of $Q2$. The remainder of the collector current of $Q1$ flows through resistor _____.

10-150
($R1$) Thus, $R1$ serves as a collector-load resistor, permitting the collector current of $Q1$ to be greater than the base current of $Q2$, and at the same time developing a voltage drop that is used as the signal input to transistor _____.

10-151
($Q2$) As in all direct-coupled circuits, any variation of bias current due to temperature instability in the first transistor is amplified by the transistors that follow. This limits the number of stages. That is, the greater the number of direct-coupled stages, the _____ is the total bias instability.

10-151
(*greater*)

VOLUME AND TONE CONTROLS
10-152
Refer to Fig. 10-152. This is an *RC*-coupled vacuum tube amplifier in which the gain or volume is controlled by resistor _____.

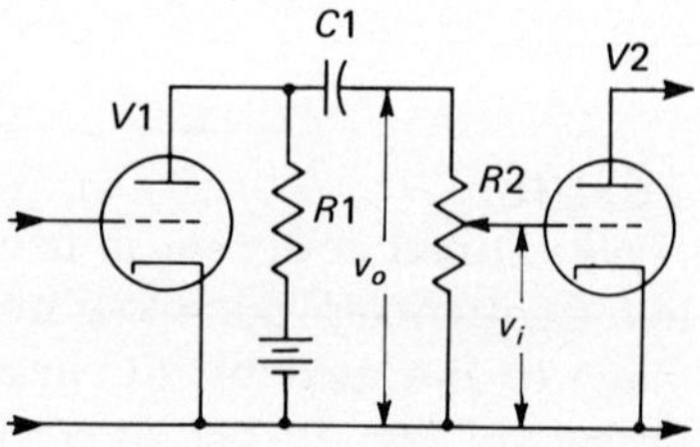

Fig. 10-152

10-153
($R2$) The source of the ac signal that reaches the grid of $V2$ is the plate circuit of $V1$. The ac output signal from the plate of $V1$ has been identified in Fig. 10-152 as _____.

10-154
(v_o) The ac input signal to the grid of $V2$ has been identified in Fig. 10-152 as _____.

10-155
(v_i) R2 is considered a *voltage divider* in Fig. 10-152. A resistive network is designated as a voltage divider when the source impedance is small compared with the load impedance. Since the plate circuit impedance of V1 is substantially lower than the grid input impedance of V2, then R2 must be considered as a _____ divider.

10-156
(*voltage*) The plate circuit impedance of V1 is considered relatively low because a reasonable signal current flows in this circuit; the grid input impedance of V2 is considered high because the current flowing in the grid-cathode circuit in normal class A operation is practically _____.

10-157
(*zero*) The signal output of V2 is controlled by the position of the movable arm of R2. When the arm is moved to the top of R2, then the input voltage to the grid of V2 becomes _____ to the output voltage from V1.

10-158
(*equal*) That is, for this position of the arm, $v_i = v_o$. This is the condition for _____ signal output voltage from the plate circuit of V2.

10-159
(*maximum*) When the movable arm of R2 is placed at the lowest point, minimum signal output voltage is obtained from V2 because, for this position of the arm, $v_i =$ _____.

10-160
(*0*) An equivalent transistor *RC*-coupled amplifier is shown in Fig. 10-160. Consider the output impedance of the source of signal (the collector circuit of Q1). The collector is always reverse-biased; hence the impedance of the collector-emitter circuit may be considered to be relatively _____ in size.

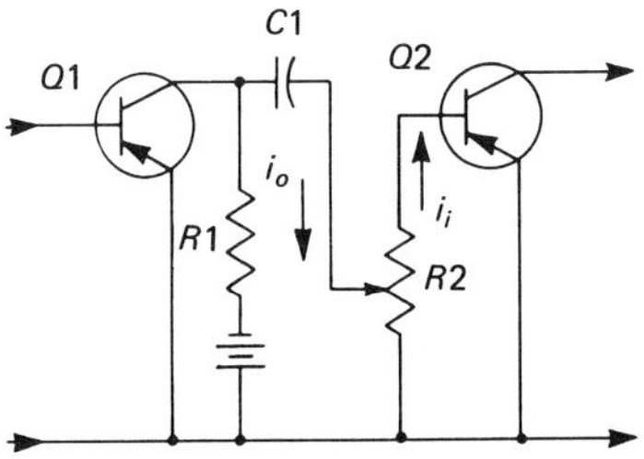

Fig. 10-160

10-161
(*large*) On the other hand, the input impedance of the base-emitter circuit of Q2 acts as the load. This circuit is forward-biased; hence its impedance must be appreciably _____ than the output impedance of the collector circuit of Q1.

AUDIO AMPLIFIERS I 269

10-162
(*lower*) When the source impedance is higher than the load impedance, the volume control then must be considered as a *current divider*. Thus, in Fig. 10-160 resistor _____ is treated as a current divider.

10-163
(*R2*) Unlike the input circuit of a vacuum tube, the base-emitter circuit of transistor Q2 draws a considerable amount of _____.

10-164
(*current*) The signal current from the collector of Q1 is symbolized in Fig. 10-160 by _____.

10-165
(i_o) If the movable contact of R2 is placed at its lowest position, all the output current i_o flows back to the emitter-collector circuit of Q1. For this condition, the signal input current, i_i, to Q2 equals _____.

10-166
(*zero*) This is the condition of _____ output from transistor Q2.

10-167
(*minimum*) As the movable contact is moved upward on R2, i_o divides so that part of it now constitutes the input current to the base-emitter circuit of Q2. For this condition, therefore, current _____ is no longer zero.

10-168
(i_i) Maximum output from Q2 is realized when the movable arm of R2 is moved to its _____ position.

10-169
(*highest* or *uppermost, etc.*) Thus, vacuum tube volume controls are generally considered as voltage dividers while transistor volume controls in the CE configuration are considered as _____ _____.

10-170
(*current dividers*) A satisfactory volume control must meet 3 requirements. The first of these is concerned with noise. A volume control must be selected and arranged in a given circuit so that it introduces as _____ noise as possible.

10-171
(*little*) The second requirement has to do with *range* of control. For a volume control to be satisfactory, it must be capable of varying the volume from maximum all the way down to _____.

10-172
(*zero*) The third requirement is that the volume control must not be *frequency-selective*. This means that as a volume control is varied from maximum toward zero, it must equally attenuate all _____ present in the system.

10-173
(*frequencies*) Let us analyze some faulty volume-control circuits first. In Fig. 10-173 we have a volume-control circuit in which the controlling element is the collector load resistor of the first stage. This resistor is identified as _____.

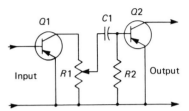

Fig. 10-173

10-174
($R1$) Since $R1$ is the collector load resistor, it is the impedance element through which the collector dc flows. Thus, whether or not a signal is present at the input, there will be a relatively heavy direct _____ flowing through $R1$.

10-175
(*current*) A movable arm (or wiper) which travels along a current-carrying resistance element is prone to pick up spurious voltages due to small but unpredictable voltage drops in the molecular structure of the resistor. Such voltage drops applied to the input of the next stage give rise to undesirable _____ as the wiper is moved.

10-176
(*noise*) Thus, the circuit of Fig. 10-173 is noisy and unsatisfactory. In Fig. 10-176, the collector dc no longer flows through the volume control. In this figure, the volume control is identified as resistor _____.

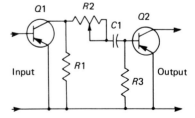

Fig. 10-176

10-177
($R2$) However, the collector current flows through resistor _____ in this case (Fig. 10-176).

10-178
(*R1*) Only the signal current now flows through the volume control *R2*. *R2* is *in series* with the base-emitter circuit of *Q2*. A series impedance element, however, cannot produce a *full range* of control. It permits maximum gain but cannot reduce the gain to _____ unless it presents an infinite impedance to the signal.

10-179
(*zero*) Another defect of the volume control circuit of Fig. 7-165 is related to *frequency response*. When the wiper is moved fully to the left, *R2* is short-circuited, and the only coupling element is capacitor _____.

10-180
(*C1*) Since the reactance of a capacitor decreases as the frequency is increased, C1 has a natural tendency to attenuate the _____ frequencies more than the _____ frequencies.

10-181
(*low, high*) Thus, at maximum volume (when the wiper is fully to the left in the diagram, Fig. 10-176) the coupling circuit has bass attenuation characteristics. As the wiper is moved toward the right, resistor _____ begins to become increasingly important as the coupling impedance.

10-182
(*R2*) When the wiper is fully to the right, the input impedance to *Q2* now becomes very large. That is, the interelement (collector-emitter) capacitance of *Q1* now looks into a very _____ input impedance.

10-183
(*large*) The shunting action of the collector-emitter capacitance now becomes significantly large. This capacitance has a much greater shunting effect on the _____ frequencies than it has on the _____ frequencies.

10-184
(*high, low*) This has the effect of attenuating the trebles more than the bass. Another way to say this is that the collector-emitter shunting capacitance has the effect of *boosting* the _____.

10-185
(*bass*) Thus we see that at maximum volume (wiper fully to the left, Fig. 10-176) the circuit tends to provide treble boost, while at smaller volume the circuit tends to produce _____ boost.

10-186
(*bass*) A volume control with frequency discrimination characteristics is unsatisfactory. Although the circuit of Fig. 10-176 does not tend to produce undesirable _____, it should not be used because it does have frequency discrimination characteristics.

10-187
(*noise*) A third unsatisfactory system for controlling volume is shown in Fig. 10-187. This circuit controls volume by increasing or decreasing the amount of resistance in the emitter circuit that is bypassed for audio frequencies by capacitor _____.

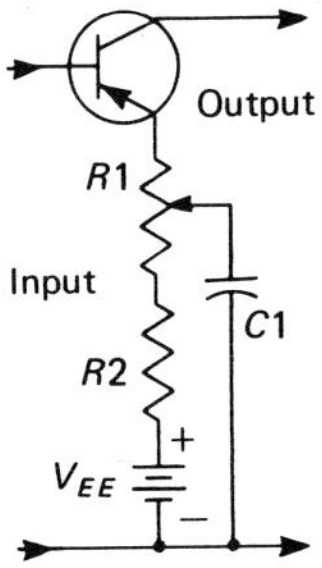

Fig. 10-187

10-188
(*C1*) When the wiper is fully up (in the diagram), the total emitter resistance is bypassed. Thus, in this condition the amount of *degeneration* is not very _____, and the volume is at a maximum.

10-189
(*great* or *large*) As the wiper is moved downward, less and less of R1 is bypassed by C1, resulting in an increase of _____.

10-190
(*degeneration*) As the degeneration increases, the volume of the output signal must _____.

10-191
(*decrease*) Hence, the circuit of Fig. 10-187 can serve to control volume. It has several unsatisfactory features, however. First, all the emitter dc flows through the volume control. This leads to excessive _____ in the output.

10-192
(*noise*) Second, unless R1 is made very large (up to 50,000 Ω), it is impossible to produce enough degeneration to reduce the volume all the way down to _____.

10-193
(*zero*) If R1 were to be made large enough to provide a full range of control, then it would be necessary to use an excessively large emitter battery voltage. The emitter battery is identified in Fig. 10-187 as _____.

AUDIO AMPLIFIERS | 273

10-194
(V_{EE}) A third defect of the degenerative control is related to frequency response. Assume first that the wiper is fully up. In this condition, capacitor C1 must bypass a relatively large resistance since all R1 and _____ are in series.

10-195
(R2) When a capacitor bypasses a large resistance (R1 and R2 in series) it will bypass both high and low frequencies relatively effectively. Thus, the amount of _____ that occurs is relatively small for both the highs and lows.

10-196
(*degeneration*) As the wiper is moved downward, however, C1 bypasses less and less resistance. When this occurs, its bypassing action becomes less and less effective for the _____ frequencies.

10-197
(*low*) But if the low frequencies are not effectively bypassed, the amount of degeneration for the low frequencies must _____.

10-198
(*increase*) Thus, the volume of the low frequencies decreases as the wiper is moved downward along R1, while the _____ frequencies are only slightly affected.

10-199
(*high*) This means that at low volume (wiper down on R1) the bass response decreases, while at high volume the bass response _____.

10-200
(*increases*) Hence, this circuit has _____-discrimination characteristics at various settings of the volume control. It is therefore unsatisfactory for this reason as well as the reasons involving noise and difficulty of obtaining zero gain.

10-201
(*frequency*) Figure 10-201 illustrates a fourth volume control circuit which, in the form shown, is unsatisfactory but which we shall later modify to correct its defects. In this circuit, the volume control is evidently resistor _____.

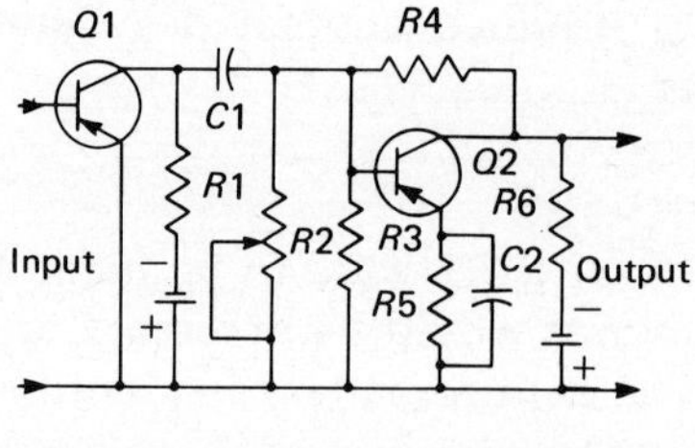

Fig. 10-201

10-202
(*R2*) Let us first analyze the circuit as shown in Fig. 10-201. The output signal of Q1 is developed across the collector load resistor identified as _____.

10-203
(*R1*) Dc blocking action and coupling from the output of Q1 to the input of Q2 is handled by the component identified as _____.

10-204
(*C1*) Base bias voltage is established by the voltage divider comprising resistors R3 and _____.

10-205
(*R4*) Resistor R5 is the emitter-_____ resistor, bypassed for ac by capacitor C2.

10-206
(*swamping*) The output signal is developed across _____ load resistor R6.

10-207
(*collector*) The first unsatisfactory feature of the circuit of Fig. 10-201 involves the shunting effect of R2 on resistor R3, and the manner in which it affects the base-emitter bias. The transistors we are dealing with in this circuit are of the _____ type.

10-208
(*PNP*) You will recall that the first letter of the type designation (P, in this case) tells us the polarity that the emitter must have with respect to the base for proper forward bias. Thus, for a PNP transistor, the emitter must be _____ with respect to the base.

10-209
(*positive*) In the Q2 collector circuit (Fig. 10-201), electron current flows up through R6 from the battery, down from collector to emitter, and down through R5 to common ground. This makes the top of resistor R5 negative with respect to ground. Since the emitter is connected to the top of R5, then the emitter is _____ with respect to ground.

10-210
(*negative*) Now consider the current through R4 and R3. The current flows through R4, down through R3, and back to the battery "+" terminal. With this direction of current flow, the top of R3 must become _____ with respect to ground.

10-211
(*negative*) Since the base of Q2 is connected to the top of R3, the base must become _____ with respect to ground because of the current flow through R4 and R3.

10-212
(*negative*) But we know that the emitter must be positive with respect to the base, or that the base must be negative with respect to the emitter. For this to occur, we see that the voltage drop across R3 which makes the base negative must be greater than the voltage drop across R _____ which makes the emitter negative.

10-213
(*5*) If the voltage drop across R3 is greater than the voltage drop across R5, then the base will be _____ with respect to the emitter and will provide the proper polarity for forward bias.

10-214
(*negative*) To establish this kind of net base-emitter bias in which the emitter is positive with respect to base, the resistance values of R5, R4, and _____ must be adjusted correctly so that the drop across R3 will more than cancel the drop across R5, thus leaving the emitter positive with respect to the base.

10-215
(*R3*) Now, noting how the volume control R2 shunts R3, we can see that R2 can short out R3 when its wiper is moved to the _____ position.

10-216
(*uppermost* or *topmost, etc.*) Thus, variation of R2 causes the base-emitter _____ to vary over a wide range, moving its operating point over a similarly wide range. This is a very undesirable action.

10-217
(*bias*) Another defect of the volume control circuit in Fig. 10-201 is that a relatively _____ direct current will flow through R2 from the collector battery of Q2.

10-218
(*large* or *heavy, etc.*) This condition would lead to excessively _____ operation of the volume control, as we have shown previously.

10-219
(*noisy*) A third defect of the circuit of Fig. 10-201 relates to frequency response. First, assume that the base-emitter resistance of $Q2$ is negligibly small (a reasonable assumption). With this assumption, it is clear that $R2$, $R3$, and _____ may all be considered to be connected in parallel with each other.

10-220
($R5$) Thus, $R2$ (the volume control) may be considered to *shunt* the emitter-swamping resistor, R _____.

10-221
(5) When $R2$ is fully in the circuit (wiper fully *down*), its effect upon $R5$ is negligible so that capacitor _____ is effective in bypassing both the low and high frequencies around $R5$.

10-222
($C2$) But as the wiper is moved upward to reduce the amount of $R2$ in the circuit, $C2$ becomes less and less effective in bypassing the low frequencies. Thus, degeneration for the _____ frequencies begins to occur.

10-223
(*low*) This _____ the low-frequency gain, causing the stage to favor the higher frequencies. This action is undesirable, since it causes emphasis of the high frequencies.

10-224
(*reduces* or *lowers, etc.*) Figure 10-224 illustrates certain modifications of the circuit of Fig. 10-201 which minimize the defects just discussed. The modifications include the addition of a new component, _____, and the rerouting of the return wire from the volume control.

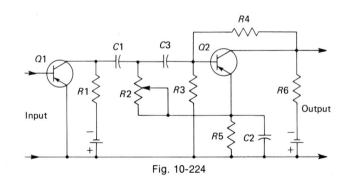

Fig. 10-224

10-225
($C3$) $C3$ serves two important functions. By producing dc isolation of $R2$ from the $Q2$ base circuit, it prevents $R2$ from affecting the bias-producing voltage drop across resistor _____.

10-226
(*R3*) The second improvement introduced by *C3* again relates to dc isolation. Prior to its addition, battery current flowed through *R4* and down through *R2*, causing a relatively large direct current in *R2*. *C3* prevents this current from flowing in *R2*, thereby ensuring that there will be no excessive ——————— from this source.

10-227
(*noise*) By returning the lower end of *R2* to ground through *R5*, *R2* no longer shunts *R5*. This prevents *R2* from reducing the bypassing effectiveness of *C2*, thereby avoiding undesirable ——————— of the lower frequencies.

10-228
(*degeneration*) Volume controls used in transformer-coupled amplifiers must meet the same requirements as discussed previously. The amplifiers shown in Figs. 10-228*a*, 10-228*b*, and 10-228*c* are all ———————-coupled types.

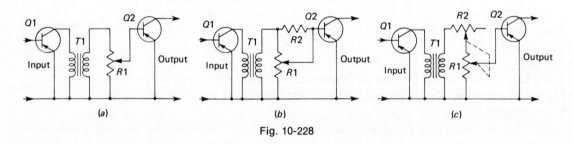

Fig. 10-228

10-229
(*transformer*) Of great importance in transformer-coupled amplifiers is the consideration of *reflected impedance*. Because of the nature of transformers, their operation is such that changes in secondary impedance are reflected into the ——————— circuit of the transformer.

10-230
(*primary*) The impedance reflected back from the secondary circuit (usually base-emitter circuit) to the primary circuit (usually the collector circuit) is given by the relation: $Z_p = (N_p/N_s)^2 Z_s$ where Z_p = the *reflected* impedance into the primary, Z_s = the secondary impedance, and N_p/N_s = the ——————— ratio of the transformer.

10-231
(*turns*) Note that the reflected impedance into the primary varies as the *square* of the turns ratio. Thus, even a small change in secondary impedance (Z_s) is likely to result in a ——————— change in reflected primary impedance Z_p.

10-232
(*large*) Every effort must be made to keep the secondary impedance changes very small. A large change in reflected impedance tends to cause unequal frequency attenuation. Thus, a large change in reflected impedance may be very detrimental to the _____ response of the amplifier.

10-233
(*frequency*) An unsatisfactory volume control circuit is shown in Fig. 10-228a. In this circuit, the volume control is identified symbolically as _____.

10-234
($R1$) When the wiper of $R1$ is at its lowest point (in the diagram), the volume is at a _____ because, for this condition, no voltage can be applied to the base-emitter circuit of $Q2$.

10-235
(*minimum*) For this position of the wiper, the only impedance across the secondary is that of $R1$, usually of the order of 1,000 Ω. As the wiper is moved upward, however, the volume control begins to shunt the input impedance of _____.

10-236
($Q2$) Since the input impedance of $Q2$ is normally about 1,000 Ω or less, the total impedance into which the secondary works at full volume (wiper up) is from 300 to 400 Ω. The impedance reflected into the _____ circuit by this variation of secondary circuit impedance has detrimental effects upon frequency response.

10-237
(*primary*) As mentioned, $R1$ has a rather low value—usually 1,000 Ω. Even at full volume, a considerable portion of the signal is lost because of the shunting effect of such a _____ resistance across the secondary winding of the transformer.

10-238
(*low*) The circuit of Fig. 10-228b is an improved volume control arrangement. The additional part in this circuit as compared with the previous one is _____.

10-239
($R2$) Usually resistor $R2$ is made equal in value to the input resistance of the transistor base-emitter circuit. With the wiper fully down (zero volume), the base-emitter circuit is shorted for ac, and resistor $R2$ is then in parallel with resistor _____.

10-240
(*R1*) Assume that both *R1* and *R2* in Fig. 10-228*b* are 1,000 Ω. Thus, the net resistance of this pair in parallel is _____ Ω.

10-241
(*500*) Thus, the impedance of the secondary circuit is 500 Ω. Now consider the wiper in position for maximum volume (fully up). For this condition, *R2* is now short-circuited, but the input impedance of the base-emitter circuit of *Q2* is now in _____ with the secondary winding of *T1*.

10-242
(*parallel*) Assume that the input resistance of *Q2* is also 1,000 Ω. Hence, the net resistance of *R1* and *Q2* in parallel is _____ Ω.

10-243
(*500*) Thus, at full volume and at zero volume the secondary impedances are _____ to each other.

10-244
(*equal*) Thus, for zero volume and maximum volume conditions, there will be no change in the _____ response of the amplifier.

10-245
(*frequency*) At the midpoint setting of the wiper, however, the secondary impedance rises about 25 percent of the value at maximum or zero volume. Thus, at the midpoint setting, the change of reflected impedance may be quite _____.

10-246
(*large*) This is not an ideal situation. It may be corrected, however, by using the arrangement shown in Fig. 10-228*c*. The dashed line indicates that the wiper arms of *R1* and _____ are mechanically ganged for simultaneous and equal movement when a single shaft is rotated.

10-247
(*R2*) In the diagram, imagine that the arm of *R2* moves to the *right* while the arm of *R1* moves *upward*. Let us start by visualizing the *R1* wiper fully down and the *R2* wiper fully to the _____.

10-248
(*left*) In this setting, *R2* is out of the circuit, the base-emitter input is shorted, and the only impedance across the secondary of *T1* is the resistance of _____.

10-249
(*R1*) Now imagine that the wiper of *R1* is moved slowly upward while that of *R2* moves slowly to the right. As more and more of the input resistance of *Q2* begins to shunt *R1* (thus reducing the secondary impedance), more and more of _____ comes into series with the secondary impedance, thereby causing it to remain essentially constant.

10-250
(*R2*) At full volume, the wiper of *R1* is fully up and the wiper of *R2* is fully to the right. In this condition, *R2* is in series with the parallel combination of *R1* and *Q2* input resistance. Thus, at full volume, the impedance across the secondary is equal to the impedances at any other settings of the dual _____ control.

10-251
(*volume*) A tone control circuit permits manual control of the frequency response of an amplifier. Any fixed low- or high-frequency compensating network can be converted to a tone control by substituting a manually variable circuit element for one of the _____ circuit elements.

10-252
(*fixed*) For example, the compensating network in Fig. 10-79 may be changed to a low-frequency-boost tone control circuit. In this drawing, the frequency compensating network consists of *C2* and _____.

10-253
(*R3*) Fixed resistor *R3* may be changed to a _____ resistor, thus converting the circuit to a low-frequency-boost tone control.

10-254
(*variable*) As another example, the high-frequency compensating network in Fig. 10-96 may be changed to a treble-boost tone control circuit. In this drawing, the frequency compensating elements are *C1*, *C2*, and _____.

10-255
(*R3*) If *R3* is changed to a variable resistor in this circuit, the circuit becomes a treble-boost _____ control.

10-255
(*tone*)

ONE-STAGE PHASE-INVERTER DRIVERS

10-256
Refer back to Fig. 10-2. As shown in this block diagram, a driver amplifier is located between the low-level preamplifier and the _____-output stage or stages.

10-257
(*power*) A driver stage may have two distinct functions. A power stage, whether single-ended or push-pull, generally requires that power be supplied to its *input* circuit in order to obtain full power output. A preamplifier is generally not expected to supply input power; a _____ stage is generally capable of supplying this power.

10-258
(*driver*) If the power-output stage is push-pull, the driver must perform a second function. A push-pull power amplifier requires _____ input signals, each 180° out of phase with the other.

10-259
(2) The stage preceding the power amplifier must be capable of supplying these two signals in phase opposition. Since the stage preceding the power amplifier is the driver, then such a driver must be able to act as a _____ inverter.

10-260
(*phase*) One common method of obtaining phase inverter action involves the use of an interstage coupling transformer having a center-tapped _____ winding.

10-261
(*secondary*) For reasons of cost, bulk, and frequency response, an interstage transformer may not be desirable. *RC* coupling is more economical, less bulky, and gives better _____ response.

10-262
(*frequency*) In the absence of a transformer having a center-tapped secondary winding, some other method must be employed to produce two signals in phase opposition suitable as input signals to a _____-_____ power amplifier stage.

10-263
(*push-pull*) Figure 10-263 illustrates an *RC*-coupled phase-inverter driver and its push-pull output stage. Transistor $Q1$ is a PNP type; hence its base-emitter circuit must be biased so that the emitter is more _____ than the base (forward bias).

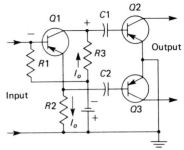

Fig. 10-263

10-264
(*positive*) Resistor R3 is the collector load resistor. Base-bias voltage is established by resistor _____.

10-265
(*R1*) Resistors R3 and R2 are made equal in value. When the input signal swing is such as to make the base more negative than it is during quiescent conditions, this aids the forward bias, causing the output current to _____.

10-266
(*increase*) The output current flows through both R3 and R2, the emitter load resistor. The increase is shown by the arrows I_o. The increased I_o through R3 causes the top of R3 to become more _____ than it was under conditions of quiescent operation, with respect to ground.

10-267
(*positive*) Similarly, the increase of I_o flowing through resistor R2 causes the top of this resistor to become more _____ than it was under conditions of quiescent operation, again with respect to ground.

10-268
(*negative*) When the input signal polarity is such as to oppose the forward bias, the output current I_o decreases below its initial value under _____ conditions of operation.

10-269
(*quiescent*) This causes the voltage drop produced by I_o across resistors R3 and _____ to decrease.

10-270
(*R2*) Such a decrease of output current will make the top of resistor R3 less positive than it was before, and the top of resistor R2 less _____ than it was before.

10-271

(*negative*) Hence the voltages at the tops of R2 and R3 swing upward and downward with respect to ground as the signal input varies. The direction of the voltage swings at each resistor are such as to make them _____° out of phase with each other.

10-272

(*180*) The signal developed across R3 is coupled to the base of Q2 through blocking capacitor C1. The signal developed across R2 is coupled to the base of Q3 through blocking capacitor _____.

10-273

(*C2*) Although R2 and R3 are equal in value, the output impedance of the collector circuit is greater than the output impedance of the emitter circuit. Thus, the signal-source impedance for Q2 is not _____ to the signal-source impedance for Q3.

10-274

(*equal*) This unbalanced condition leads to distortion at high signal levels. Figure 10-274 illustrates one method that may be used to equalize the signal-source _____ for Q2 and Q3.

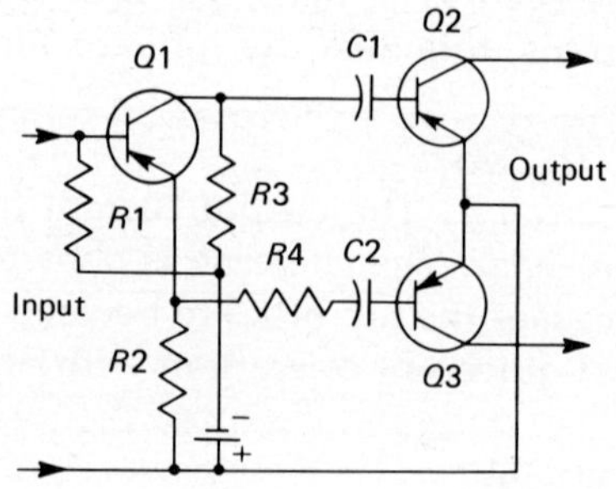

Fig. 10-274

10-275

(*impedances*) Resistors R2 and R4 are selected so that both Q2 and Q3 "look back" into equal signal-source impedances. The presence of R4 in series with the signal, however, must cause some loss of signal _____ as a result of the drop across this resistor.

10-276

(*voltage*) In the unbalanced impedance circuit of Fig. 10-263, R2 and R3 were made equal. To make up for the signal voltage loss across R4, a slightly greater signal voltage must now appear across resistor _____ than previously.

10-277

(*R2*) Hence, resistor R2 is made _____ in value than R3 so that the same output current, I_o, can produce a somewhat larger voltage drop across R2 than R3.

10-278
(*larger*) Note that R2 is not bypassed for signal frequencies. If it were, there could be no signal _____ drop across it for excitation of Q3.

10-279
(*voltage*) With R2 unbypassed, there is a relatively large negative feedback voltage developed across it. That is, R2 produces a _____ amount of degeneration.

10-280
(*large*) When the degeneration of an amplifier is large, its gain is reduced. This in turn calls for a much _____ signal input if maximum output is to be obtained.

10-280
(*larger*)

TWO-STAGE PHASE INVERTERS
10-281
Figure 10-281 illustrates a two-stage phase-inverter system. In this diagram, the driver-phase inverters are Q1 and Q2. Push-pull output is obtained from the transistors identified as _____ and _____.

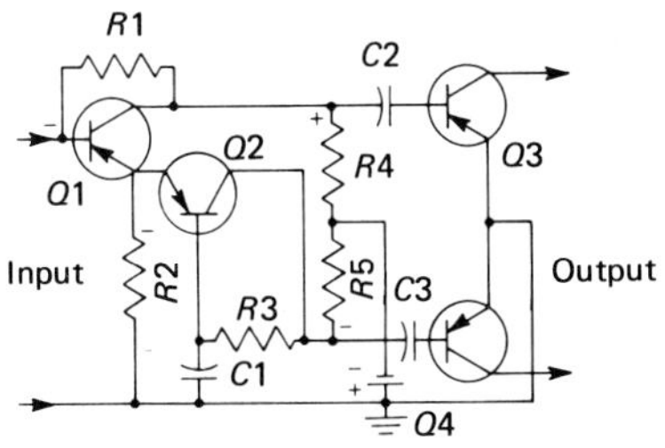

Fig. 10-281

10-282
(*Q3, Q4*) Note the position of the common-ground lead to help establish the configurations used. Since the emitter of Q1 returns to this common-ground lead, the configuration used for Q1 is common emitter. Since the base of Q2 returns to ground through a low ac reactance C1, the configuration for Q2 is common _____.

10-283
(*base*) To see how phase inversion is accomplished, let us start by considering that the input lead to the base of Q1 is negative-going as a result of signal input. You will recall that a common-emitter configuration produces a _____° phase reversal from input to output.

10-284
(*180*) Thus, if the base is made negative-going, then the collector output is _____-going.

AUDIO AMPLIFIERS | **285**

10-285
(*positive*) This positive signal is fed to the base of transistor _____ through blocking capacitor C2.

10-286
(*Q3*) Let us bear in mind, then, that a negative-going signal produces a positive-going signal at the input to Q3. Let us complete the analysis of the Q1 circuit before going on to Q2. The base bias on Q1 is established by resistor _____.

10-287
(*R1*) The output signal from Q1 is developed across the collector load resistor. This resistor is identified as _____ in Fig. 10-281.

10-288
(*R4*) R2 is the emitter load resistor for Q1. The same signal that produces the output in the collector circuit also causes a small signal to appear across the emitter load resistor R2. This signal voltage is applied directly to the _____ of Q2.

10-289
(*emitter*) The same signal that causes the top of R4 to be positive-going causes the top of R2 to be _____-going.

10-290
(*negative*) The reason is that as a negative-going signal *increases* the forward bias on Q1, the emitter current must increase. This current flows downward through R2, making the top of R2 _____ with respect to the bottom.

10-291
(*negative*) Thus, a _____-going signal is coupled to the emitter of the CB configuration of Q2.

10-292
(*negative*) You should remember that a CB configuration produces no phase inversion. That is, the _____ signal is in the same phase as the output signal.

10-293
(*input*) The collector load resistor for Q2 is resistor R5 (trace the circuit!). Since the CB amplifier produces no phase inversion, then a negative-going input signal from the top of R2 must cause a _____-going signal voltage across R5.

10-294
(*negative*) Thus, the bottom of R5 becomes more _____ than its top.

10-295
(*negative*) This negative output voltage is coupled to the base of Q4 via blocking capacitor _____.

10-296
(C3) The negative-going signal, as we emphasized previously, produces a positive-going signal at the base of Q3. Now we see that the same negative-going signal produces a _____-going voltage at the base of Q4.

10-297
(*negative*) Thus, the push-pull output stage receives its respective input signals 180° apart. Hence, the circuit has accomplished the required _____ inversion.

10-298
(*phase*) Resistor R3, to complete the analysis, provides the required base bias for Q2, while _____ keeps the base of Q2 at ac ground potential.

10-299
(C1) If the incoming signal swings positive rather than negative (as it does on the next half-cycle of input), the ac polarities indicated at R2, R4, and R5 would simply _____.

10-300
(*reverse*) Regardless of the swing direction of the input signal, therefore, the output transistors Q3 and Q4 always receive their respective input voltages _____° out of phase.

10-301
(180) A phase inverter may also be designed around a pair of CE configurations as in Fig. 10-301. Again assume a negative-going signal applied to the base of Q1. Since this is a CE amplifier, the voltage at the collector of Q1 must be a _____-going signal because of the 180° phase inversion inherent in the CE configuration.

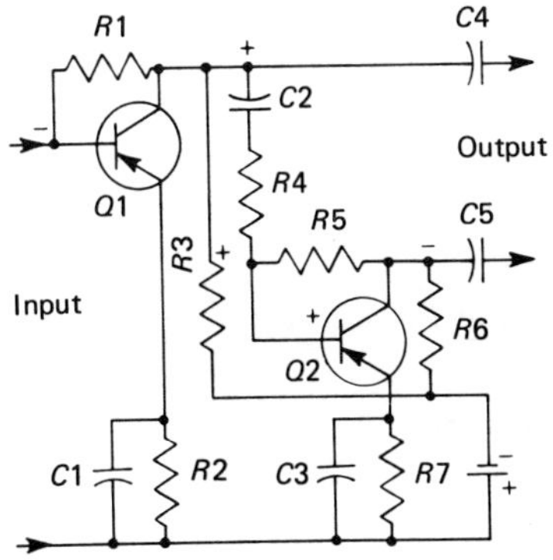

Fig. 10-301

10-302

(*positive*) Part of this signal is coupled to one of the output transistors (not shown) through C4. Since a capacitor does not invert phase, then the polarity of the voltage fed to the transistor through C4 must be _____-going.

10-303

(*positive*) Let us note that a part of the collector output of Q1 is also fed to the base of Q2 through C2 and R4 (latter to be discussed later). Thus, the positive-going output of Q1 appears as a _____-going input signal at the base of Q2.

10-304

(*positive*) Since a CE configuration as used for Q2 causes a 180° phase inversion, a positive-going signal at the base must result in a _____-going signal at the top of R6, the collector load resistor.

10-305

(*negative*) The negative-going voltage at the top of R6 is coupled to the output transistor (not shown) through C5. Thus, for this output transistor, the base input signal will be _____-going.

10-306

(*negative*) We have shown that Q1 causes a positive-going signal to appear at the input of the top output transistor and that Q2 causes a negative-going signal to appear at the input of the lower output transistor. Thus, the double CE amplifier in Fig. 10-301 produces a _____° inversion of phase.

10-307

(*180*) The remainder of the circuit is analyzed as follows: Base bias for Q1 is provided by the component identified as _____.

10-308

(*R1*) The emitter-swamping resistor for Q1 is R2. The emitter-swamping resistor for Q2 is _____.

10-309

(*R7*) Degeneration in the Q2 emitter circuit is minimized by C3. Degeneration in the Q1 emitter circuit is minimized by _____.

10-310

(*C1*) The collector load resistor for Q1 is R3. The collector load resistor for Q2 is _____.

10-311
(*R6*) Resistor *R5* provides _____ _____ for *Q2*.

10-312
(*base bias*) The function of *R4* may be explained as follows: In order to obtain equal driving power from *Q1* and *Q2* for their respective output transistors, the input signal to *Q1* should be _____ to the input signal to *Q2*.

10-313
(*equal*) The signal for *Q1* comes from the preamplifier at some given level. The output of *Q1*, however, is larger due to amplification. If this amplified output were to be fed directly to the base of *Q2* through *C2*, the input to *Q2* would be much _____ in magnitude than the input to *Q1*.

10-314
(*larger*) This is undesirable for good balance. Thus, the signal fed to the base of *Q2* must be attenuated until its magnitude is equal to that of the signal fed to the base of *Q1*. This attenuation is accomplished by the voltage drop in resistor _____ .

10-314
(*R4*)

CRITERION CHECK TEST

The following 3 questions refer to the block diagram of an audio amplifier below.

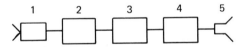

____10-1 Block 1 represents a (*a*) power amplifier, (*b*) driver, (*c*) preamplifier, (*d*) microphone.

____10-2 Block 4 represents a (*a*) power amplifier, (*b*) driver, (*c*) preamplifier, (*d*) microphone.

____10-3 The preamplifier is identified as block (*a*) 1, (*b*) 2, (*c*) 3, (*d*) 4.

____10-4 At which stage is it most necessary to control the signal-to-noise ratio? (*a*) Preamplifier stage, (*b*) driver stage, (*c*) output stage, (*d*) power stage.

____10-5 The noise factor of an ideal amplifier expressed in decibels is (*a*) 0, (*b*) 0.1, (*c*) 1, (*d*) 10.

____10-6 Choose the correct statement: (*a*) The smaller the signal-to-noise ratio, the better the signal, (*b*) all signals have some noise, (*c*) the larger the noise factor, the better the amplifier, (*d*) one decibel is equal to 10 volts.

_____10-7 In Fig. 10-32, if $V_C = 3$ V and $I_E = 0.25$ mA, the noise factor is (a) 15 dB, (b) 20 dB, (c) 25 dB, (d) 30 dB.

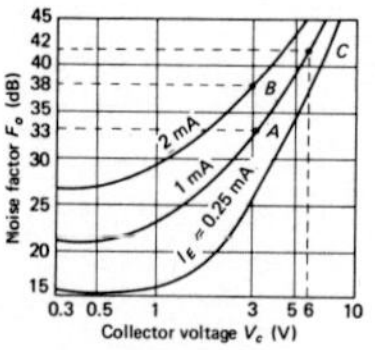

Fig. 10-32 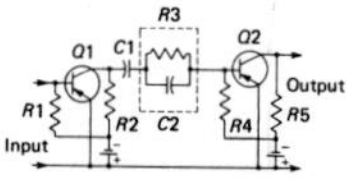   Fig. 10-96 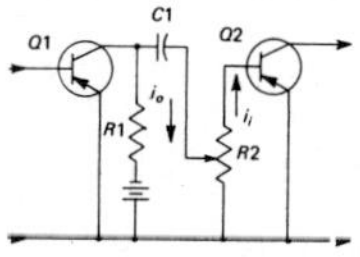   Fig. 10-160 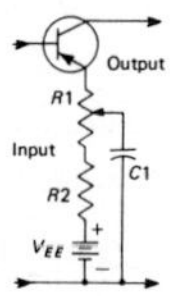   Fig. 10-187

_____10-8 Which frequency produces the highest noise factor in a transistor? (a) dc, (b) 500 Hz, (c) 1,000 Hz, (d) 10,000 Hz.

_____10-9 If a carbon microphone has a resistance of 100 Ω, then the best transistor configuration to use as the input of a preamp is (a) common base, (b) common emitter, (c) common collector, (d) common plate.

_____10-10 If a portable voltmeter required an input resistance of 10 MΩ, which device would you choose? (a) A PNP transistor in the common-collector mode, (b) a PNP transistor in the common-emitter mode, (c) a vacuum tube, (d) a FET.

_____10-11 In obtaining the frequency response curve of an amplifier, the (a) amplifier output is kept constant, (b) generator frequency is held constant, (c) generator output level is kept constant, (d) amplifier power supply is varied.

_____10-12 An amplifier with a frequency response of 50 Hz to 8,000 Hz (a) is considered a high fidelity amplifier, (b) could be used as a high fidelity preamplifier, (c) can reproduce intelligible speech, (d) can be used as a dc amplifier.

The following 3 questions refer to Fig. 10-96.

_____10-13 The resistor which provides forward bias for Q1 is (a) R1, (b) R2, (c) R3, (d) R4.

_____10-14 The element which decouples dc between Q1 and Q2 is (a) C1, (b) C2, (c) C3, (d) C4.

_____10-15 The elements which provide high-frequency compensation are (a) R2 and C1, (b) R1 and C1, (c) R3 and C2, (d) R4 and R5.

_____10-16 In an RC coupling scheme, C must be large enough to (a) pass dc between stages, (b) give good bias stability, (c) not attenuate the low frequencies, (d) dissipate high power.

_____10-17 An advantage of the RC coupling scheme is (a) economy, (b) excellent frequency response, (c) high efficiency, (d) good impedance matching.

_____10-18 A typical RC coupling capacitor for a transistor circuit is (a) 100 pF, (b) 0.001 μF, (c) 0.1 μF, (d) 10 μF.

_____10-19 Transformer coupling provides high efficiency because the (a) dc resistance is low, (b) collector voltage is stepped up, (c) collector voltage is stepped down, (d) flux linkages are incomplete.

_____10-20 When a multistage transistor amplifier is to amplify dc signals, then one must use (a) transformer coupling, (b) RC coupling, (c) direct coupling, (d) link coupling.

_____10-21 To avoid a high noise figure, the volume control should (a) not have large direct currents flowing through it, (b) have a value less than 10,000 Ω, (c) have a value greater than 10,000 Ω, (d) be much smaller than the collector resistance.

___10-22 A weakness of the volume control in Fig. 10-187 is that (a) direct current flows through R1, (b) degeneration is provided at all frequencies, (c) V_{EE} provides only forward bias, (d) V_{EE} provides only collector bias.

___10-23 In Fig. 10-224, C3 was introduced to (a) prevent direct current flowing in R2, (b) emphasize the bass over the treble, (c) emphasize the treble over the bass, (d) block ac from reaching the base of Q2.

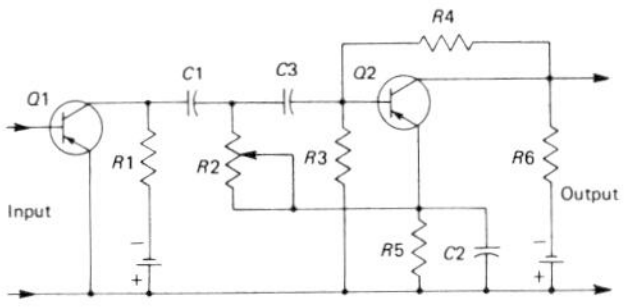

Fig. 10-224

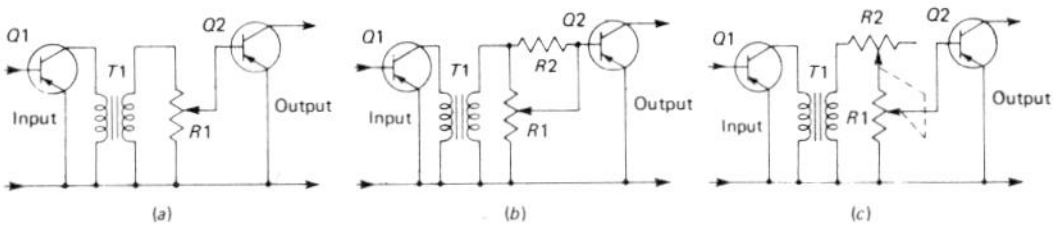

Fig. 10-228

___10-24 In Fig. 10-224, R2 is returned to ground through (a) R4C3, (b) C3C1, (c) R5C2, (d) C3R6.

___10-25 In Fig. 10-228a, when the wiper arm of R1 is all the way down, the impedance in the secondary of T1 is (a) R1, (b) R1 in parallel with the input of Q2, (c) 0, (d) the input of Q2.

___10-26 The purpose of the ganged resistors in Fig. 10-228c is to (a) provide a constant dc bias, (b) minimize power losses, (c) achieve a constant impedance, (d) provide a low secondary impedance.

The following 3 questions refer to Fig. 10-263.

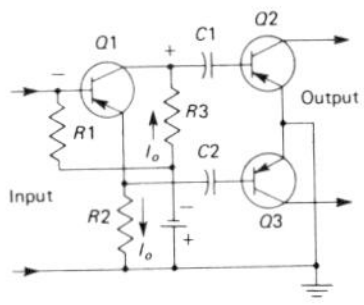

Fig. 10-263

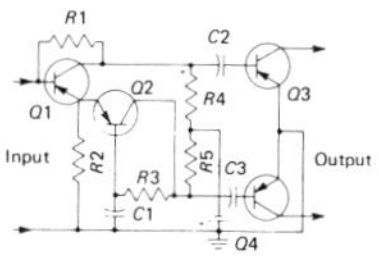

Fig. 10-281

___10-27 When the base of Q1 has a negative ac signal, then (a) the base of Q2 is positive, the base of Q3 is negative; (b) the base of Q2 is positive, the base of Q3 is positive; (c) the base of Q2 is negative, the base of Q3 is positive; (d) the base of Q2 is negative, the base of Q3 is negative.

___10-28 Q2 and Q3 are in the (a) CE configuration, (b) CB configuration, (c) CC configuration, (d) common-cathode configuration.

___10-29 When the base of Q1 is negative, (a) the collector of Q1 is negative, (b) the base of Q2 is negative, (c) the collector of Q2 is negative, (d) the collector of Q3 is negative.

The following 3 questions relate to Fig. 10-281.

___10-30 Transistor Q2 is in the (a) CB configuration, (b) CE configuration, (c) CC configuration, (d) common-drain configuration.

___10-31 When the base of Q1 goes positive (a) the collector of Q2 goes positive, (b) the emitter of Q1 goes negative, (c) the base of Q3 goes positive, (d) the collector of Q3 goes negative.

___10-32 Forward bias for Q1 is provided by (a) R1 and R5, (b) R5 and R3, (c) R4 and R1, (d) R4 and R5.

audio amplifiers II

OBJECTIVES

(1) Demonstrate that distortion occurs in an unbiased transistor push-pull amplifier. **(2)** Demonstrate that forward bias eliminates crossover distortion in a transformer-coupled push-pull amplifier. **(3)** Modify a push-pull circuit to avoid operation in the class C region. **(4)** Discuss the basic principles of a complementary-symmetry amplifier. **(5)** Draw a basic complementary-symmetry 2-transistor circuit and trace the output current for a complete cycle of input. **(6)** Sketch and describe a complementary-symmetry circuit with transformer-coupled input and output, a direct-coupled complementary-symmetry CE driver and output stage circuit, and a complementary-symmetry circuit whose driver stage uses the CC configuration. **(7)** Discuss the linearity of the current gain for a transistor as a function of collector current. **(8)** Analyze a compound-connected transistor pair by depicting such a pair and by tracing the dc current flow through the 2 transistors. **(9)** Compute the short-circuit forward-current amplification factor, h_{fb}, and h_{fe} for the compound pair, and compare h_{fe} for the pair with h_{fe} for a single transistor. **(10)** Draw a general 4-element bridge circuit. **(11)** State the condition for "balancing" a bridge. **(12)** Analyze a 2-transistor bridge circuit. **(13)** Discuss a bridge circuit with 2 NPN and 2 PNP transistors. **(14)** Depict a 4-transistor bridge circuit and compare with the 2-transistor circuit with respect to input and battery supply requirements. **(15)** Trace the current flow through the bridge due to an ac input. **(16)** Discuss the bridge circuit operation under class A conditions. **(17)** Depict a 4-PNP-transistor bridge circuit using 2 transistors in the CC mode and 2 in the CE mode. **(18)** Describe the use of a transformer and the use of dropping resistors to provide different levels of drive to the different configurations (CC and CE). **(19)** List at least 5 advantages of the bridge-type circuit.

POWER AMPLIFIERS, PART 1

11-1
Refer to Fig. 10-2, the block diagram of a complete audio amplifier system. We have discussed several typical preamplifiers and drivers. The remaining amplifier stage to which we shall give our attention now is the _____ amplifier.

11-2
(*power*) Amplifiers may be single-ended or push-pull. If an amplifier is single-ended, it normally uses a single transistor in that stage. A push-pull amplifier makes use of _____ transistors.

11-3
(*2*) In Fig. 11-3, 2 transistors are employed in a single power amplifier stage. Both of these transistors contribute to the gain of the same stage. This is evidently a _____-_____ circuit since 2 transistors are involved in the same stage.

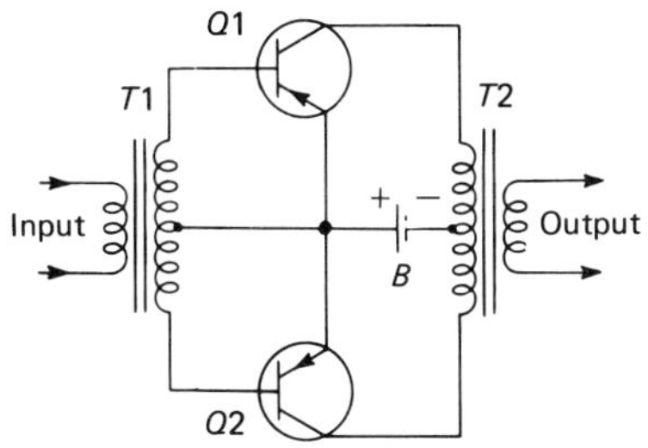

Fig. 11-3

11-4
(*push-pull*) Battery B supplies only the collector-emitter voltage. Since there is no other battery, nor is there any network that would enable B to supply base-emitter bias, the system will operate under conditions of zero _____ difference between emitter and base when there is no input signal.

11-5
(*voltage*) This condition is called *zero bias*. There is no forward bias on either transistor. Assuming that I_{CBO} (leakage current) is negligible, the collector current with zero signal may then be considered to be _____.

11-6
(*zero*) Thus, very high efficiency is obtained because neither transistor conducts during the period when the signal is zero. With a signal applied to the primary of the input transformer, a signal of the same frequency will appear across the _____ of the input transformer.

11-7
(*secondary*) During signal input, the base of one transistor will be positive-going, while the base of the other transistor will be _____-going.

11-8
(*negative*) At the instant that the base of Q1 is driven negative, the base of Q2 will be driven _____.

11-9
(*positive*) For a PNP transistor, a negative base voltage is a *forward* voltage. That is, as the base is driven negative with respect to the emitter, the collector current of the transistor _____.

11-10
(*increases*) A positive base voltage (PNP) is a *reverse* voltage. If the collector current of a transistor is zero for no-signal conditions, applying a positive voltage to the base will not change the _____ current.

11-11
(*collector*) Thus, assume that the base of Q1 in Fig. 11-3 is driven negative while the base of Q2 is driven positive. For this condition, only transistor _____ will conduct in its collector circuit.

11-12
(*Q1*) When the signal reverses, then the base of Q2 will be made negative-going while the base of Q1 is positive-going. For this condition, only transistor _____ will conduct.

11-13
(*Q2*) This shows that each transistor conducts on alternate half-cycles of the input signal. The output signal is a composite of the signals from both transistors; the combination of the 2 signals occurs in the primary of the output transformer, identified as _____ (Fig. 11-3).

11-14
(*T2*) To obtain a picture of the total output waveform we consider the dynamic transfer characteristic for the amplifier. Figure 11-14 shows the characteristic for one of the transistors. The dynamic transfer characteristic is a graph showing the relationship between collector current and _____ _____.

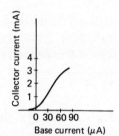

Fig. 11-14

11-15
(*base current*) Assuming that both transistors have identical characteristics, the total output characteristic can be obtained by placing two characteristic curves back-to-back as in Fig. 11-15. Note that the zero line of each curve is lined up vertically to reflect the fact that the bias current is _____ for no-signal conditions.

Fig. 11-15

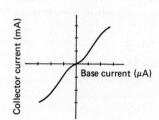

11-16
(*zero*) The curves are said to be *combined* when their zero lines are thus lined up. In Fig. 11-16, the effect of an input _____ current upon the output collector current is shown.

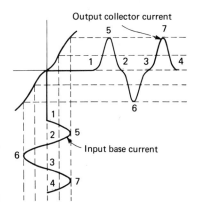

Fig. 11-16

11-17
(*base*) As may be seen from the curve on the vertical axis, the input base current is sinusoidal. It is projected upward on the combined characteristic, then to the right to produce the waveform of the _____ collector current.

11-18
(*output*) Severe distortion occurs at the *crossover points*. The crossover points are 1, 2, 3, and _____.

11-19
(4) For faithful reproduction of the input signal it is necessary to eliminate or minimize the distortion at the _____ points.

11-19
(*crossover*)

POWER AMPLIFIERS, PART 2
11-20
To appreciate how crossover distortion is minimized, refer first to Fig. 11-20. This circuit differs from the push-pull amplifier discussed in the last section in that it contains 2 additional components: $R1$ and _____.

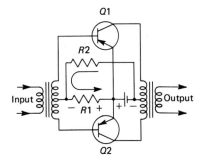

Fig. 11-20

AUDIO AMPLIFIERS II **295**

11-21
(*R2*) *R2* and *R1* form a voltage divider through which current from the battery flows. The voltage drop across *R1* has the polarity shown. This polarity is such as to make the emitters of both transistors more _____ than the bases.

11-22
(*positive*) When the emitter of a PNP transistor is more positive than the base, the transistor is then operating under conditions of forward _____.

11-23
(*bias*) Thus the *R2R1* voltage divider provides a small amount of forward bias for both transistors. The dynamic transfer characteristics in *uncombined* form are shown in Fig. 11-23. In this form, the 2 curves have been aligned so that zero base current for one is immediately above _____ base current for the other.

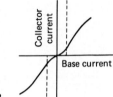

Fig. 11-23

11-24
(*zero*) This alignment is incorrect since, with forward bias applied, the base current of neither transistor is equal to _____.

11-25
(*zero*) The base current in one transistor for zero-signal conditions is shown by the dashed line above the horizontal axis. The base current in the other transistor for zero-signal conditions is shown by the dashed line _____ the horizontal axis (Fig. 11-23).

11-26
(*below*) To combine the characteristics properly, the top dashed line must be aligned with the _____ dashed line, as shown in Fig. 11-26.

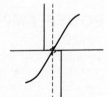

Fig. 11-26

11-27
(*bottom* or *lower*) When this is properly done, as in Fig. 11-26, the combined push-pull dynamic characteristic follows a reasonably _____ line as it crosses the horizontal and vertical axes at the origin.

11-28

(*straight*) Then, as indicated in Fig. 11-28, a projected input base current sinusoid gives rise to a projected output _____ current sinusoid of the same waveshape as the input.

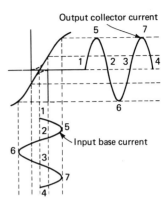

Fig. 11-28

11-29

(*collector*) Thus, by applying a small forward bias to the push-pull amplifier, _____ distortion is virtually eliminated.

11-30

(*crossover*) Figure 11-30 shows the effect of adding a bypass capacitor across R1. This capacitor is identified in the schematic as _____ .

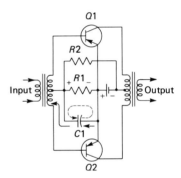

Fig. 11-30

11-31

(C1) With a signal applied to the base of the lower transistor Q2, capacitor C1 would tend to charge through the _____-emitter junction of Q2, as shown by the solid arrows.

11-32

(*base*) This action would occur during the time that Q2 was conducting because of a negative-going half-cycle on its base (forward voltage). During the positive half-cycle, C1 would tend to discharge through _____, as shown by the dashed arrow.

11-33

(R1) This discharge current would develop a voltage drop across R1 having the polarity indicated in Fig. 11-30. This voltage drop would tend to make the base _____ with respect to emitter (for each transistor).

11-34
(*positive*) This is *reverse* bias which would tend to drive both transistors beyond cutoff into the class C region. Class C operation would be very undesirable because an audio amplifier operating in class C would produce severe _____.

11-35
(*distortion*) Thus, the use of a bypass capacitor in the circuit of Fig. 11-30 is forbidden. When the capacitor is not used, the transistors operate in class _____ since they are virtually at cutoff when there is no signal.

11-36
(*B*) Class B operation at audio frequencies is permissible provided that _____ transistors are used in a push-pull circuit.

11-37
(*2*) During class B operation, one transistor conducts for one-_____ the input cycle while the other is nonconducting except for the small amount of collector current due to the forward bias used to minimize crossover distortion.

11-38
(*half*) In ordinary *RC* coupling to a class B push-pull stage, a certain undesirable effect occurs. Refer to Fig. 11-38. Assume that the input signal is, at a particular moment, driving the base of *Q1* in a negative direction. Since a _____ going signal has a forward effect, *Q1* will conduct.

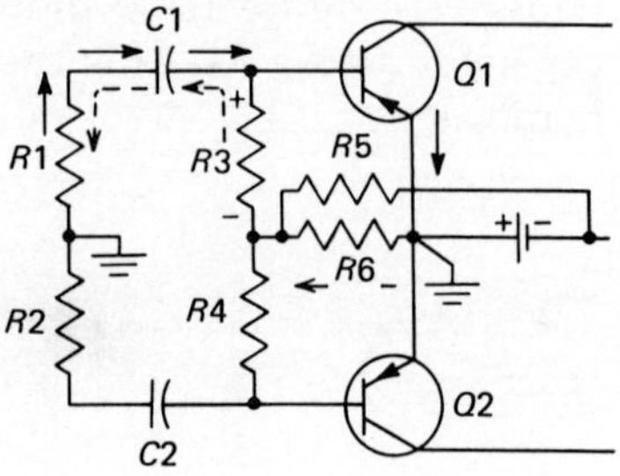

Fig. 11-38

11-39
(*negative*) Electrons leave the righthand plate of *C1*, enter the base-emitter junction of *Q1*, flow down to the emitter ground connection, up from ground to the connection between *R1* and *R2*, up through _____, and back to the lefthand plate of *C1*.

11-40
(*R1*) This path is shown by the _____-line arrows in Fig. 11-38.

11-41
(*solid*) Resistor *R1* is the output resistance of the previous stage and has a relatively low value. Since the base-emitter junction also has a low resistance, the entire charging path for *C1* is low-resistance. This means that the time for charging *C1* is _____. Even if the output resistance of the previous stage were large, there is the problem of the nonlinear *RC* time constant due to the transistor's rectification-like characteristic.

11-42
(*short* or *small*, etc.) In a PNP transistor, the path from emitter to base for electron flow is virtually an open circuit. For C1 to be able to discharge through the emitter-base junction, electrons would have to flow from emitter to base. Hence, C1 cannot _____ through this junction.

11-43
(*discharge*) The only discharge path it has is shown by the dashed arrows in Fig. 11-38. Electrons move out of the lefthand plate of C1, down through R1 to ground, up from ground at the emitter ground point, to the left through R6, and up through _____ to the righthand plate of C1.

11-44
(R3) This discharge is slow because C1 must discharge through resistor R3. Normally, R3 is made quite _____ to avoid shunting signal currents around the base-emitter junction of Q1.

11-45
(*large*) The discharge current through R3 develops a voltage across R3 having the polarity shown in Fig. 11-38. This polarity causes the base to become _____ with respect to the emitter.

11-46
(*positive*) This is *reverse* bias, causing the operating point of the transistor to move from class B toward class _____ as previously described.

11-47
(C) Since class C operation causes severe _____, this condition of emitter-base rectification of the signal to produce a reverse bias must be corrected.

11-48
(*distortion*) Rectified reverse bias can be avoided by making the circuit and component changes shown in Fig. 11-48. In this circuit, resistors R3 and R4 have been replaced by diodes CR1 and _____.

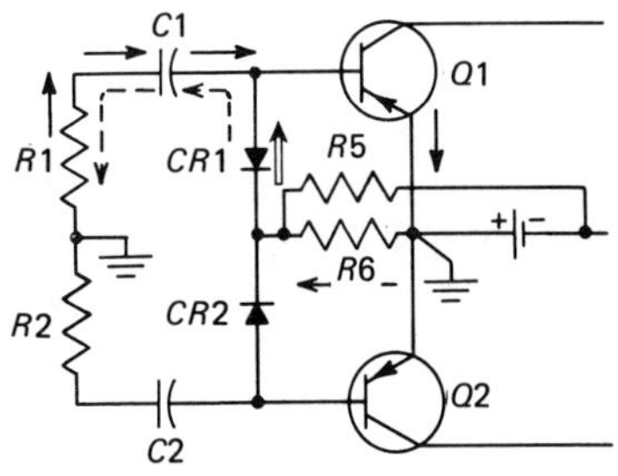

Fig. 11-48

11-49

(*CR2*) The electron conduction direction of *CR1* is upward in Fig. 11-48. This _____ conduction direction is shown by the double arrow in the drawing.

11-50

(*upward*) The signal is applied to the base of *Q1* through *C1*. Electrons flowing out of the righthand plate of *C1* must follow the same charge path as before (solid arrows), because the conduction direction of *CR1* is such that it will not permit _____ to flow downward through it.

11-51

(*electrons* or *current*) Hence, during the conduction cycle of *Q1*, *CR1* does not shunt out the signal and the circuit operation is normal. The discharge path is also the same as before, except that electrons now flow through *CR1* to the right-hand plate of *C1* instead of having to flow through a large-valued _____ as they did before.

11-52

(*resistor*) *CR1* is fully conductive for the discharge electrons. As such, it represents an extremely _____ resistance.

11-53

(*low*) *C1* is now discharging, therefore, through a very low-valued resistance. Hence, its discharge time is correspondingly _____.

11-54

(*short* or *small, etc.*) Thus, *C1* charges *and* discharges rapidly in the circuit of Fig. 11-48. This makes it impossible for a reverse-bias charge to build up on *C1*. Also, since *CR1* has negligible resistance in the discharge direction, the voltage developed across it is very _____.

11-55

(*small*) Hence, the base always has a small forward bias because of the *R5-R6* voltage divider, and the circuit can operate in class _____ without developing a reverse bias that could drive it into class C.

11-55
(*B*)

PRINCIPLES OF COMPLEMENTARY SYMMETRY

11-56
In an electron tube, the electrons can flow in only one direction during normal operation. That is, normally electrons flow from the _____ to the plate of the tube.

11-57
(*cathode*) Since there is only one kind of electron tube (*kind* as used here means direction of electron flow in the envelope), the plate of the tube is always connected to a source of potential that is more _____ than the potential terminal connected to the cathode.

11-58
(*positive*) There are 2 kinds of transistors: PNP and NPN. Figure 11-58 illustrates the directions of the electron flow in each electrode lead for both types. For example, the collector current I_C in the PNP transistor flows toward this electrode in the external lead, while in the NPN type the electron flow for I_C is directed _____ from the transistor.

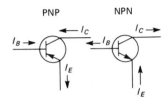

Fig. 11-58

11-59
(*away*) A similar situation prevails for the other electrodes. That is, the direction of electron flow in each lead of the PNP type is _____ from that in each lead of the NPN type.

11-60
(*opposite*) Figure 11-60 shows 2 transistors in a circuit that combines a PNP and an NPN transistor. In this circuit, Q1 is a(n) _____ type and Q2 is a(n) _____ type.

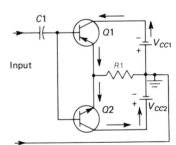

Fig. 11-60

11-61
(*PNP, NPN*) The dc electron path indicated by the arrows in Fig. 11-60 is completed through the collector-_____ junctions of the transistors.

11-62
(*emitter*) When 2 transistors, one PNP and the other NPN, are connected in a single stage in this manner, the circuit is known as *complementary symmetry*. A complementary-symmetry circuit cannot be set up with vacuum-tube amplifiers because there is only _____ kind of vacuum tube.

11-63
(*one*) Note that the signal is fed to the base of each transistor through blocking capacitor C1. No attempt is made to invert the phase of the input signal to one transistor as we normally find in push-pull amplifiers. Thus, the base of each transistor is fed the same _____ of the signal simultaneously.

11-64
(*phase*) Assume that this is a positive-going input half-cycle. A PNP transistor is forward-biased by a negative-going signal and reverse-biased by a positive-going signal. Thus, with this assumed signal phase, Q1 is _____-biased.

11-65
(*reverse*) But Q2 is an NPN transistor. Hence application of a positive-going signal to this transistor produces a condition of _____ bias.

11-66
(*forward*) Thus, a single positive-going half-cycle causes transistor Q1 to be nonconducting while the same positive-going half-cycle causes Q2 to be _____.

11-67
(*conducting*) The opposite effect occurs when the negative-going half-cycle arrives. This negative-going half-cycle reverse-biases transistor _____.

11-68
(*Q2*) At the same time, the negative-going half-cycle _____-biases Q1.

11-69
(*forward*) For the negative-going half-cycle Q1 is conducting and Q2 is _____.

11-70
(*nonconducting*) Thus, when one transistor is conducting, the other is nonconducting because the signal that forward-biases one transistor, _____-biases the other transistor.

302 A PROGRAMMED COURSE IN BASIC ELECTRONICS

11-71
(*reverse*) Consider Fig. 11-71 which is a simplification of the output circuit of Fig. 11-60. The internal emitter-collector circuit of Q1 is represented by R1; the internal emitter-collector circuit of Q2 is represented by ―――――――.

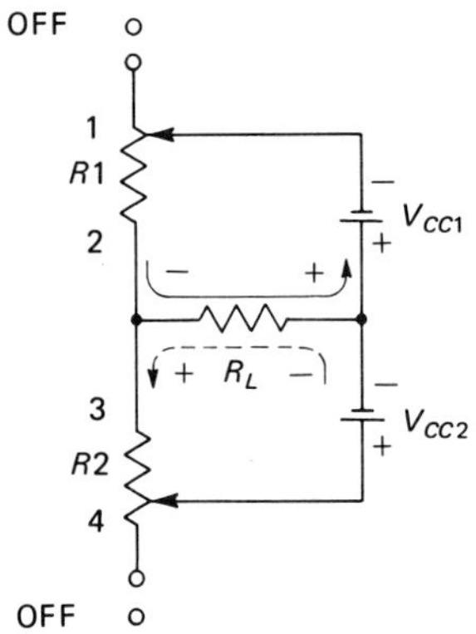

Fig. 11-71

11-72
(*R2*) Note that no provision is made in the circuit of Fig. 11-60 for emitter-base bias. Hence, this bias is zero, and both transistors will operate in class ―――――.

11-73
(*B*) With no input signal, therefore, the collector current of both transistors may be considered zero. Comparing this condition with a setting of the wipers of R1 and R2, we see that these wipers must be moved to the position marked ――――――, a position representing the nonconducting state for both transistors.

11-74
(*Off*) For the Off setting of R1 and R2, the current flowing through the transistors is zero; also, the current flowing through the load R_L is ―――――.

11-75
(*zero*) Now assume that the incoming signal goes positive. For a positive-going half-cycle of signal, transistor ――――― conducts.

11-76
(*Q2*) As this occurs, transistor ――――― remains nonconducting.

11-77
(*Q1*) Since Q1 remains nonconducting, variable resistor ――――― may be considered to remain in the Off position.

11-78
(*R1*) At the same time, the arm of resistor R2 may be considered to move from the Off position since Q2 is conducting. Thus, the arm of R2 moves from point 4 toward point ―――――.

11-79
(3) With the arm of R2 between point 4 and point 3, current flows through the series circuit consisting of battery V_{CC2}, resistor R_L, variable resistor R2, and back to battery _____.

11-80
(V_{CC2}) The magnitude of the current flowing in this series circuit depends upon the size of the incoming signal. For increasing signal (forward bias), the wiper of R2 is considered to move closer to point 3; as the signal becomes weaker, the wiper moves closer to point _____.

11-81
(4) The direction of this current flow in Fig. 11-71 is indicated by the dashed arrow. This current produces the voltage drop in R_L as shown, making the left side of R_L _____ with respect to the right.

11-82
(*positive*) When the input signal goes negative, transistor _____ begins to conduct.

11-83
(Q1) At this time, transistor Q2 becomes _____.

11-84
(*nonconducting*) Current now flows from battery V_{CC1}, through R1 (which leaves the Off position), through _____ back to the battery as shown by the solid arrow in Fig. 11-71.

11-85
(R_L) The current through R_L produces a voltage drop having a polarity _____ that produced by the current through R_L that flowed when Q2 was conducting.

11-86
(*opposite*) Thus ac voltage corresponding to the signal frequency appears across the load because of the alternate conduction and _____ of transistors Q1 and Q2.

11-87
(*nonconduction*) As previously mentioned, the transistors in Fig. 11-60 work in class B. Let us see what would happen if a small amount of forward bias were applied to each transistor. First, this would bring the transistors out of class B into class _____.

11-88
(*A*) For class A, the collector _____ is not cut off at any time, in either transistor.

11-89
(*current*) This means that R1 and R2 (the equivalent "collector-emitter" circuits in Fig. 11-71) will never be in the _____ position, even when there is zero signal input.

11-90
(*Off*) There is a complete circuit now formed involving both batteries and both variable resistors. That is, current will flow out of the negative terminal of V_{CC2}, into the positive terminal of V_{CC1}, through R1, through R2, and back to the positive terminal of _____.

11-91
(V_{CC2}) No current flows through R_L because the circuit is now behaving as a balanced bridge in which V_{CC1} and V_{CC2} are two corresponding arms, and resistors R1 and R2 are the other corresponding _____.

11-92
(*arms*) When an input signal appears, the wiper of R1 may be considered to move one way and the wiper of R2 the other way, depending upon whether the input half-cycle is _____-going or negative-going.

11-93
(*positive*) This upsets the bridge balance and current flows through R_L in the direction of either the solid arrow or the dashed arrow (Fig. 11-71), depending upon which of the transistors is conducting and which is _____ for that half-cycle.

11-94
(*nonconducting*) Upon reversal of the input signal polarity, the current in R_L also reverses. Thus, ac flows through R_L in accordance with the signal input, but no _____ flows through R_L, since in the absence of signal the bridge is balanced.

11-95
(*dc*) If this circuit is used for audio power amplification a loudspeaker can be used in place of R_L. The voice coil will not be offset by any current other than signal current because _____ does not flow through the load at any time.

11-96
(*dc*) The advantages of complementary-symmetry circuits may be listed as follows: There is no need for a phase-inverter stage or a center-tapped input transformer because the signal is fed to the bases of Q1 and Q2 in the same _____ .

11-97
(*phase*) The input coupling capacitor (C1, Fig. 11-60) charges through one transistor during the positive half-cycle of the input signal and _____ through the other transistor during the succeeding half-cycle.

11-98
(*discharges*) Thus, no reverse-bias voltage can be built up to cause distortion as previously explained. This eliminates the need for discharge diodes with capacitance coupling as is required in conventional class _____ push-pull amplifiers.

11-99
(*B*) The parallel connection of the output circuit with respect to the load eliminates the need for a tapped primary transformer in the _____ circuit.

11-99
(*output*)

COMPLEMENTARY-SYMMETRY CIRCUITS
11-100
As an example of a typical complementary-symmetry circuit, study the circuit of Fig. 11-100. The first thing to note is that the input signal is coupled to the base of each transistor via an input transformer identified as _____ in the figure.

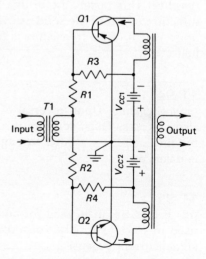

Fig. 11-100

11-101
(*T1*) As we have previously pointed out, the signal fed to the base of one transistor is in the same phase as the signal fed to the _____ of the other transistor.

11-102
(*base*) In most push-pull circuits of conventional design, an input transformer having a center-tapped secondary is required to supply the 180° phase difference between input signals. In this circuit, no _____ _____ in the secondary winding is needed because there is no need for a 180° phase difference.

11-103
(*center tap*) Forward bias for transistor Q1 is supplied by the voltage divider consisting of resistors _____ and R1.

11-104
(*R3*) Forward bias for transistor Q2 is supplied by the voltage divider comprising resistors R4 and _____.

11-105
(*R2*) Since both transistors are _____-biased, this circuit is not designed for class B operation.

11-106
(*forward*) Collector current flows during the entire input cycle and during idling time. If the bias is properly adjusted for these conditions, the transistors will operate in class _____.

11-107
(*A*) Note the grounded electrodes in Fig. 11-100. This indicates that this is a common-_____ configuration.

11-108
(*emitter*) Since Q1 is a PNP transistor, its collector must be maintained _____ with respect to its emitter. This polarity is established by battery V_{CC1}.

11-109
(*negative*) In contrast, Q2 is an NPN transistor; hence its collector must be kept _____ with respect to its emitter. This polarity is established by V_{CC2}.

11-110
(*positive*) In order to keep the polarities correct, therefore, it is necessary to use a split-primary output transformer. This is a transformer having _____ separate primary windings as shown in Fig. 11-100.

11-111
(2) In the circuit of Fig. 11-111, Q1 and Q2 make up a complementary-symmetry driver, while Q3 and Q4 constitute a complementary-symmetry _____ stage.

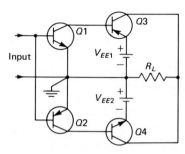

Fig. 11-111

11-112
(*output*) Since neither RC coupling nor transformer coupling is used, there is good economy of parts. The stage consisting of Q1 and Q2 is _____ coupled to the output stage.

11-113
(*directly*) Consider first the action of Q1 and Q3. Q1 is an NPN transistor, while Q3 is a(n) _____ type.

11-114
(*PNP*) Assume that the input signal to Q1 happens to be positive-going. Since a positive-going signal is a forward voltage for an NPN transistor, when the signal is positive-going, the collector current of Q1 must _____.

11-115
(*increase*) Thus, Q1 conducts with a positive-going signal. But in a common-emitter configuration, there is a 180° phase reversal from input to output. Hence, when the base goes positive, the collector goes _____.

11-116
(*negative*) Thus, the collector of Q1 goes negative when Q1 conducts. Since the collector of Q1 is directly coupled to the base of Q3, then the base of Q3 must go in the _____ direction.

11-117
(*negative*) Q3, however, is a PNP transistor. A negative-going voltage applied to the base of a PNP transistor is a forward voltage. Hence, the collector current of Q3 must _____.

11-118
(*increase*) Thus, when the input signal is positive-going, Q1 conducts and causes Q3 to conduct with consequent amplification in cascade. The collector current of Q3 is supplied by battery _____ .

11-119
(V_{EE1}) Let us check the bias voltages for Q3 and Q1 in this circuit. Starting from the negative terminal of V_{EE1} we find that the emitter-collector junction of Q1 is in series with the _____-emitter junction of Q3 and battery V_{EE1}.

11-120
(*base*) Since the base of Q3 is connected to the "—" terminal of V_{EE1} through the emitter-collector junction of Q1, then the base of Q3 is _____ with respect to its emitter. This is correct for forward bias for Q3.

11-121
(*negative*) Also, the positive terminal of V_{EE1} is connected to the collector of Q1 through the resistance of the base-emitter junction of Q3. Hence, the collector of Q1 is _____ with respect to its emitter. This is correct for the reverse bias needed in the emitter-collector circuit.

11-122
(*positive*) When the signal becomes negative-going, Q1 and Q3 become quiescent while Q2 and _____ begin to conduct in exactly the same fashion as previously described for Q1 and Q3.

11-123
(*Q4*) Bias voltages for Q2 and Q4 are obtained just as described previously for Q1 and Q3. When Q3 conducts, its collector current flows through R_L from left to right. When Q4 conducts, its collector current flows through R_L from _____ to _____ .

11-124
(*right, left*) Thus, the current in R_L is an _____ current, changing its direction of flow in accordance with the input signal variations.

11-125
(*alternating*) In Fig. 11-111, note that the emitters of both output transistors are common to both input and output circuits. Hence, Q3 and Q4 are wired in the common-_____ configuration.

11-126
(*emitter*) A similar *common-collector* circuit is given in Fig. 11-126. That this is the circuit configuration is apparent from the fact that the collectors of both Q3 and Q4 are _____ to the input and output circuit of the respective transistors.

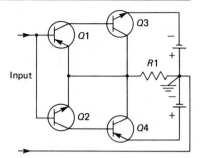

Fig. 11-126

11-127
(*common*) Although the common-collector configuration has less voltage, current, and power gain than the common-_____ configuration, it is often used because it has a higher input resistance.

11-127
(*emitter*)

COMPOUND-CONNECTED TRANSISTORS

11-128
It will be recalled from discussions in previous sections that the short-circuit forward-current-amplification factor for the CB configuration is h_{fb}; for the CE configuration is h_{fe}; and for CC configuration is _____.

11-129
(h_{fc}) From the equations previously developed it was shown that the short-circuit forward-current-amplification factor is the ratio of the output current to the _____ current measured with the output short-circuited.

11-130
(*input*) Reference to these equations will also show that if h_{fb} is high for a given transistor, the current gains in the other configurations will also be high. That is, if h_{fb} is large, h_{fc} and _____ will also be large.

11-131
(h_{fe}) Refer to the curve for a single transistor in Fig. 11-131. This curve is a plot of the variation of collector current (dependent variable) versus the _____ current (independent variable).

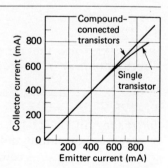

Fig. 11-131

310 A PROGRAMMED COURSE IN BASIC ELECTRONICS

11-132
(*emitter*) The straight-line portion of this curve, which shows a constant ratio of collector current to emitter current, represents that region of operation in which h_{fb} for the single transistor also remains _____ .

11-133
(*constant*) Thus, h_{fb} remains constant up to about 400 mA of emitter current. For larger emitter currents, the collector current increases _____ rapidly than it did over the linear portion of the curve.

11-134
(*less*) This indicates that the value of _____ also decreases for larger values of emitter current.

11-135
(h_{fb}) A special method of connecting two transistors is often used to prevent the dropoff of collector current (or reduction of h_{fb}) at high emitter currents. As shown in the upper curve (Fig. 11-131), when 2 transistors are connected in this special way, they are said to be _____-connected.

11-136
(*compound*) As shown by the upper curve in Fig. 11-131, the value of _____ remains relatively constant over the entire range of emitter currents.

11-137
(h_{fb}) Two compound-connected transistors are shown in Fig. 11-137 inside the dashed lines. (Two transistors connected in this configuration are called a *Darlington pair*.) Note first that the base of Q1 is connected to the _____ of Q2.

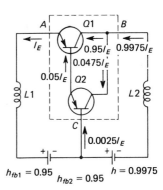

Fig. 11-137

11-138
(*emitter*) Note also that the _____ of Q1 and Q2 are connected directly together.

11-139
(*collectors*) Using Fig. 11-137, we will now show that compound-connected transistors can provide a larger h_{fb} than a single transistor. Assume first that the short-circuit forward-current-amplification for each transistor is 0.95. That is, assume that h_{fb} for each transistor = _____.

11-140
(*0.95*) By definition, $h_{fb} = I_{C1}/I_E$ where I_{C1} is the collector current of transistor Q1. This equation should now be solved for I_{C1}. This gives us the form: I_{C1} = _____.

11-141
($h_{fb}I_E$) Since h_{fb} is assumed to be 0.95 for both transistors, this figure may be substituted in the equation above. This results in the equation: _____.

11-142
($I_{C1} = 0.95I_E$) Note the arrow near Q1 that shows this current in Fig. 11-137. The base current I_{B1} of transistor Q1 equals its emitter current I_E minus its collector current, $0.95I_E$. Thus, $I_{B1} = I_E - 0.95I_E =$ _____.

11-143
($0.05I_E$) Note the arrow near the base of Q1 that denotes this current. The base current of Q1 *is* the emitter current of Q2, since these currents flow in the common lead between these elements. Thus, the emitter current of Q2 is equal to _____.

11-144
($0.05I_E$) Remembering that collector current is $I_C = h_{fb}I_E$ in general, then the collector current of Q2 is given by the relation $I_{C2} = h_{fb} \times 0.05I_E$. We have assumed that h_{fb} for both transistors equals 0.95. Thus, $I_{C2} = 0.95 \times 0.05I_E =$ _____.

11-145
($0.0475I_E$) Note the arrow near the collector lead of Q2 that denotes this current. The total output current at point B (I_O) is the sum of the individual collector currents of Q1 and Q2. Hence, $I_O = 0.95I_E + 0.0475I_E =$ _____.

11-146
($0.9975I_E$) The 2 compound-connected transistors can be considered as a single transistor having an emitter at point A, a base at point C, and a _____ at point B.

11-147
(*collector*) Considered as a single unit, the short-circuit forward-current-amplification factor is given by the relation: $h_{fb} = I_o/I_E$. Substituting the value of I_o just obtained, we have $h_{fb} = 0.9975 I_E/I_E =$ _____ .

11-148
(*0.9975*) Thus we see that compound connection has led to a substantial increase of short-circuit forward-current-amplification factor. The original assumed value for h_{fb} was _____ while the value obtained from the compound-connected unit is 0.9975.

11-149
(*0.95*) To appreciate the importance of this increase in current amplification factor, refer to Fig. 7-131. In this table, find the formula which will enable you to convert h_{fb} to h_{fe}. This formula will help us find the current amplification factor for the CE circuit and reads as follows: $h_{fe} =$ _____ .

11-150
$\left(\dfrac{-h_{fb}}{1 + h_{fb}}\right)$ Now let us find the h_{fe} of a transistor having an $h_{fb} = 0.95$ using this formula and then compare it with the h_{fe} obtained for the compound connection. For a transistor whose $h_{fb} = -0.95$, $h_{fe} = 0.95/(1 - 0.95) = 0.95/0.05 =$ _____ .

11-151
(*19*) Thus, for a single unit the forward short-circuit current amplification factor in the CE circuit would be 19. Now let us repeat this process for a transistor with an $h_{fb} = -0.9975$ as obtained for the compound connection. This yields

$$h_{fe} = \frac{0.9975}{1 - 0.9975} = \frac{0.9975}{0.0025} = \underline{}$$

11-152
(*399*) Thus, by using 2 transistors in compound connection, we have assembled a unit which has a substantially higher _____ than either of the 2 original transistors.

11-153
(h_{fe}) To simplify the arithmetic, we originally assumed equal values of h_{fb} for each transistor. It is easily shown, however, that the advantages of compound connection are not restricted to the use of transistors having _____ values of h_{fb}.

11-153
(*equal*)

TWO-TRANSISTOR BRIDGE CONNECTION

11-154
Certain definite advantages may be gained by connecting transistor amplifiers in bridge circuits. First let us review the basic bridge concept. Figure 11-154 illustrates a fundamental bridge circuit containing 4 circuit elements identified as W, X, Y, and _____.

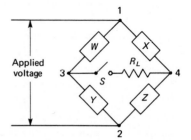

Fig. 11-154

11-155
(Z) The portions of the bridge containing the circuit elements are known as the "arms" of the bridge. Either ac or dc voltage may be applied across points 1 and 2, while the load may be connected across points 3 and _____.

11-156
(4) The arms of the bridge may contain any types of circuit elements such as resistors, capacitors, inductors, transistors, vacuum tubes, or batteries. For example, the bridge circuit in Fig. 11-156 contains batteries in 2 arms and _____ in the other 2 arms.

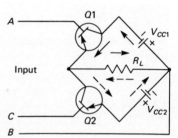

Fig. 11-156

11-157
(*transistors*) Returning to Fig. 11-154, assume that switch S is open. For this condition, the circuit component identified as _____ may be considered absent from the circuit.

11-158
(R_L) With the branch containing R_L open, the remainder of the circuit may be considered to contain 2 parallel branches. The left branch contains elements W and Y in series; the right branch contains X and _____ in series.

11-159
(Z) As in any parallel circuit, there will be separately determinable currents flowing in the 2 branches. These currents may or may not be equal. In any case, the voltage drop across element W will be $I_W R_W$, and the voltage drop across element X will be _____.

314 A PROGRAMMED COURSE IN BASIC ELECTRONICS

11-160
($I_X R_X$) If R_X and R_W are correctly adjusted with respect to the rest of the circuit, it is possible to make the voltage drops across these elements equal to each other. If the voltage drops across W and X are made equal to each other, then the *difference of potential* between points 3 and 4 will be _____.

11-161
(*zero*) Thus, if the switch S is closed under these conditions, that is, when the voltage from point 3 to point 4 is zero, no current will flow through component _____.

11-162
(R_L) The bridge is now said to be "balanced." If any one of the arms is now changed in resistance, the bridge becomes unbalanced and the current through R_L will no longer be _____.

11-163
(*zero*) Now consider the transistor circuit in Fig. 11-156. Suppose that $Q1$ is biased so that a small collector current flows from V_{CC1} through $Q1$ and to the right through R_L. This current is indicated by the _____ arrows in Fig. 11-156.

11-164
(*solid*) Similarly, suppose that $Q2$ is biased so that a small collector current flows from V_{CC2}, to the left through R_L, and through $Q2$ back to the battery. This current is indicated by the _____ arrows in Fig. 11-156.

11-165
(*dashed*) If the bias is adjusted so that the current through $Q1$ equals the current through $Q2$, then the *net* current through R_L would have to be equal to _____.

11-166
(*zero*) Since the current through R_L is zero when the bias on $Q1$ and $Q2$ is adjusted for equal emitter-collector currents, the bridge may be said to be in the _____ state.

11-167
(*balanced*) That is, the bridge is balanced for dc, since the voltages and currents present are due entirely to the batteries in the bridge arms. Therefore, with or without signal input to the points A, B, C, the direct current through R_L is _____.

11-168
(*zero*) Now picture an ac signal applied across points A and B, and a second identical signal fed to points C and B 180° out of phase with the first. Say at a given instant that the signal fed to A is positive-going. Then at the same instant, the signal fed to C will be _____-going.

11-169
(*negative*) Q1 and Q2 are PNP transistors; hence a negative-going signal produces a *forward* effect and a positive-going signal produces a _____ effect.

11-170
(*reverse*) Thus, the positive-going signal applied to A will reduce the current indicated by the solid arrows while the negative-going signal applied to C will _____ the current indicated by the dashed arrows.

11-171
(*increase*) With the solid and dashed arrows no longer equal in magnitude, the current through R_L will no longer be _____.

11-172
(*zero*) When the input signal reverses and the base of Q1 goes negative while the base of Q2 goes positive, the current indicated by the dashed arrows will become _____ than the current indicated by the solid arrows.

11-173
(*smaller*) This will result in a net current through R_L that will have a direction _____ that of the current under the previous set of conditions.

11-174
(*opposite*) Thus, a signal current will flow in R_L, even though the dc through R_L is still equal to _____.

11-175
(*zero*) Since this signal current consists of net collector current, it will be an amplified reproduction of the original signal applied out of phase across A–B and _____.

11-176
(*C–B*) The advantage of bridge connection over other types of circuits is that the dc in load resistor R_L is _____.

11-177
(*zero*) If R_L is replaced by a speaker voice coil, the distortion will be minimized. In a normal circuit the speaker voice coil is *offset* by the _____ current flowing through it. This causes distortion.

11-178
(*direct*) Another advantage of bridge connection is the elimination of a center-tapped output transformer or voice coil. In a standard push-pull amplifier, using two similar transistors (both PNP or both NPN), it would be necessary to use a _____-_____ voice coil if the speaker were connected in place of the load resistor.

11-179
(*center-tapped*) Thus, bridge connection provides two distinct advantages over more conventional output circuits. It makes use of simpler components such as untapped coils and/or transformers, and there is no danger of _____ current flowing through the load.

11-179
(*direct*)

OTHER BRIDGE ARRANGEMENTS
11-180
Figure 11-180 illustrates a circuit in which transistors appear in all 4 _____ of the bridge.

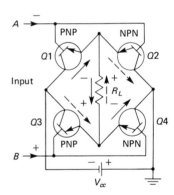

Fig. 11-180

11-181
(*arms*) It is important to note several other essential differences between the circuit of Fig. 11-180 and previous bridge circuits. In Fig. 11-180, 2 of the transistors are PNP types, while the other 2 are _____ types.

11-182
(*NPN*) In addition, the circuit of Fig. 11-180 is a 2-terminal input type, while that of the previous figure (Fig. 11-156) has 3 input terminals. Thus, the amplifier of Fig. 11-180 does not require 2 input signals _____ ° out of phase.

11-183
(*180*) This is a distinct advantage. Another advantage is apparent from a study of the dc power sources in each case. The 4-transistor circuit requires only 1 _____ rather than 2 as in the case of the circuit of Fig. 11-156.

11-184
(*battery* or *dc source*, *etc.*) Let us analyze the operation of the circuit in Fig. 11-180 with the help of the solid and dashed arrows. Assume class B operation is to be used. Thus we are assuming that, under quiescent conditions, all the emitter-collector currents are _____.

11-185
(*zero*) If the current flowing through the transistors is zero, then there is no complete circuit across the battery and there can be no direct current flowing through the load _____.

11-186
(R_L) Assume that an input signal causes point *A* to become negative with respect to point *B*. Note that point *A* is directly connected to the base of one PNP transistor, Q1, and to the base of one _____ transistor, Q2.

11-187
(*NPN*) A negative-going voltage like that at point *A* is a *forward-bias* voltage for a PNP transistor and a _____-bias voltage for an NPN transistor.

11-188
(*reverse*) Hence, Q1 will tend to conduct while Q2 will remain cut off. Now consider the positive-going voltage applied to point *B*. Point *B* is connected to the base of one PNP transistor Q3 and to the _____ of the NPN transistor Q4.

11-189
(*base*) A positive-going voltage at point *B* is *forward bias* for an NPN transistor and *reverse bias* for a _____ transistor.

11-190
(*PNP*) Thus, Q4 tends to _____ while Q3 remains in the cutoff condition.

318 A PROGRAMMED COURSE IN BASIC ELECTRONICS

11-191
(*conduct*) For the polarity of input signal given, we have shown that transistors _____ and Q4 will tend to conduct.

11-192
(*Q1*) A flow of current will therefore occur through these transistors along the following path: from the "−" terminal of the battery V_{CC}, through the collector-emitter circuit of Q1, down through _____, through the emitter-collector circuit of Q4, and back to the "+" terminal of the battery.

11-193
(R_L) This current path is indicated by the _____ arrows in Fig. 11-180.

11-194
(*solid*) This current flow develops a voltage across R_L having the polarity indicated. That is, because of the input signal described (negative at *A* with respect to *B*), the top of R_L becomes _____ with respect to the bottom.

11-195
(*negative*) Now consider the result when the input signal reverses polarity, becoming positive at point *A* with respect to point *B*. In this case, transistors Q1 and _____ become nonconducting.

11-196
(*Q4*) For reasons similar to those presented previously, transistors Q2 and Q3 begin to _____. A current flows in their emitter-collector circuits from battery V_{CC}.

11-197
(*conduct*) The current flowing through V_{CC} and transistors Q2 and Q3 is indicated by the _____ arrows in Fig. 11-180.

11-198
(*dashed*) As shown, therefore, the current flowing through R_L has a direction such that the top of the resistor becomes _____ with respect to the bottom.

11-199
(*positive*) For a complete input cycle, an ac voltage drop appears across the load resistor. This voltage drop has the same waveform and frequency as the _____ signal, provided that no distortion occurs.

11-200

(*input*) As in the 2-transistor bridge discussed in the last section, the transistors in Fig. 11-180 may be biased for class A operation as well as for class B. In this case, the quiescent currents through the transistors are not _____, since all the transistors conduct to a small extent at all times.

11-201

(*zero*) However, as previously shown, the current flowing through R_L as a result of the conduction of Q1 and Q4 is _____ to the current flowing through R_L as a result of the conduction through Q2 and Q3.

11-202

(*equal*) Hence, the net current during the quiescent period through R_L must be _____.

11-203

(*zero*) An ac input signal under these conditions would cause one pair of transistors (such as Q1 and Q4) to conduct a larger current than the other pair. This would _____ the bridge and give a net signal voltage drop across R_L.

11-204

(*unbalance*) Thus, the 4-transistor bridge circuit will operate in either class A or class B. In either case, the quiescent dc through R_L is _____.

11-205

(*zero*) As in all amplifiers, class B operation provides better power efficiency, but class A gives less _____.

11-206

(*distortion*) Another 4-transistor bridge circuit is shown in Fig. 11-206. In this circuit, all the transistors are _____ types.

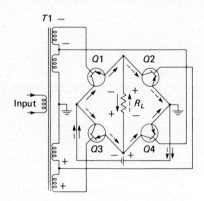

Fig. 11-206

320 A PROGRAMMED COURSE IN BASIC ELECTRONICS

11-207
(*PNP*) As in the previous cases, for class B operation, none of the transistors conducts and the current through R_L is _____.

11-208
(*zero*) Or, for class A operation, the transistors draw equal currents for quiescent conditions and the current through R_L is zero because the bridge is _____.

11-209
(*balanced*) In this circuit, an input transformer having 5 leads (3 taps) is required. Let us see why. Trace the solid arrows through Q1 and note that the output current into the load is taken from the *emitter*, making this a CC configuration. A second transistor (dashed arrows) whose emitter supplies the output current is _____.

11-210
(*Q3*) Hence, both Q1 and Q3 are connected in the _____ configuration.

11-211
(*CC*) On the other hand, transistors Q2 and Q4 are connected so that their *collectors* join R_L directly. Thus, the output current is taken from the collectors of Q2 and Q4. The input signal is fed to the base of each of these transistors. Hence, the configuration used for Q2 and Q4 is _____.

11-212
(*CE* or *common emitter*) A CC amplifier has a relatively high input impedance; a CE amplifier has a relatively _____ input impedance.

11-213
(*low*) When the input impedance is high, the input voltage must be relatively large to obtain a given output current; when the input impedance is low, a relatively _____ input voltage will provide the same output current.

11-214
(*small*) For satisfactory bridge operation, both transistors of a mated pair (e.g., Q1 and Q4) must draw the same output current. However, Q1 is in CC configuration and Q2 is in CE configuration. Hence, if the same input voltage is applied to each of these, the output current in each will not be the _____.

11-215
(*same*) During the half of the input cycle when Q1 and Q4 are driven into conduction (solid arrows), a larger voltage must be fed to _____ since it has a higher input impedance, if we want the output currents in both transistors to be equal.

AUDIO AMPLIFIERS II **321**

11-216
(Q1) Similarly, when the input signal reverses it is necessary to feed a larger input voltage to transistor _____, since it, too, is wired in the CC configuration (dashed arrows).

11-217
(Q3) Thus, the base of Q1 is wired to the top of T1, while the base of _____ is wired to the first tap down from the top, thus providing a smaller input voltage for Q4 than for Q1.

11-218
(Q4) Similarly, the base of Q3 is wired to the bottom of the transformer, while the base of Q2 is wired to the first tap up from the bottom in order to feed a _____ input voltage to Q3.

11-219
(*larger*) Figure 11-219 shows how the 5-lead (3-tap) transformer requirements may be eliminated. The input signal is applied only to the bases of transistors Q1 and _____.

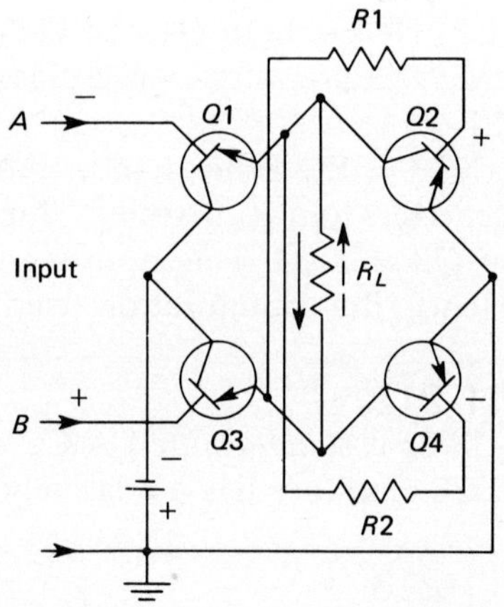

Fig. 11-219

11-220
(Q3) A portion of the output of Q1 is coupled to the base of Q4 through limiting resistor _____.

11-221
(R2) Similarly, a portion of the output of transistor Q3 is coupled to the base of transistor _____ through limiting resistor R1.

11-222
(Q2) The quiescent conditions for this circuit are the same as those for the circuits previously discussed. That is, the bridge is balanced for either class A or class B operation so that the dc through the load is _____.

322 A PROGRAMMED COURSE IN BASIC ELECTRONICS

11-223
(*zero*) Assume that point A goes instantaneously negative with respect to ground and that point B goes instantaneously positive with respect to ground. These are all PNP transistors. Of the 2 transistors Q1 and Q3, _____ will conduct and _____ will not conduct.

11-224
(*Q1, Q3*) Since Q1 is connected in the CC configuration, there will be no phase reversal between its input and output. That is, if a negative-going signal is fed to its base, a _____-going signal will appear at its emitter.

11-225
(*negative*) Thus, Q1 couples a _____-going voltage to the base of Q4 through R2 for this condition.

11-226
(*negative*) It will be remembered that Q4 is connected in the _____ configuration.

11-227
(*CE*) Hence, Q4 should be fed a smaller signal voltage than Q1 was fed initially if its collector current is to _____ that of Q1.

11-228
(*equal*) The decrease in signal voltage fed to the base of Q4 is obtained as a voltage drop across resistor _____.

11-229
(*R2*) Exactly the same situation prevails when Q3 and Q2 conduct on the next half of the input cycle. That is, Q3 feeds a small portion of its output voltage (with no phase change) to the base of Q2 through R1 when point A goes _____ and point B goes _____.

11-230
(*positive, negative*) When Q1 and Q4 are conducting, the current in the load flows in the direction of the _____ arrow; when Q3 and Q2 are conducting, the current in the load flows in the direction of the _____ arrow.

11-231
(*solid, dashed*) Bridge-connected amplifiers have many advantages. First, as we have seen, dc does not pass through the _____ when the bridge is correctly balanced.

11-232
(*load*) Second, although a bridge amplifier provides push-pull output, no center tap is required on the loudspeaker _____ coil.

11-233
(*voice*) Each mated pair of transistors (Q1-Q4 or Q2-Q3) can develop a peak voltage with reference to ground which may approach the supply voltage. That is, if the battery is a 9-V type, each mated pair can develop an output signal voltage that approaches _____ V.

11-234
(*9*) Since the output voltage developed across the load is the vector sum of the magnitudes of the 2 output half-cycles (one from each mated pair), then a bridge amplifier can develop a "peak-to-peak" output voltage that approaches _____ the supply voltage. With a 9-V battery, the peak-to-peak output may approach 18 V.

11-235
(*twice*) Third, since the transistors of each mated pair are in series with each other during the conduction interval, each transistor experiences a voltage that is less than the supply voltage. Thus, if a 9-V battery is used, no one of the transistors will have applied across it a voltage that approaches _____ V.

11-236
(*9*) Fourth, a bridge circuit requires only _____ battery, whereas other types of push-pull circuits demand 2 batteries in some cases.

11-237
(*1*) Each transistor of a mated pair handles only half the power present in that particular half-cycle. This means that a 4-transistor bridge amplifier can handle _____ the power of a conventional push-pull amplifier without overheating the transistors.

11-237
(*twice*)

CRITERION CHECK TEST

The following 3 questions refer to Fig. 11-3.

_____11-1 The dc bias conditions are: (*a*) Q1 and Q2 have reverse bias on their bases, (*b*) Q1 and Q2 have forward bias on their bases, (*c*) Q1 and Q2 have no bias on their bases, (*d*) Q1 has forward bias, Q2 reverse bias.

_____11-2 A function of T1 is to (*a*) provide forward bias to Q1 and Q2, (*b*) provide 2 signals 180° out of phase to Q1 and Q2, (*c*) couple the collectors to the output, (*d*) match the impedance of the base of Q1 to the impedance of the collector of Q2.

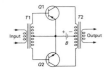

Fig. 11-3

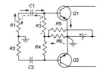

Fig. 11-38

Fig. 11-60

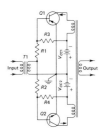

Fig. 11-100

_____11-3 When an ac signal forward-biases Q1, then (a) Q2 receives no bias, (b) Q2 receives a reverse bias, (c) Q1 does not conduct, (d) Q2 conducts.

_____11-4 Crossover distortion is eliminated in a push-pull amplifier by (a) using a transformer with a large stepdown ratio, (b) using a transformer with a large stepup ratio, (c) providing a small forward bias to the transistors, (d) supplying both transistors with in-phase signals.

_____11-5 In Fig. 11-38, C2 discharges through (a) R1, R6, and R3; (b) R4 and R6; (c) R2 and R4; (d) R2, R6, and R4.

_____11-6 In Fig. 11-38, (a) R6 would be much larger than R5, (b) R5 would be much larger than R6, (c) R5 would be about the same magnitude as R6, (d) relative magnitudes of R5 and R6 cannot be determined from the drawing.

_____11-7 A complementary-symmetry amplifier has (a) 1 PNP and 1 NPN transistor, (b) 2 PNP transistors, (c) 2 NPN transistors, (d) 2 N-channel FETs.

_____11-8 In the circuit of Fig. 11-60, during the negative half-cycle of an input signal fed through C1, (a) Q1 is cut off, (b) Q2 is cut off, (c) Q1 and Q2 are both conducting, (d) Q1 and Q2 are both cut off.

_____11-9 In the circuit of Fig. 11-60, during the positive half-cycle of an input signal fed through C1, (a) electrons flow through R_L from right to left, (b) electrons flow through R_L from left to right, (c) holes flow through R_L from left to right, (d) holes flow through R_L from right to left.

_____11-10 In the circuit of Fig. 11-100, the primary winding of the output transformer must be split because (a) direct currents must flow at no signal, (b) the collectors of the transistors need opposite polarity voltages, (c) opposite biases are needed by the transistors, (d) the CE configuration is used.

_____11-11 In the circuit of Fig. 11-100, the bias for Q1 is (a) the total voltage drop across R1 and R2, (b) half the voltage drop across R1, (c) V_{CC_1}, (d) the voltage drop across R1.

_____11-12 In Fig. 11-100, with a negative-going signal at the junction of R1 and R2, (a) more current will flow in the lower half of the primary winding of the output transformer, (b) more current will flow in the upper half of the primary winding of the output transformer, (c) the current in both halves of the primary winding will be equal, (d) Q1 will be cut off.

_____11-13 The short-circuit forward-current-amplification factor is the ratio of (a) the output current to the input current with the output shorted, (b) the output current to the input current with the input shorted, (c) the input current to the output current with the output shorted, (d) the input current to the output current with the input shorted.

_____11-14 In the basic bridge circuit of Fig. 11-154, if the voltage is applied at points 1 and 2, the bridge is balanced when (a) the voltage between points 3 and 4 is zero, (b) the voltage drop across W is equal to the voltage drop across Y, (c) the impedance of W is equal to the impedance of X, (d) the current through W is equal to the current through Z.

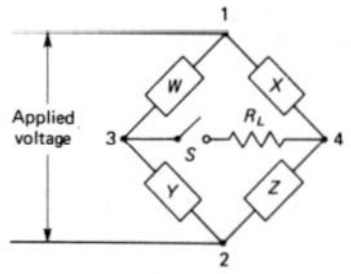

Fig. 11-154

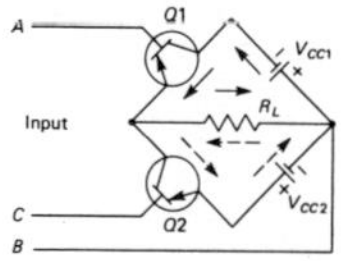

Fig. 11-156

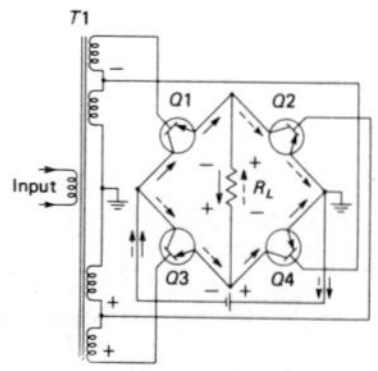

Fig. 11-206

_____11-15 In the bridge circuit of Fig. 11-156, if the negative input signal is at A while the positive input signal is at C, the current through R_L would be (a) ac, (b) from right to left, (c) zero, (d) from left to right.

_____11-16 An advantage of the bridge circuit of Fig. 11-156 over an ordinary push-pull amplifier is (a) fewer transistors are needed, (b) higher efficiency, (c) no center-tapped output transformer is needed, (d) the transistors handle less power.

_____11-17 When the bridge circuit of Fig. 11-156 is balanced, (a) the voltage across R_L is zero, (b) no current flows through Q1, (c) no current flows through Q2, (d) no current flows through Q1 and Q2.

_____11-18 In the circuit of Fig. 11-206, Q1 is given a higher voltage signal than Q4 because (a) Q1 drives Q4, (b) the input impedance of Q1 is higher than that of Q4, (c) Q1 must start the current flow through the load resistor and therefore needs a larger driving voltage, (d) Q1 is a PNP type, while Q4 is an NPN type.

_____11-19 In Fig. 11-206, as the signal at the top of the secondary of T1 goes positive, (a) all the transistors conduct equally well, (b) Q3 and Q1 conduct more heavily, (c) Q3 and Q2 conduct more heavily, (d) all the transistors are cut off.

_____11-20 One advantage of a 4-transistor bridge over a conventional push-pull amplifier is that (a) each transistor dissipates half of the power of a push-pull transistor, (b) each transistor dissipates twice the power of a push-pull transistor, (c) 4 transistors are used instead of 2, (d) the bridge is a power amplifier.

_____11-21 For a 4-transistor bridge circuit, if a peak-to-peak voltage of 36 V is needed, then each transistor should have a V_{CE} rating of at least (a) 9 V, (b) 18 V, (c) 36 V, (d) 72 V.

12 wideband amplifiers

OBJECTIVES

(1) Depict at least 4 recurrent waveforms commonly found in electronics. **(2)** Contrast the current response of a sinusoidal and nonsinusoidal voltage for various loads. **(3)** Give an argument for using the sinusoid as the fundamental waveform in electronics. **(4)** Determine the harmonics for a given complex waveform. **(5)** Discuss the behavior of a complex waveform when it is passed through a narrowband and a wideband amplifier. **(6)** Indicate the kind of amplifier needed to pass pulses used in radar and television work. **(7)** Indicate the components in an *RC*-coupled transistor amplifier which limit its frequency response. **(8)** Define *phase delay* of a sinusoidal wave and relate it to time delay. **(9)** Tell the effect on a complex wave of phase distortion. **(10)** State the condition on the phase delay which will produce no phase distortion. **(11)** Depict and discuss a transistor circuit which uses a coil for high-frequency shunt compensation, one which uses a coil for high-frequency series compensation, and one which uses two coils in a combination peaking configuration. **(12)** Indicate those components in an *RC*-coupled amplifier which degrade low-frequency performance. **(13)** Discuss the phase distortion caused by the *RC*-coupling network. **(14)** Depict and describe a 2-stage transistor circuit which has a low-frequency *RC* compensating network. **(15)** State the amplitude and phase conditions which must be met by the harmonics of a square wave. **(16)** Depict several examples of square waves which suffer from frequency and/or phase distortion.

REVIEW OF RECURRENT WAVEFORMS

12-1
Figure 12-1 illustrates 4 different types of waveforms that may be encountered in electronics. Type (*a*) is a *sinusoid* or sine wave, (*b*) is a *triangular* wave, (*c*) is a *sawtooth* wave, and (*d*) is a *square* or _____ wave.

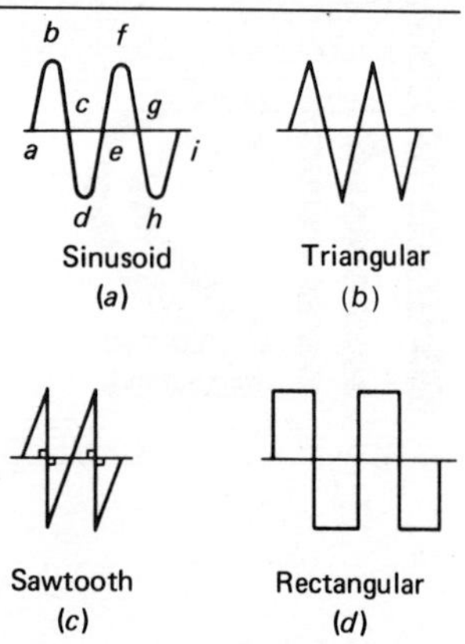

Fig. 12-1
Sinusoid (*a*)
Triangular (*b*)
Sawtooth (*c*)
Rectangular (*d*)

12-2
(*rectangular*) All these waveforms are called *recurrent waves* because each cycle has the _____ shape as the preceding cycle and the cycle that follows.

12-3
(*same*) In (*a*), Fig. 12-1, we have shown a total of _____ complete cycles.

12-4
(*2*) The first cycle executes the variations that carry it through points *a*, *b*, *c*, *d*, and _____.

12-5
(*e*) The second cycle in (*a*), Fig. 12-1, goes through the variations identified as *e*, *f*, _____, _____, and _____.

12-6
(*g*, *h*, and *i*) The second cycle is identical with the first; hence the waveform in (*a*) is a _____ waveform.

12-7
(*recurrent*) In (*b*), the second cycle is identical with the first, too. Hence (*b*) is a recurrent waveform. This is also true of (*c*) and (*d*) since in each case each _____ is exactly the same as its neighbors on either side.

12-8
(*cycle*) Figure 12-8 shows a recurrent waveform, in this case a sine wave, in the form of a voltage *v*. If this voltage waveform is impressed across a pure resistance, a current flows through the resistance. In Fig. 12-8, the resulting current is symbolized by _____.

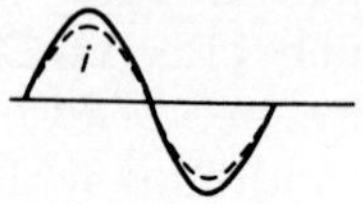

Fig. 12-8

328 A PROGRAMMED COURSE IN BASIC ELECTRONICS

12-9
(*i*) When a sinusoidal voltage is impressed across a pure resistance, the resulting current is in the _____ phase as the voltage as shown in Fig. 12-8.

12-10
(*same*) But what is of greater importance at this moment is that both the current and the voltage have identical _____ forms.

12-11
(*wave*) That is, when a voltage sinusoid is impressed across a pure resistance, the current is also a _____ waveform.

12-12
(*sinusoidal*) When a sinusoidal voltage is impressed across a pure inductance, as shown in Fig. 12-12, the waveform of the current lags behind the impressed voltage by _____ electrical degrees.

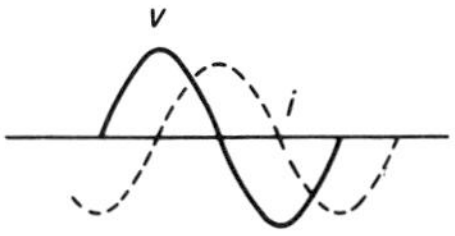

Fig. 12-12

12-13
(*90*) But, what is more important, the current waveform is *sinusoidal* if the voltage waveform is _____.

12-14
(*sinusoidal*) Figure 12-14 illustrates the *v* and *i* waveforms when the voltage is a _____ wave and is impressed across a pure capacitance.

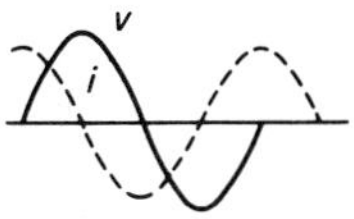

Fig. 12-14

12-15
(*sine*) Here we see that the current *i* _____ the voltage *v* by 90°.

12-16
(*leads*) Again, it is even more important to note in this case that if the applied voltage is sinusoidal across the capacitance, then the resulting _____ will also be sinusoidal.

12-17
(*current*) We can summarize these findings thus: If the voltage is a pure sinusoid, the current that flows through the load will be sinusoidal regardless of the nature of the load. This is true whether the load is resistive, inductive, or _____.

WIDEBAND AMPLIFIERS 329

12-18

(*capacitive*) Consider the nonsinusoidal voltage *v* in Fig. 12-18. As indicated, the nonsinusoidal voltage in this figure is being applied to a _____ load.

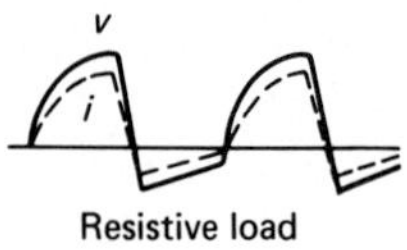

Fig. 12-18   Resistive load

12-19

(*resistive*) For this condition, the current waveform *i* is the same as the _____ waveform *v*.

12-20

(*voltage*) Now, however, conditions change radically when the load becomes inductive or capacitive. Figure 12-20 illustrates the conditions for an _____ load.

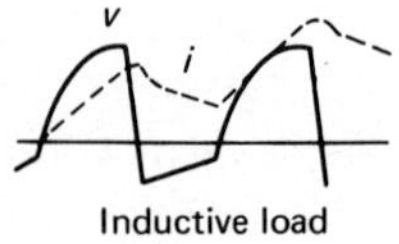

Fig. 12-20   Inductive load

12-21

(*inductive*) The applied voltage *v* is nonsinusoidal; the resulting current is also non_____.

12-22

(*sinusoidal*) But note that the waveform of the current *i* (Fig. 12-20) does not even remotely resemble the waveform of the applied _____.

12-23

(*voltage*) That is, a *sinusoidal* voltage applied to an inductive load yields a *sinusoidal* current; however, a nonsinusoidal voltage applied to an inductive load yields a current that is completely dissimilar in _____ from the voltage that causes the current.

12-24

(*waveform* or *shape, etc.*) Finally, consider the waveforms shown in Fig. 12-24. A _____ voltage *v* is being applied to a capacitive load to produce the current waveform indicated as *i*. The current is also nonsinusoidal.

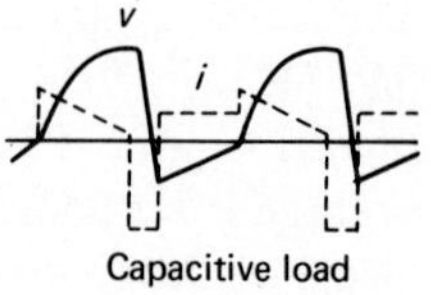

Fig. 12-24   Capacitive load

12-25

(*nonsinusoidal*) As in the case of the inductive load and nonsinusoidal voltage, the waveform of the current is completely different from the waveform of the voltage when the load is _____.

12-26
(*capacitive*) Thus, only a sinusoidal voltage will produce a current having the same _____ as itself, regardless of whether the load is resistive, inductive, or capacitive.

12-27
(*waveform* or *shape, etc.*) That is, any nonsinusoidal voltage may produce a waveform of current the same as its own if the load is resistive, but it will *not* do so if the load is either capacitive or _____.

12-28
(*inductive*) Assume that the fundamental frequency of each of the waveforms in Fig. 12-1 is the same; let us call this frequency f. If the sine wave has a frequency of, say, 100 Hz, then the frequency of the rectangular wave is assumed to be _____ Hz.

12-29
(*100*) We are also assuming that the frequency of each of the other waveforms — the triangular wave and the _____ wave — is 100 Hz.

12-30
(*sawtooth*) You will recall that a "harmonic" is an integral multiple of a fundamental frequency. The second harmonic of 100 Hz is 200 Hz. The third harmonic of 100 Hz is _____ Hz, etc.

12-31
(*300*) It may be shown that any of the waveforms in Fig. 12-1 may be built up from *any one of the other waveforms*. Suppose we select the triangular wave as the "standard" waveform; we can show that a sine wave can be built up by taking the fundamental frequency of the _____ waveform and adding to it the right number of its own harmonics.

12-32
(*standard* or *triangular*) That is, we could build up a 100-Hz *sine* wave by adding to a 100-Hz fundamental *triangular* wave a large number of _____ of the 100-Hz triangular wave.

12-33
(*harmonics*) If we chose the rectangular wave as the standard waveform, we could build up a triangular wave of the same fundamental frequency by adding the proper number of harmonics to the fundamental frequency of the _____ wave.

12-34

(*rectangular*) If, on the other hand, the sine wave is selected as the standard or basic waveform, we could build up a triangular wave, a sawtooth wave, a rectangular wave, or any other non_____ waveform merely by adding the right number of harmonics to the fundamental sine wave.

12-35

(*sinusoidal*) Which waveform shall we adopt as the *standard waveform?* Suppose we arbitrarily adopt the triangular waveform as standard. Then, the waveform shown in _____ of Fig. 12-1 will be considered as a single frequency from which all other waveforms may be constructed by harmonic addition.

12-36

(*b*) If our standard triangular waveform is now applied as a voltage to a pure resistor, the current waveform will also be _____, as previously discussed.

12-37

(*triangular*) But, if the standard triangular voltage is applied to either a capacitor or an inductor or any combination thereof, the current will have some waveform other than _____.

12-38

(*triangular*) Thus, if a *single-frequency* standard waveform (triangular) of voltage is used on a reactive circuit, the _____ resulting from this voltage does *not have a standard, single-frequency waveform.*

12-39

(*current*) Instead, the current resulting from a single-frequency fundamental triangular voltage is a *new* waveform consisting of many frequencies in addition to the fundamental, each of these frequencies being a _____ of the standard waveform.

12-40

(*harmonic*) The same would hold for any nonsinusoidal waveform. That is, any *nonsinusoidal waveform considered as the standard single-frequency voltage will produce, in reactive loads, a current consisting of many* different _____, each of which is a harmonic of the arbitrarily selected standard frequency.

12-41

(*frequencies*) As may be well imagined, the analysis of waveforms is substantially simplified by choosing a standard waveform of voltage such that it will *reproduce* itself as a current regardless of the resistive or reactive nature of the _____ to which the voltage is applied.

332 A PROGRAMMED COURSE IN BASIC ELECTRONICS

12-42
(*load*) That is, it is highly desirable to use a standard voltage which, in single-frequency form, will give rise to a _____-frequency current (i.e., a current of the same waveform as the voltage) in any type of load.

12-43
(*single*) As shown in Figs. 12-8, 12-12, and 12-14, the only waveform that will produce a single-frequency current from a single-frequency voltage in R, L, and C loads is that of a _____.

12-44
(*sinusoid* or *sine wave*) That is, only a sinusoidal voltage produces a sinusoidal current in any type of _____.

12-45
(*load*) For this reason, the sine wave has been adopted universally as the fundamental waveform. Thus, it is understood in all electrical literature that when a single frequency f is spoken of, it means a _____ wave with f Hz.

12-46
(*sine*) Following this concept through, we will now recognize any *nonsinusoidal* waveform as consisting of a sine wave of the same fundamental frequency *plus* the proper number, kinds, and amplitudes of sinusoidal _____ that have been added to the fundamental sine wave.

12-47
(*harmonics*) For example, consider the complex voltage that appears in Fig. 12-47. By inspection, it is at once apparent that this waveform is non_____.

Fig. 12-47

12-48
(*sinusoidal*) Referring now to Fig. 12-48, we see 2 pure sinusoids drawn on the same axis. The one having the larger amplitude is called the _____ sinusoid.

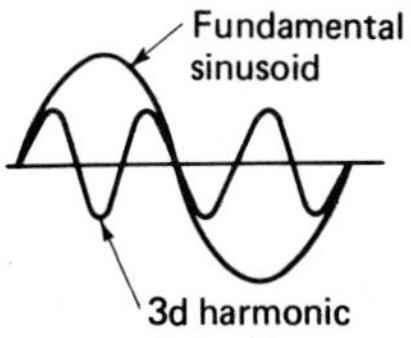

Fig. 12-48 — Fundamental sinusoid, 3d harmonic sinusoid

12-49
(*fundamental*) The sinusoid of smaller amplitude in Fig. 12-48 has a frequency of 3 times the fundamental and is therefore the _____ harmonic of the fundamental sinusoid.

12-50

(*third*) If the 2 sinusoids of Fig. 12-48 are added algebraically point by point, we obtain the waveform of Fig. 12-47. Thus, the complex wave in Fig. 12-47 consists of a fundamental sinusoid plus the third _____ of this fundamental.

12-51

(*harmonic*) As another example, refer to Fig. 12-51. The waveform in part B is a complex wave. That is, it is non_____.

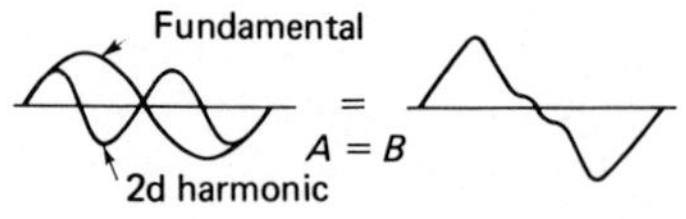

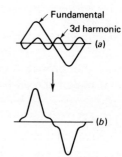

Fig. 12-51

12-52

(*sinusoidal*) If the two waveforms in part A (both sinusoidal) are added together, however, we obtain the complex waveform of part B. Thus, the waveform in part B is obtained by adding the _____ harmonic to the fundamental sinusoid.

12-53

(*second*) As a final example, refer to Fig. 12-53. Here again we are dealing with the fundamental and the _____ harmonic in an additive situation.

Fig. 12-53

12-54

(*third*) The difference between Figs. 12-53a and 12-48 is one of phase relationships. That is, the third harmonic is being added to the fundamental sine wave in both cases, but the _____ of the harmonic is shifted by 180° in one case as compared with the other.

12-55

(*phase*) The complex wave that results from the addition of fundamental and third harmonic in Fig. 12-53 is therefore completely _____ in appearance from that of Fig. 12-47.

12-56

(*different*) Summarizing: In all waveform analyses, the sine waveform is taken as the single-_____ standard waveform.

12-57
(*frequency*) Summarizing: The sine waveform is selected for this purpose because it is the only voltage waveform that produces an identical current waveform in any type of load, regardless of whether the load is resistive or _____.

12-58
(*reactive*) Summarizing: When harmonics of a fundamental sine wave are added to the fundamental, the resulting waveform is called a _____ wave.

12-59
(*complex* or *nonsinusoidal*) Summarizing: Any complex but recurring wave may be built up by adding the proper number, magnitudes, and types of _____ to the fundamental sine waveform.

12-60
(*harmonics*) Summarizing: Conversely, any periodic _____ waveform can be shown to be made up of a fundamental sine wave plus a definite number of harmonics of this sine wave of the correct magnitudes and phases.

12-61
(*complex* or *nonsinusoidal*) Summarizing: Thus, only a sine waveform can be considered a single frequency. Most complex, periodic waves must be considered to consist of many _____.

12-61
(*frequencies*)

BEHAVIOR OF WIDEBAND AMPLIFIERS
12-62
In many amplifier systems, the amplifiers are called upon to amplify signals having sawtooth waveform, trapezoidal waveform, rectangular waveform, etc. Such waveforms, as we have seen, are complex or non_____.

12-63
(*sinusoidal*) Nonsinusoidal waveforms consist of a fundamental component plus a number of _____.

12-64
(*harmonics*) If such a signal is passed through an amplifier that is not capable of providing equal gain for all the harmonics as well as the fundamental, then the output signal will not be an exact reproduction of the _____ signal.

12-65

(*input*) In general, an amplifier having a poor high-frequency response will not be able to amplify the harmonics to the same degree as it amplifies the _____ frequency, which, of course, is a lower frequency.

12-66

(*fundamental*) As illustrated in Fig. 12-66, an amplifier that "loses" all but the fundamental and the lower-order harmonics is called a _____ band amplifier.

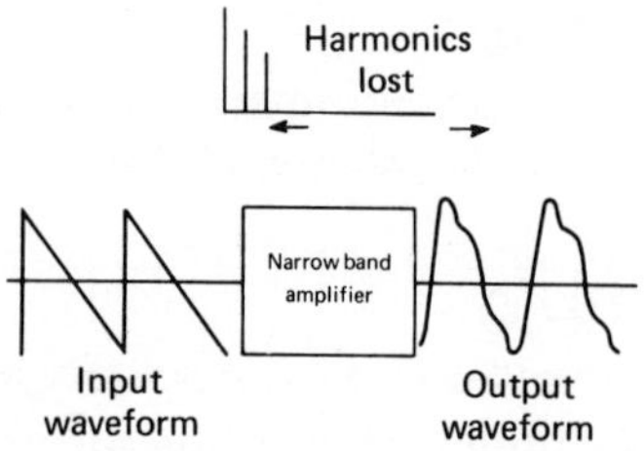

Fig. 12-66

12-67

(*narrow*) In Fig. 12-66, the input wave has the characteristic waveform that we have referred to as a _____ wave.

12-68

(*sawtooth*) Since a sawtooth wave is nonsinusoidal, it is composed of a _____ frequency plus a number of harmonics of the fundamental frequency.

12-69

(*fundamental*) When such a waveform is passed through a narrowband amplifier, many of the higher-order _____ are lost in that they receive little or no amplification.

12-70

(*harmonics*) The output waveform, therefore, is lacking many of the components that were present in the _____ waveform.

12-71

(*input*) Hence, the _____ waveform is highly distorted in that it is not an exact reproduction of the input waveform.

12-72

(*output*) An amplifier that is capable of providing equal amplification for the fundamental frequency and all the harmonics is illustrated in Fig. 12-72. Such an amplifier is known as a _____ amplifier.

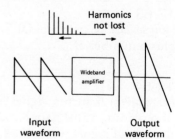

Fig. 12-72

12-73
(*wideband*) Compare the output and input waveforms in Fig. 12-72. Since there is little difference between the waveforms of the output and input, such an amplifier causes very little _____.

12-74
(*distortion*) The sawtooth signals in the sweep circuits of the oscilloscope require amplification in many cases. To amplify such signals, a _____ amplifier would be required.

12-75
(*wideband*) Many television and radar circuits are required to amplify rectangular waveforms (pulses). Such waveforms are also non_____.

12-76
(*sinusoidal*) Hence, such amplifiers must also have the amplifying characteristics associated with wideband amplifiers since, if they do not, the output waveforms will not be exact _____ of the input waveforms.

12-77
(*reproductions*) Coupling systems contribute to the bandpass characteristics of amplifiers. Refer to Fig. 12-77a and 12-77b. The response curve in Fig. 12-77a shows equal amplification for a relatively small range of frequencies; hence the amplifier producing this response would be thought of as a _____ band amplifier.

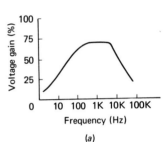

(a)

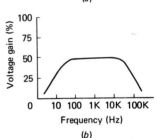

(b)

Fig. 12-77

12-78
(*narrow*) The curve given in Fig. 12-77a is that of a transformer-coupled amplifier. The response curve in Fig. 12-77b has lower gain but a considerably _____ bandpass than that of Fig. 12-77a.

12-79
(*larger* or *wider*, etc.) On a comparative basis, the amplifier whose response is shown in Fig. 12-77b may be considered to have better wideband characteristics. This curve was obtained from an *RC*-coupled system. Thus, _____ coupling gives better promise than _____ coupling for the design of wideband amplifiers.

12-80
(*RC, transformer*) It should be noted that even in the *RC*-coupled response curve, there is a dropoff of gain at both the _____ - and high-frequency ends of the range covered.

12-81
(*low*) Let us see what the frequency-limiting elements in an *RC*-coupled amplifier are. Refer to Fig. 12-81. Transistor $Q1$ works into a load impedance R_L that has a definite capacitance; in addition stray output capacitances are present. All these are lumped together in the diagram and are shown as a capacitor identified as _____.

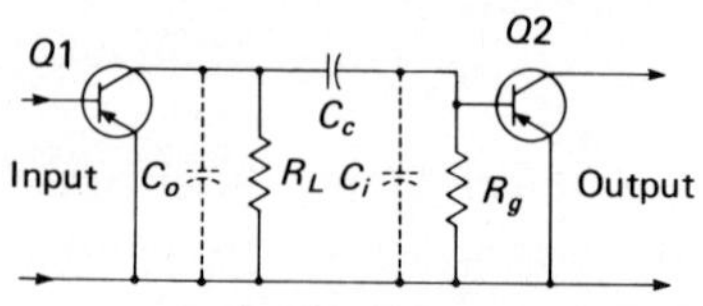

Fig. 12-81

12-82
(C_o) The coupling circuit works into an input impedance that has capacitance. This, and the stray input capacitances, is shown as a lumped capacitance identified in the figure as _____.

12-83
(C_i) The reactance of any capacitor _____ as the frequency increases.

12-84
(*decreases*) Thus, as the frequency rises, the reactances of C_o and C_i decrease and cause shunting or bypassing of the higher frequencies. This causes the _____-frequency dropoff shown in Fig. 12-77b.

12-85
(*high*) In order for C_c to couple the low frequencies into the next stage with minimum loss, the time constant of C_c and R_g must be *long* in comparison with the _____ frequency to be amplified.

12-86
(*lowest*) On the other hand, if the time constant is too long, there is danger of instability. By restricting the time constant to a "safe" value, the _____-frequency response suffers.

12-87
(*low*) This effect is seen as a _____ of gain at the low-frequency end of the range in Fig. 12-77b.

12-87
(*loss* or *dropoff* or *reduction, etc.*)

THE MEANING AND CAUSE OF PHASE DISTORTION
12-88
Phase distortion is a type of distortion that results from a nonuniform time delay as various components of a complex wave pass through the amplifier. In Fig. 12-88, wave b is said to be lagging wave _____ since it is reaching its various phases at later times.

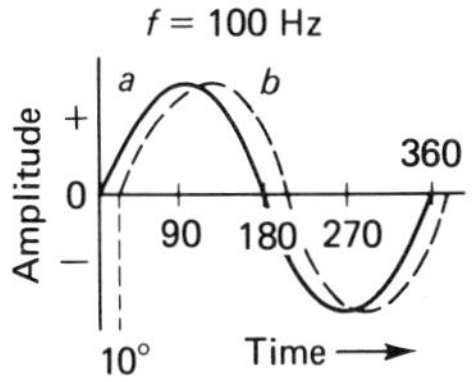

Fig. 12-88

12-89
(*a*) The phase delay suffered by wave b may be stated in terms of an angle, as in Fig. 12-88. Since all points on wave b are _____ ° behind equivalent points on wave a, then wave b is said to be delayed in phase by _____ °, as indicated in the figure.

12-90
(*10, 10*) Phase delay is equivalent to *time delay*. To state phase delay in time units such as seconds (s), milliseconds (ms), or microseconds (μs), we must take into account the frequency of the waves. The frequency of the wave in Fig. 12-88 is _____ Hz.

12-91
(*100*) Since the frequency is 100 Hz we can determine the "period" by taking the reciprocal of the frequency. Thus, the period of this wave is _____ s.

12-92
($\frac{1}{100}$) It requires $\frac{1}{100}$ s to complete one cycle of 360°. Thus, how long does it take to complete only 10° of the full cycle? This time is, of course, $\frac{10}{360} \times \frac{1}{100}$. When reduced to a decimal figure, the answer is _____ s.

12-93
(*0.000278*) This number—0.000278 s—stated in μs is _____.

12-94
(*278*) Thus, a phase delay of 10° in a wave of frequency 100 Hz amounts to 278 μs of *time delay*. Now let us determine the time delay for the same phase in degrees when the frequency of the wave is 1,000 Hz. In this case the period is _____ s.

12-95
($\frac{1}{1,000}$) Thus the time delay is $\frac{10}{360} \times \frac{1}{1,000} =$ _____ μs.

12-96
(*27.8*) Note that the time delay goes down by a factor of 10 (from 278 μs to 27.8 μs) when the frequency goes _____ by a factor of 10 (from 100 to 1,000 Hz).

12-97
(*up*) In other words, if the frequency were 10,000 Hz a 10° phase shift would be the equivalent of 2.78 μs. If the frequency were 100,000 Hz, the phase delay would be the equivalent of _____ μs.

12-98
(*0.278*) Thus, a given angular phase delay at a low frequency like 100 Hz causes a relatively *large* time delay between the 2 waves; at the high frequencies, however, the same angular phase delay causes a relatively _____ time delay.

12-99
(*small*) This means that even a small phase shift at low frequencies can cause a serious time difference between two waves that are initially in a given phase relationship; on the other hand, the same phase delay is not nearly so serious at _____ frequencies.

12-100
(*high*) Now let us study what happens to a complex wave as a result of phase delays that might occur in an amplifier. First, refer to Fig. 12-100. In (*a*) of this figure, we have drawn in a dashed line at the _____ ° angle for use in the forthcoming discussion.

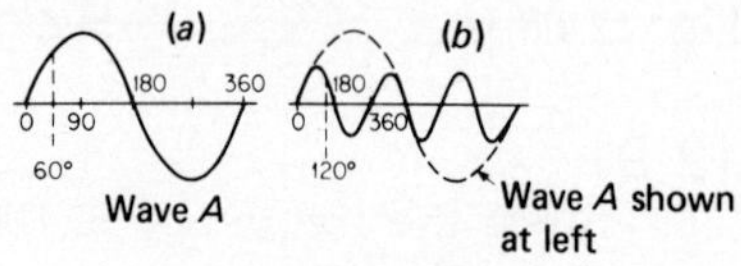

Fig. 12-100

12-101
(*60*) Let us consider wave *A* in Fig. 12-100 as the fundamental frequency. Now refer to (*b*) of Fig. 12-100. This is the third harmonic of the wave shown in (*a*) since _____ full cycles of the shorter wave fit into the same time interval as 1 cycle of the fundamental.

340 A PROGRAMMED COURSE IN BASIC ELECTRONICS

12-102
(3) On the third harmonic wave we have drawn in a dashed line indicating the position of the _____ ° angle of this wave. We shall use this marker in the forthcoming discussion, too.

12-103
(120) Now refer to Fig. 12-103. In (a), we illustrate the fundamental and third harmonic drawn on the same axes. Let us assume that the initial and desired phase relationship between this fundamental and _____ harmonic are as shown on the left in (a), Fig. 12-103.

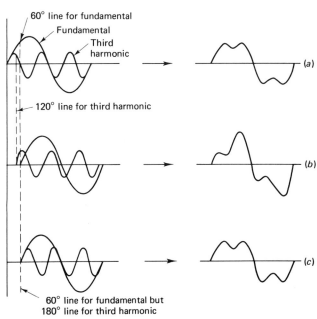

Fig. 12-103

12-104
(*third*) When these waveforms are added, the complex wave on the right in part (a) is obtained. Remember, this is the desired waveform. It is fed to an amplifier. If the amplifier is properly designed, the output waveform will be exactly the _____ as this desired input waveform.

12-105
(*same*) Suppose, however, that the amplifier causes a phase delay in the fundamental frequency. This is shown in (b), Fig. 12-103. The fundamental *output* waveform starts later in time than the input waveform; that is, it starts at the 60° line as shown. Hence, the phase delay in the fundamental is _____ ° of its cycle.

12-106
(*60*) But, suppose the same amplifier produces a phase shift of 120° in the third harmonic. Thus, as shown in (b), Fig. 12-103, the third harmonic output wave starts 120° later than its input wave. Note how the third harmonic wave is shown starting on the _____ line for its cycle in (b).

12-107
(*120°*) Now the fundamental and third harmonic waveforms have a phase relationship that is completely different from the input waveform, resulting in a complex wave [right side, (b), Fig. 12-103] that is also completely _____ from the input waveform.

12-108
(*different*) This, of course, signifies that distortion is being produced by the amplifier since it cannot handle waves of different frequencies without producing a detrimental _____ delay.

12-109
(*phase* or *time*) Now note how the symmetry of the complex wave is restored by having the phase delays in the fundamental and harmonic *differ by proper amounts*. In (c), we show the fundamental again delayed by _____ °.

12-110
(*60*) But this time we show the third harmonic delayed by *180° of its cycle*. Note that the third harmonic has a frequency that is 3 times the fundamental; note also that the phase delay in the harmonic is _____ times the phase delay in the fundamental (that is, 180° as compared with 60°).

12-111
(*3*) By exactly the same procedure applied to *all other harmonics* we can show that distortion will not occur if the *phase delay is proportional to the frequency*. That is, distortion will not occur if the delay for the *second* harmonic is *twice* that of the fundamental, for the *fourth* harmonic is _____ times that of the fundamental, etc.

12-112
(*4*) To clarify this point, let us translate phase delay into time delay as in this example: Consider a 1,000-Hz signal as a fundamental carrying its second, fourth, and sixth harmonics. The harmonic frequencies, therefore, are 2,000 Hz, 4,000 Hz, and _____ Hz, respectively.

12-113
(*6,000*) Assume that the fundamental is delayed by 60°. As we saw previously, this is equivalent to a time delay of: $\frac{60}{360} \times \frac{1}{1,000} =$ _____ _____ (leave as fraction).

12-114
($\frac{1}{6,000}$ s) If the delay is made proportional to the frequency, then the second harmonic should be delayed by _____ ° since its frequency is twice that of the fundamental.

12-115
(*120*) The time delay in this case is therefore: $\frac{120}{360} \times \frac{1}{2,000} =$ _____ s (leave as fraction).

12-116
($\frac{1}{6,000}$) Similarly if the phase delay is proportional to the frequency, then for the fourth harmonic, the delay should be _____ °

12-117
(*240*) Again, the same delay as obtained from: $\frac{240}{360} \times \frac{1}{4,000} =$ _____ s (leave as fraction).

12-118
($\frac{1}{6,000}$) Finally the sixth harmonic, following the same pattern, gives us the figures: _____ $= \frac{1}{6,000}$ s.

12-119
($\frac{360}{360} \times \frac{1}{6,000}$) It is therefore evident that when the *phase delay is proportional to the frequency* of the component in the complex wave, there will be no distortion because the time delay for each component will be _____ .

12-120
(*the same* or *constant* or *uniform, etc.*) Time delay does no harm, nor is it undesirable, provided that such delay is _____ for all components of the complex wave.

12-121
(*the same* or *constant* or *uniform, etc.*) The only effect of uniform time delay is to shift the entire wave to a later time. But, since all components would retain their proper phase with respect to all others, there would be no _____ of the waveform.

12-122
(*distortion*) Hence, for uniform time delay and no phase distortion, it is necessary for the phase delay of each component of a complex wave to be _____ to the frequency of that component.

12-122
(*proportional*)

FREQUENCY COMPENSATION IN WIDEBAND TRANSISTOR AMPLIFIERS

12-123

To convert the *RC*-coupled circuit of Fig. 12-123 into a *wideband* amplifier, it is necessary to compensate for low-frequency dropoff, high-frequency dropoff, and component phase shifts that are not proportional to the _____ of the component.

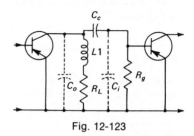

Fig. 12-123

12-124

(*frequency*) Let us first consider *high-frequency* compensation. This type of compensation is intended to minimize the loss of gain of the amplifier at the _____-frequency end of its range.

12-125

(*high*) Referring to Fig. 12-123, you will recall that loss of gain at the high frequencies is largely due to the presence of C_o, the output capacitance of the first stage, and _____, the input capacitance to the second stage.

12-126

(C_i) Note and compare Fig. 12-123 with Fig. 12-81. The circuit of Fig. 12-123 contains an additional component that was not present in the other circuit. This additional component is identified as _____.

12-127

($L1$) The inductance of $L1$ is selected so that it forms a broadly resonant circuit with C_o and C_i, especially at the higher frequencies. Without $L1$ in the circuit, you will remember, C_o and C_i tend to shunt out the higher frequencies, because capacitive reactance _____ as the frequency increases.

12-128

(*decreases*) With $L1$ in the circuit, however, resonance over a broad range is established at the higher frequencies. As a result, the shunting impedance remains relatively _____, since the impedance of a parallel resonant circuit (even a broadly tuned one) is relatively high.

12-129

(*high*) This tends to flatten out the response curve at the high-frequency end of the range, thus minimizing _____-frequency dropoff.

344 A PROGRAMMED COURSE IN BASIC ELECTRONICS

12-130
(*high*) Compensation such as that shown in Fig. 12-123 is known as *shunt compensation*. In shunt compensation, the compensating device ($L1$ in this case) is connected in _____ with the load resistor R_L as shown in the figure.

12-131
(*series*) This arrangement is called _____ compensation because the combination of $L1$ and R_L is connected in parallel with the components for which compensation is being made, in this case C_o and C_i.

12-132
(*shunt*) Figure 12-132 illustrates a second form of compensation known as *series compensation*. To understand this system, we must first recognize that C_c is a large capacitor (10 μF or more). A capacitor of this large value has a very _____ capacitive reactance at the high frequencies.

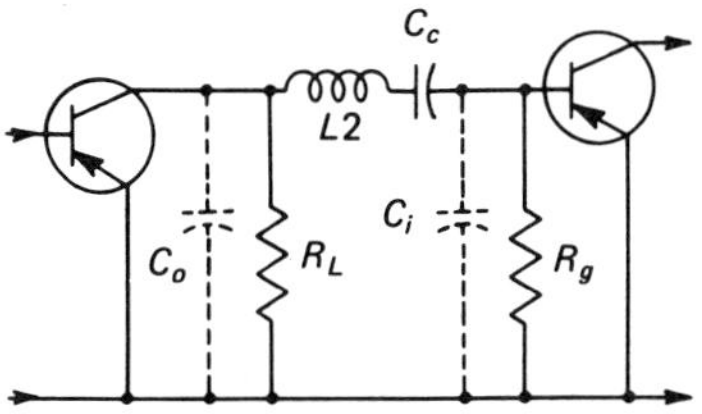

Fig. 12-132

12-133
(*small* or *low*, etc.) For this reason we may consider C_c as a short circuit for the high frequencies and thus eliminate it from consideration. $L2$ may thus be taken to be in _____ with the input capacitance C_i.

12-134
(*series*) Again, the value of $L2$ is chosen so that this inductor and C_i form a *series* resonant circuit at the high frequencies. That is, as the high frequencies are approached, the series circuit containing $L2$ and C_i approaches the condition of _____.

12-135
(*resonance*) As a series circuit approaches resonance, its impedance begins to _____.

12-136
(*drop* or *decrease*) Thus, the current flowing through the $L2C_i$ combination must _____ as resonance is approached.

12-137
(*increase*) An increased current in C_i means that the signal voltage appearing across the capacitor must become larger; at the same time, however, the decreasing capacitive _____ of C_o reduces the voltage of the signal across the load resistor.

12-138
(*reactance*) The compensation action is now clear. As the signal voltage across R_L decreases with rising frequency because of the shunting action of C_o, the signal voltage drop across C_i _____ as a result of the series resonant condition.

12-139
(*increases*) Shunt peaking is satisfactory if the maximum frequency limit is not too high and there are few amplifier stages. Series peaking provides about 50 percent more high-frequency gain than shunt peaking, however. For this reason, _____ peaking is preferred in wideband amplifiers having stricter requirements relative to high-frequency gain.

12-140
(*series*) The circuit arrangement shown in Fig. 12-140 is often called *combination peaking* because it utilizes both a shunt and a _____ peaking coil.

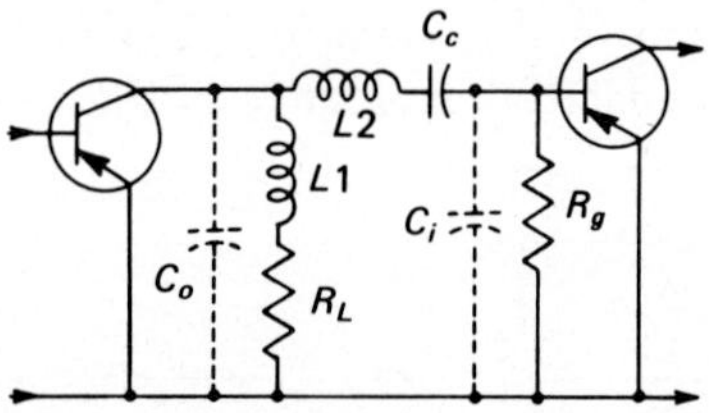

Fig. 12-140

12-141
(*series*) The coil identified as _____ is the shunt peaking coil, and the coil identified as _____ is the series peaking coil.

12-142
(*L1, L2*) When properly designed, the high-frequency gain of the _____ peaking circuit shown in Fig. 12-140 is approximately 80 percent greater than that of the series peaking arrangement.

12-143
(*combination*) Now let us consider the low-frequency response of the RC-coupled transistor amplifiers. The capacitance of C_o is quite small; hence at low frequencies its capacitive reactance is relatively _____.

12-144
(*large*) Since its reactance is large at low frequencies, then at these frequencies its shunting effect is relatively _____.

12-145
(*small*) Actually, the shunting effect of C_o at low frequencies is negligible. The same is true of the input capacitance to the following stage, _____.

12-146
(C_i) At low frequencies, therefore, the effects on frequency response of C_o and C_i can be ignored. We must look to the coupling capacitor, identified as _____ in Fig. 12-140, and the input resistor R_g as the controlling elements for low-frequency response.

12-147
(C_c) Capacitor C_c can be given a reasonably large capacitance. In Fig. 12-147, a typical value is shown. It is _____ μF.

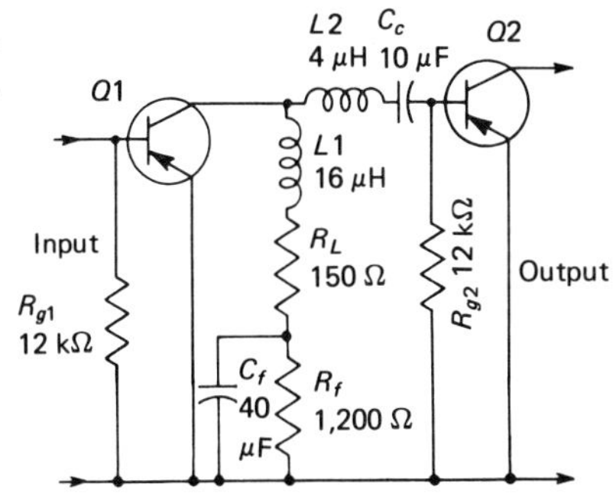

Fig. 12-147

12-148
(*10*) Notwithstanding this large value for C_c, as the frequency goes lower and lower, its capacitive reactance becomes increasingly _____.

12-149
(*larger*) C_c and R_g form a signal voltage divider for the amplified output of the previous transistor. Only the voltage that appears across _____ is applied to the base and emitter of the second transistor for further amplification.

12-150
(R_g) At high frequencies, the reactance of C_c becomes very small. This means that the voltage drop across C_c at high frequencies must also be very _____.

12-151
(*small*) Thus, at high frequencies virtually all the signal voltage drop must appear across _____ and appear as the signal applied to the base-emitter circuit.

12-152
(R_g) It follows that, at high frequencies, the amplifier response is not affected to a serious extent by C_c and R_g. As the reactance of C_c rises for the low frequencies, however, an increasing _____ drop appears across this component.

12-153
(*voltage*) As more of the signal voltage appears across C_c, the voltage drop across R_g must necessarily _____.

WIDEBAND AMPLIFIERS 347

12-154
(*decrease*) This accounts for the dropoff of response of this coupling system at the _____ frequencies.

12-155
(*low*) Even more serious, however, is the _____ distortion brought about by the large time delay introduced by the coupling circuit for the low frequencies. Let us investigate this.

12-156
(*phase*) Consider the collector output signal from the first transistor as an ac generator driving the $R_g C_c$ circuit. Since the coupling circuit is capacitive, the current must _____ the generator voltage by a definite angle.

12-157
(*lead*) The angle of lead is that angle whose tangent is defined by the capacitive reactance divided by the resistance. In standard symbols, the phase angle $\phi = \arctan X_C/$_____.

12-158
(*R*) The current flowing through R_g is, of course, what produces the signal voltage drop across this resistor. Thus, the signal voltage across R_g has the same phase as the current and _____ the input voltage by the same angle.

12-159
(*leads*) It is evident, therefore, that the phase of the signal applied to the base of the second transistor is different from that of the input signal to the coupling network. Also, as the frequency decreases, X_C becomes *larger*, and the leading phase angle must also _____.

12-160
(*increase*) Since $\phi = \arctan X_C/R$, in order to decrease phase distortion caused by the leading phase angle just described, it is necessary to make X_C as *small* as possible and R_g as _____ as possible.

12-161
(*large*) But, to make X_C as small as possible even at low frequencies, it is necessary to make C_c as _____ in capacitance as possible.

12-162
(*large*) Thus, to establish the best time constant for C_c and R_g, we must make both components the _____ possible values without introducing other problems resulting from this selection.

12-163
(*largest*) One problem that often results from the use of extremely large values of C_c and R_g is "motorboating" or relaxation _____ that arises as a result of an excessively large time constant.

12-164
(*oscillation*) To avoid motorboating, C_c and R_g must be selected carefully and must not be allowed to exceed the limits that would cause motorboating. To compensate for the phase and frequency distortion imposed by this limitation, a special compensating network is added (Fig. 12-164) consisting of capacitor C_f and resistor _____.

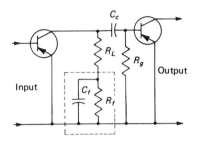

Fig. 12-164

12-165
(R_f) Let us see the effect of this filter network (C_f and R_f) at *high* frequencies. Capacitor C_f is a large capacitor. As indicated in the typical circuit of Fig. 12-147, a suitable value for this capacitor would be _____ μF.

12-166
(*40*) At the high frequencies, the reactance of C_f will be very _____.

12-167
(*small*) Since C_f is connected in parallel with R_f and has a negligible reactance at the high frequencies, the net impedance of the parallel combination is extremely _____ at the high frequencies.

12-168
(*small*) Thus, the $C_f R_f$ combination has a negligibly small effect at high frequencies and can be ignored. At the lower frequencies, however, the reactance of C_f begins to assume a _____ value.

12-169
(*greater* or *larger, etc.*) At very low frequencies, the reactance of this capacitor becomes high enough for us to consider it as an open circuit. Hence, its bypassing action is negligible and the net impedance of this $C_f R_f$ combination becomes approximately equal to the resistance of _____ alone.

12-170
(R_f) Thus, R_f can now be considered as the only component in series with R_L, thus making the total load impedance equal to the sum of _____ and _____.

WIDEBAND AMPLIFIERS

12-171
(R_f, R_L) Now let us compare the high- and low-frequency gains. At high frequencies, R_f is short-circuited out by the bypassing action of _____.

12-172
(C_f) In the typical circuit, Fig. 12-147, the total load resistance would then be equal to R_L alone. That is, the resistive component of the load at the high frequencies in this typical circuit is _____ Ω.

12-173
(*150*) As we have shown, the total load resistance at low frequencies is the sum of R_L and R_f. Thus, in the typical circuit, the total load resistance at low frequencies would be _____ Ω.

12-174
(*1,350*) It is easily shown that the voltage and power gain of a transistor amplifier will rise when the load resistance is increased from 150 to 1,350 Ω. Thus, in the typical circuit, the low-frequency amplification is _____ by the presence of the compensation network.

12-175
(*increased*) This tends to compensate for the low-frequency dropoff caused by the increasing reactance of C_c at the _____ frequencies.

12-176
(*low*) Analysis also shows that the phase of the amplifier output is shifted in the *lagging* direction because C_f is a *shunt* capacitive reactance forming a part of the collector _____ impedance.

12-177
(*load*) Taken as a whole, the effect of the $C_f R_f$ combination is just opposite to the distortion effects produced by $R_g C_c$. Thus, when the compensation components are carefully selected, the network can successfully minimize the low-frequency phase _____ caused by the coupling components.

12-177
(*distortion*)

SQUARE-WAVE ANALYSIS OF FREQUENCY AND PHASE RESPONSE

12-178

The low-frequency response of a wideband amplifier can be quickly evaluated by observing its output waveform when a _____-wave input waveform is used, as shown in Fig. 12-178.

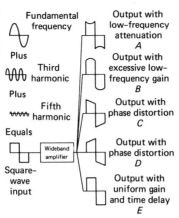

Description of a Square Wave

1. A *perfect square wave* consists of the sum of a fundamental sine wave plus all the *odd* harmonics of this sine wave related to the fundamental as follows:

   a. The amplitude of each harmonic is inversely proportional to its harmonic number.
   b. The phase of the fundamental and each harmonic is the same at 0°, and for the fundamental wave, the same at 180°.

2. A square wave whose shape approaches a perfect square wave may be formed by summing up the fundamental sine wave and the first five or six odd harmonics.

Fig. 12-178

12-179

(*square*) A perfect square wave is obtained by adding all the _____ harmonics of a given fundamental sine wave to the fundamental in the proper phase and with the proper amplitude.

12-180

(*odd*) The proper amplitude is realized when the amplitude of each odd harmonic is inversely proportional to the harmonic number. For instance, the third harmonic should be $\frac{1}{3}$ the amplitude of the _____ waveform.

12-181

(*fundamental*) Similarly, the fifth harmonic should be $\frac{1}{5}$ the amplitude of the fundamental; the seventh harmonic should be _____ the amplitude of the fundamental; etc.

12-182

($\frac{1}{7}$) Obviously, the higher-order harmonics contribute less and less to the final waveform because their amplitudes are very _____ according to this inverse relationship.

12-183

(*small*) Thus, a reasonably good square wave may be formed by combining the fundamental with the first few odd harmonics, as shown in Fig. 12-178. Another requirement for the formation of a square wave is that the _____ of the harmonics and fundamental all be the same at the 0 and 180° phases of fundamental wave.

12-184
(*phase*) The output waveforms for various forms of low-frequency distortion are shown in Fig. 12-178A–E. When there is too much low-frequency dropoff (insufficient compensation), the waveform has the appearance as shown in _____ of Fig. 12-178.

12-185
(*A*) An amplifier that is overcompensated at low frequencies yields an output waveform like that of _____ of Fig. 12-178.

12-186
(*B*) When the flat top of the output square wave slants to the right or left, the indicated trouble is _____ distortion.

12-187
(*phase*) The phase distortion shown in C is caused by an excessive leading angle due to the lack of phase compensation; the opposite is shown in D. In this output waveform, we have a case of overcompensation in which the phase angle has become a _____ one.

12-188
(*lagging*) The output waveform shown in E indicates faithful reproduction of the input waveform. That is, the gain is _____, and the phase delay for each component is proportional to the frequency of the component.

12-188
(*uniform*)

CRITERION CHECK TEST

____12-1 Choose the sawtooth waveform from the figure below. (*a*) 1, (*b*) 2, (*c*) 3, (*d*) 4.

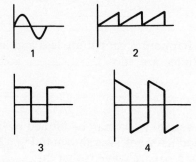

____12-2 When a rectangular voltage waveform is applied to a capacitor, then the current waveform is (*a*) a sinusoid, (*b*) a rectangular waveform, (*c*) a sawtooth, (*d*) none of the above.

_____ 12-3 What is a synonym for the first harmonic? (a) Fundamental, (b) period, (c) phase shift, (d) time delay.

_____ 12-4 The main difference between Figs. 12-48 and 12-53 is that (a) a different number of harmonics is added, (b) the fundamentals have a different phase, (c) the third harmonics have a different phase, (d) the third harmonics have different amplitudes.

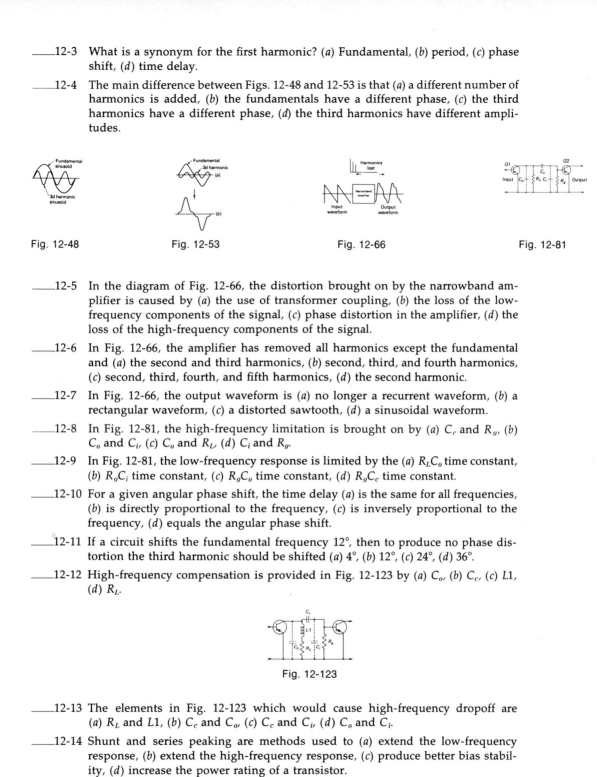

Fig. 12-48   Fig. 12-53   Fig. 12-66   Fig. 12-81

_____ 12-5 In the diagram of Fig. 12-66, the distortion brought on by the narrowband amplifier is caused by (a) the use of transformer coupling, (b) the loss of the low-frequency components of the signal, (c) phase distortion in the amplifier, (d) the loss of the high-frequency components of the signal.

_____ 12-6 In Fig. 12-66, the amplifier has removed all harmonics except the fundamental and (a) the second and third harmonics, (b) second, third, and fourth harmonics, (c) second, third, fourth, and fifth harmonics, (d) the second harmonic.

_____ 12-7 In Fig. 12-66, the output waveform is (a) no longer a recurrent waveform, (b) a rectangular waveform, (c) a distorted sawtooth, (d) a sinusoidal waveform.

_____ 12-8 In Fig. 12-81, the high-frequency limitation is brought on by (a) C_c and R_g, (b) C_o and C_i, (c) C_o and R_L, (d) C_i and R_g.

_____ 12-9 In Fig. 12-81, the low-frequency response is limited by the (a) $R_L C_o$ time constant, (b) $R_g C_i$ time constant, (c) $R_g C_o$ time constant, (d) $R_g C_c$ time constant.

_____ 12-10 For a given angular phase shift, the time delay (a) is the same for all frequencies, (b) is directly proportional to the frequency, (c) is inversely proportional to the frequency, (d) equals the angular phase shift.

_____ 12-11 If a circuit shifts the fundamental frequency 12°, then to produce no phase distortion the third harmonic should be shifted (a) 4°, (b) 12°, (c) 24°, (d) 36°.

_____ 12-12 High-frequency compensation is provided in Fig. 12-123 by (a) C_o, (b) C_c, (c) L1, (d) R_L.

Fig. 12-123

_____ 12-13 The elements in Fig. 12-123 which would cause high-frequency dropoff are (a) R_L and L1, (b) C_c and C_o, (c) C_c and C_i, (d) C_o and C_i.

_____ 12-14 Shunt and series peaking are methods used to (a) extend the low-frequency response, (b) extend the high-frequency response, (c) produce better bias stability, (d) increase the power rating of a transistor.

The following 6 questions refer to Fig. 12-147.

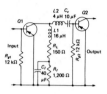

Fig. 12-147

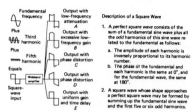

Fig. 12-178

_____12-15 High-frequency compensation is provided by (a) only shunt peaking, (b) only series peaking, (c) combination peaking, (d) field-effect transistors.

_____12-16 The element degrading low-frequency performance is (a) L1, (b) L2, (c) C_f, (d) C_c.

_____12-17 At low frequencies, the collector load resistance of Q1 is about (a) 150 Ω, (b) 1,350 Ω, (c) 12 kΩ, (d) 6 kΩ.

_____12-18 At midband frequency, the collector load resistance of Q1 is about (a) 150 Ω, (b) 1,350 Ω, (c) 12 kΩ, (d) 6 kΩ.

_____12-19 If a square wave is fed to the input, and the output is as shown in Fig. 12-178 A, then to obtain a better square wave one could make (a) R_f larger, (b) R_L larger, (c) C_c smaller, (d) R_{g1} smaller.

_____12-20 If a square wave is fed to the input, and the output is as shown in Fig. 12-178 B, then to obtain a better square wave one could make (a) C_c smaller, (b) R_f larger, (c) R_L smaller, (d) L1 smaller.

13 radio-frequency amplifiers

OBJECTIVES

(1) List the bands and corresponding frequencies found in the RF spectrum.
(2) Give the formula for and compute the resonant frequency of a tuned circuit.
(3) Indicate a method for varying a tuned network to bring in different radio stations. (4) Depict and describe a 2-stage vacuum tube RF amplifier which uses a *single-tuned* circuit for interstage coupling. (5) Depict and describe a double-tuned RF amplifier. (6) State the criteria for a tuned circuit to have good selectivity. (7) Recall the definition of Q and draw a set of selectivity curves for several values of Q. (8) Define and label the passband limits on a selectivity curve. (9) Draw selectivity curves for amplifiers with multiple tuned circuits. (10) Discuss the bandpass requirements of an amplifier designed to receive a modulated RF signal. (11) Depict and describe an amplitude-modulated RF carrier wave and its sidebands. (12) Relate the bandpass characteristics of an RF amplifier to the carrier frequency and sidebands. (13) Determine the bandpass characteristic of a receiving system, given the range of modulating frequencies. (14) Compare a double-humped bandpass characteristic with a single-humped bandpass characteristic. (15) Insert resistors to broaden the bandwidth of a double-tuned circuit. (16) Define the coefficient of coupling, k, for two coils. (17) Define the conditions of perfect coupling and critical coupling. (18) Depict selectivity curves for tuned circuits which are overcoupled, tight-coupled, critically coupled, and loosely coupled. (19) State which condition of coupling gives the best broadband response. (20) State a method used in radio and TV sets to adjust the coil coupling. (21) Broaden the bandwidth of a double-tuned circuit by varying the coil coupling.

FREQUENCY SPECTRUM

The term *radio frequencies* refers to that portion of the electromagnetic spectrum which contains the frequencies used for communication by wireless. There is no essential difference between audio frequencies (AF) and radio frequencies (RF) other than frequency; just as it is possible to have sinusoidal waveforms in AF, it is possible to have sinusoidal waveforms in RF. As we know, the AF spectrum is arbitrarily assigned the range from 20 to 20,000 Hz. The RF spectrum is vastly more extensive in terms of frequency limits, as we will see in the work that immediately follows.

13-1
Radio waves used for radio, television, radar, and other transmissions may also be amplified by vacuum tubes when they are generated or received. AF is often used as an abbreviation for audio frequencies, so _____ is used as an abbreviation for radio frequencies.

13-2
(*RF*) Before studying RF amplification, we must know what range of frequencies the radio-frequency spectrum comprises. This is shown in Fig. 13-2 with the frequencies given in kilohertz (abbrev. kHz) along the top and in megahertz (abbrev. MHz) along the bottom. Since 1 kHz contains 1,000 Hz and 1 MHz contains 1,000,000 Hz, then 1 MHz contains _____ kHz.

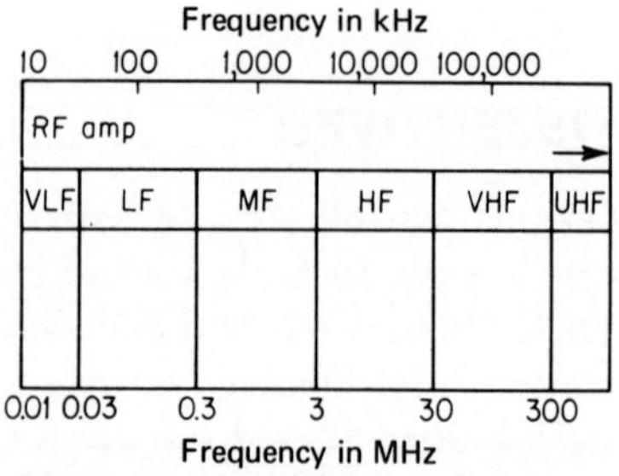

Fig. 13-2

13-3
(*1,000*) Thus, 3 MHz means the same thing as 3,000 kHz. Also, 30 MHz means the same thing as 30,000 kHz. Thus, a frequency of 300 MHz is the same as a frequency of _____ kHz.

13-4
(*300,000*) A frequency of 100,000 Hz stated in MHz would be 0.1 MHz. Similarly, a frequency of 10,000,000 Hz is the same as 10 MHz. Thus, 1,000,000,000 Hz is the same frequency as _____ MHz.

13-5
(*1,000*) In Fig. 13-2 we arbitrarily choose 10 kHz as the low end of the radio-frequency spectrum. Thus, the low end of the arbitrary RF spectrum starts at _____ MHz.

13-6
(*0.01*) By common agreement, authorities now demarcate the lowest frequency *band* in the RF spectrum as the very low frequencies (abbrev. VLF). This band extends from 10 to _____ kHz.

13-7
(*30*) The next band is the low-frequency band (abbrev. LF). The lower limit of this band is at 30 kHz and the upper limit at _____ kHz.

13-8
(*300*) Next comes the medium-frequency band (abbrev. MF). This band extends from 300 to _____ kHz.

13-9
(*3,000*) The broadcast radio band (standard broadcast AM) ranges from about 500 to 1,600 kHz. Thus, standard broadcast, according to the classification in Fig. 13-2, falls in the _____ band.

13-10
(*MF*) Extending from 3 to 30 MHz we find the high-frequency or _____ band.

13-11
(*HF*) The very high-frequency band (abbrev. VHF) has a lower limit of _____ MHz and an upper limit of 300 MHz.

13-12
(*30*) Next comes the ultra-high-frequency band (abbrev. UHF) which starts at _____ MHz on its lower end and extends up to 3,000 MHz on its upper end (not shown in Fig. 13-2).

13-13
(*300*) There are 2 additional bands in the microwave region. The super-high-frequency band (abbrev. SHF) starts at the upper limit of the UHF band, or _____ MHz, and extends to 30 gigahertz (kilomegahertz) (abbrev. GHz). At this point the extremely high-frequency band (abbrev. EHF) starts. Its upper limit is 300 GHz. These bands are not shown in the figure.

13-14
(*3,000*) The standard frequency-modulation (abbrev. FM) stations occupy a range from 88 to 108 MHz. This range falls within the _____ band.

13-15
(*VHF*) Standard television broadcast, from channel 2 through 13, extends from 54 to 216 MHz. Thus, the television band falls within the _____ band.

13-16
(*VHF*) Aeronautical and government stations are found on frequencies such as 4,400 MHz. This falls in the _____ band.

13-16
(*SHF*)

RADIO-FREQUENCY AMPLIFIERS

RF TUNED CIRCUIT

Audio-frequency amplifiers normally are required to amplify with equal or nearly equal gain all the frequencies within the AF spectrum. In a few isolated cases, RF amplifiers may be required to do a similar job on a portion of the RF spectrum, but for the most part, this is not true. In general, one uses an RF amplifier to amplify a single frequency or a *small band* of frequencies. Thus, the RF amplifier is usually equipped with a frequency-selective circuit which determines the frequency or band of frequencies to be fed to the amplifier.

13-17
We start our study of RF amplifiers with a circuit typical of radio receivers (Fig. 13-17). The receiving antenna intercepts many frequencies, only a few of which are shown. For a radio receiver to operate satisfactorily, only _____ of these frequencies should be reproduced by the set.

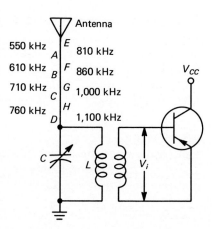

Fig. 13-17

13-18
(*1*) Connected between the antenna and ground is a combination of inductance and variable capacitance. These units are symbolized, respectively, by L and _____ in Fig. 13-17.

13-19
(*C*) With respect to each other, L and C are connected in _____.

13-20
(*parallel*) Assume that the inductance of L is a specific value and that the capacitance of variable capacitor C is set at 200 pF. For this particular value of L and C, there is only one _____ for which this LC circuit will be in resonance.

13-21
(*frequency*) Let us say that $L = 127$ microhenrys (abbrev. μH) and $C = 200$ pF. The frequency at which these values will resonate can be found from the equation $f_r = 1/(2\pi$ _____ $)$.

358 A PROGRAMMED COURSE IN BASIC ELECTRONICS

13-22
($\sqrt{LC}$) In this equation, f_r is in cycles per second, L is in henrys (H), and C is in _____.

13-23
(*farads*) Substituting the values of L and C in this equation gives us a resonant frequency of 1,000 kHz or 1,000,000 Hz. In Fig. 13-17, the station identified by the letter _____ is transmitting a radio wave of this frequency.

13-24
(G) For all the frequencies beside 1,000 kHz, this circuit is *not* resonant. It is resonant only for a frequency of 1,000 kHz. As applied to the resonant frequency, a parallel resonant circuit develops the _____ impedance possible.

13-25
(*highest*) Therefore, the input circuit shown in Fig. 13-17 has the highest impedance between antenna and ground for a frequency of _____ kHz.

13-26
(*1,000*) For all the other frequencies intercepted by the antenna, the impedance of the LC circuit is substantially _____.

13-27
(*lower* or *less*) Since voltage drop depends upon the product of IZ ($V = IZ$), and since Z is the highest at resonance, the frequency for which the greatest voltage drop will appear across the LC circuit is _____ kHz.

13-28
(*1,000*) By the same reasoning, the voltage drops produced across the LC circuit by signals having frequencies other than 1,000 kHz will be considerably _____.

13-29
(*smaller*) If the resonant voltage drop at 1,000 kHz is called v_i, then as shown in Fig. 13-17, this voltage is applied between the _____ and emitter of the transistor.

13-30
(*base*) The purpose of the transformer T is to match the _____ impedance of the parallel resonant circuit to the _____ impedance of the transistor.

RADIO-FREQUENCY AMPLIFIERS

13-31
(*high, low*) All the voltage drops for frequencies other than the resonant frequency will be much smaller than v_i and may be ignored. Thus, the transistor will amplify only a radio signal having a frequency of 1,000 kHz for the values of C and L given. If C is now reduced, the resonant frequency of the LC circuit will go _____.

13-32
(*up*) In this case, the next of the incident waves to be tuned in would be the one having a frequency of _____ _____ in Fig. 13-17.

13-33
(*1,100 kHz*) If C were to be increased to a value higher than 200 pF, the next station to be tuned in would be station F because its frequency is _____ than that of station G.

13-33
(*lower*)

RF COUPLING

The performance of an RF amplifier is partly dependent upon the effectiveness of the coupling arrangement. The two coupling methods discussed in the work that immediately follows are referred to as a *single-tuned* and *double-tuned* circuit, respectively.

13-34
One possible coupling method between RF amplifier stages is given in Fig. 13-34. In this system, the desired signal induced in the antenna is first tuned in by the combination of _____ and _____.

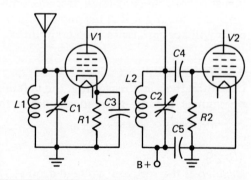

Fig. 13-34

13-35
(*L1, C1*) The selected RF voltage is then applied between the grid and _____ of the pentode V1.

13-36
(*cathode*) The selected RF voltage is first amplified by tube _____.

13-37
(V1) The amplified voltage then appears as a drop across the combination of _____ and _____.

13-38
(L2, C2) This voltage drop is coupled to the grid and cathode of V2 through capacitors C5 and _____.

13-39
(C4) Grid bias for V1 is provided by _____.

13-40
(R1) The choice of components for producing the bias for V1 will depend upon the class of operation of V1. If R1 is selected so that plate current flows throughout the entire input cycle, then V1 will be operating in class _____.

13-41
(A) A preferred method of coupling two RF amplifiers is shown in Fig. 13-41. In this drawing, the resonant collector load for Q1 consists of components _____ and _____.

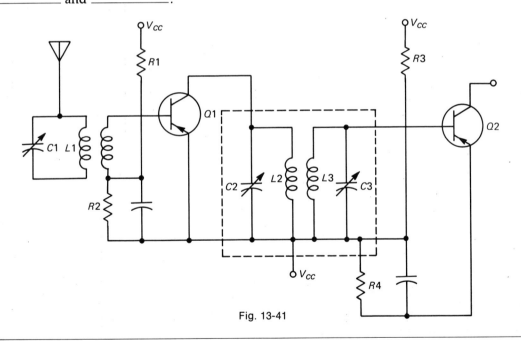

Fig. 13-41

13-42
(C2, L2) Of the various signals intercepted by the antenna, the one that is desired is selected by components _____ and _____.

13-43
(C1, L1) Since only the signal tuned in by C1 and L1 is to be amplified by Q1, the frequency to which C2 and L2 are resonant must be the _____ as the frequency of resonance of C1 and L1.

13-44
(*same*) Therefore, the impedance for the selected signal developed by C2 and L2 is _____.

13-45
(*high*) In the parallel resonant circuit comprising C2 and L2, a signal current circulates through the coil L2. Since L3 is placed very close to L2, a _____ will be induced across L3.

13-46
(*voltage*) The frequency of this induced voltage will be the same as the frequency of the current flowing in the resonant circuit containing C2 and _____.

13-47
(*L2*) In order that the voltage induced across L3 and C3 be *high*, the impedance of the L3C3 circuit at this frequency must be _____.

13-48
(*high*) This means that L3 and C3 must be tuned to the same frequency as that to which L1C1 and _____ are both tuned.

13-49
(*L2C2*) The components enclosed in the dashed-line box in Fig. 13-41 together make up a double-tuned RF *transformer*. This is called a transformer because energy is transferred from the output of Q1 to the input of Q2 by the method of electromagnetic _____.

13-50
(*induction*) Forward bias is provided for Q1 by resistors R1 and R2, and for Q2 by resistors _____ and _____.

13-51
(*R3, R4*) RF amplifiers used in receiving circuits such as those shown in Fig. 13-41 are generally biased in class A. Thus, the _____ current of Q1 and Q2 flows during the entire input cycle.

13-51
(*collector*)

SELECTIVITY

In your study of the phenomenon of resonance, you will recall that you learned the meaning of selectivity. A resonant system has good selectivity if it can separate closely adjacent frequencies. The degree to which an LC circuit is selective is determined in large part by the *figure of merit* or Q of the coil. If a coil with zero resistance could be wound, its Q would be infinite; but since all coils have some resistance, a coil of infinite Q could never be built. We define and measure Q as the ratio of the inductive reactance of the coil to the resistance of the coil, or $Q = X_L/R = 2\pi fL/R$. A comparison of the selectivity of given amplifiers is usually made by means of selectivity curves. If the selectivity curve is narrow, high, and sharply peaked, this is an indication of good selectivity. Poor selectivity is indicated when the curve is broad and flat.

13-52
If a tuned circuit in an RF amplifier is capable of discriminating between closely adjacent frequencies, it is said to have good _____.

13-53
(*selectivity*) The selectivity of a single tuned circuit depends largely upon the amount of pure resistance present in the circuit. The figure of merit of a coil is the ratio of its X_L to its R. This ratio is called the _____ of the coil.

13-54
(*Q*) In order for the selectivity of a tuned circuit to be good, the Q of the coil must be relatively _____.

13-55
(*large*) In Fig. 13-55, three selectivity curves are shown for three different Qs. The best selectivity characteristic is displayed by curve _____.

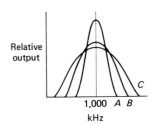

Fig. 13-55

13-56
(*A*) The worst selectivity characteristic of the curves in Fig. 13-55 is displayed by curve _____.

13-57
(*C*) The circuit responsible for curve _____ has the highest Q of the three.

13-58
(*A*) Reviewing the meaning of *passband*, we refer to Fig. 13-58. The half-power point in this diagram is the horizontal line marked _____ percent.

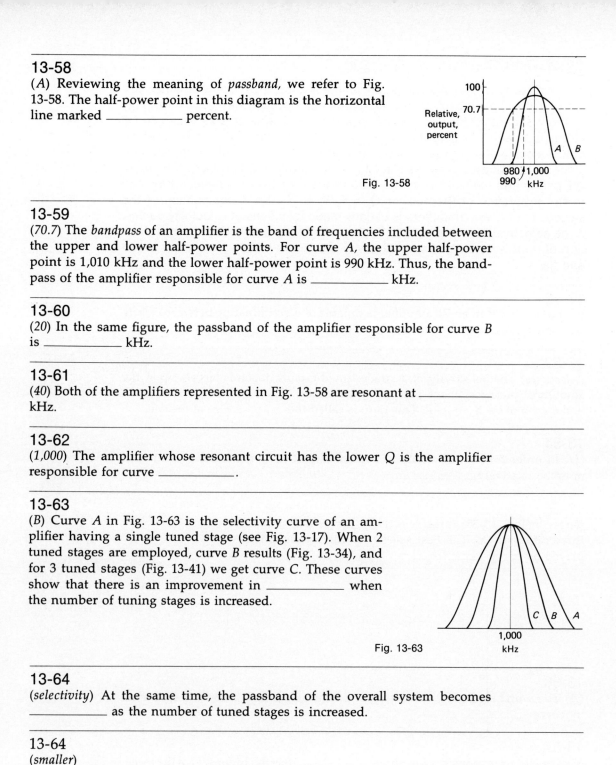

Fig. 13-58

13-59
(*70.7*) The *bandpass* of an amplifier is the band of frequencies included between the upper and lower half-power points. For curve *A*, the upper half-power point is 1,010 kHz and the lower half-power point is 990 kHz. Thus, the bandpass of the amplifier responsible for curve *A* is _____ kHz.

13-60
(*20*) In the same figure, the passband of the amplifier responsible for curve *B* is _____ kHz.

13-61
(*40*) Both of the amplifiers represented in Fig. 13-58 are resonant at _____ kHz.

13-62
(*1,000*) The amplifier whose resonant circuit has the lower *Q* is the amplifier responsible for curve _____.

13-63
(*B*) Curve *A* in Fig. 13-63 is the selectivity curve of an amplifier having a single tuned stage (see Fig. 13-17). When 2 tuned stages are employed, curve *B* results (Fig. 13-34), and for 3 tuned stages (Fig. 13-41) we get curve *C*. These curves show that there is an improvement in _____ when the number of tuning stages is increased.

Fig. 13-63

13-64
(*selectivity*) At the same time, the passband of the overall system becomes _____ as the number of tuned stages is increased.

13-64
(*smaller*)

BANDPASS CHARACTERISTICS

In this section of our work we shall see that the highest achievable selectivity is not always the most desirable condition. In radio and television reception, as well as in many other phases of electronics, we must be able to adjust the bandpass of an amplifier to meet the conditions. There are situations in which very narrow bandpass characteristics are needed, while in other circumstances the amplifier must be designed with wide bandpass characteristics.

13-65
On the basis of information presented thus far, it would seem that the passband of an RF amplifier in a radio receiver should be as _____ as possible in order that the amplifier be able to discriminate between stations that are close to each other in frequency.

13-66
(*small* or *narrow*) This is quite true for single-frequency transmission. For example, a continuous RF wave is shown in Fig. 13-66. This is a single-frequency transmission. Therefore, to avoid interference between this signal and closely adjacent ones, the passband of the receiver should be as _____ as possible.

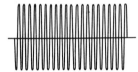

Fig. 13-66

13-67
(*narrow*) The continuous wave of Fig. 13-66 cannot convey information of any kind since it is not varying. It can be made to convey information, however, if it is broken up into a series of discontinuous emissions (Fig. 13-67) forming a code pattern of dots and dashes. This is still a _____-frequency transmission, however.

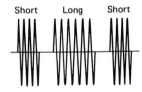

Fig. 13-67

13-68
(*single*) Thus, the same rule applies to this emission as to the continuous wave. To separate the coded emission from other closely adjacent frequencies, the _____ of the receiver should be as narrow as possible.

13-69
(*passband*) In Fig. 13-69, 2 waveforms are shown. One of these is an audio-frequency wave of, say, 2,000 Hz, and the other is an RF wave of, say, 1,000 kHz. Purely on the basis of diagrammatic representation, you can tell that wave *A* is the _____-frequency wave.

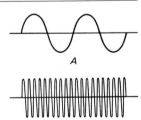

Fig. 13-69

RADIO-FREQUENCY AMPLIFIERS

13-70
(*audio*) Since the frequency of wave B is so much greater than the frequency of wave A in Fig. 13-69, it is clear that wave B is intended to represent the _____-frequency wave.

13-71
(*radio*) The process of *modulation* consists of superimposing the audio wave upon the radio wave. The resulting wave pattern is given in Fig. 13-71. In this wave, the *envelope* of the RF wave is a line connecting all the peak voltages of the _____-frequency wave.

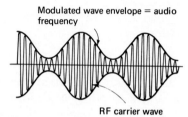

Fig. 13-71

13-72
(*radio*) The frequency of variation of the *envelope* is the same as the _____ frequency that was superimposed on the RF wave.

13-73
(*audio*) In the modulated wave, the radio-frequency portion merely transports the audio in a kind of piggyback system. For this reason, the RF portion is called the RF _____ wave (see Fig. 13-71).

13-74
(*carrier*) When the modulated wave is analyzed, it is found that its waveform consists of 3 different frequencies (Fig. 13-74). These are the carrier frequency (assumed 1,000 kHz); an *upper side frequency*, abbreviated USF; and a *lower side frequency*, abbreviated _____.

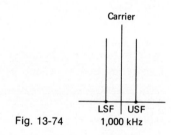

Fig. 13-74

13-75
(*LSF*) If the modulating or audio frequency is 2,000 Hz (*or 2 kHz*), then it is found that the LSF is 2 kHz lower than the carrier in frequency, and the USF is _____ kHz higher than the carrier in frequency.

13-76
(*2*) Thus, the frequency of the LSF is $1,000 - 2 = 998$ kHz. By the same reasoning, the frequency of the USF is _____ kHz.

13-77
(*1,002*) Hence, when an RF carrier of 1,000 kHz is modulated by a pure audio sine-wave tone of 2,000 Hz, the emission consists of a total of _____ different radio-frequency waves.

13-78
(*3*) For faithful reproduction of this modulated wave (or the modulation contained therein), the RF amplifier in the receiver should be capable of passing and amplifying the LSF of 998 kHz, the carrier of 1,000 kHz, and the USF of _____ kHz.

13-79
(*1,002*) Should the tuned circuits in the receiver be very sharply selective, then when the set is tuned to the carrier of 1,000 kHz, it is possible that either or both of the _____ frequencies will not be passed.

13-80
(*side*) Since the side frequencies contain the audio information, an RF amplifier with too much selectivity to pass the side frequencies will tend to distort or remove the _____ portion of the modulated wave.

13-81
(*audio*) For the modulated wave under discussion, the *bandwidth* required for faithful reproduction of the audio component extends from 998 kHz at the LSF up to _____ kHz at the USF.

13-82
(*1,002*) Thus, the passband of the amplifier must be at least _____ kHz wide.

13-83
(*4*) When speech or music is used as the modulating source, every frequency present in the speech or music produces its own side frequency. As shown in Fig. 13-83, many side frequencies are present, forming *sidebands*. The lower sideband is labeled LSB, and the upper sideband is labeled _____.

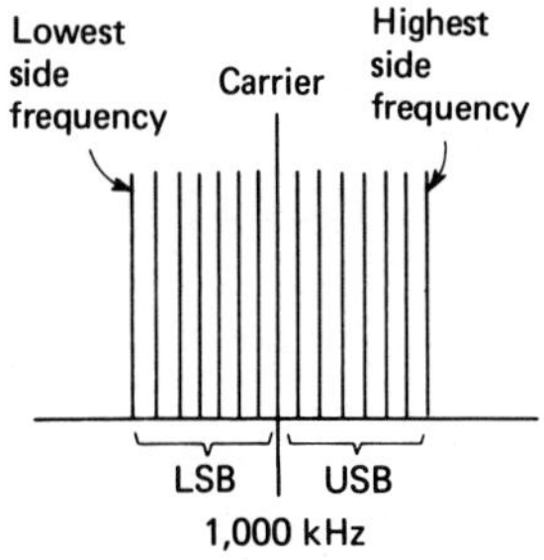

Fig. 13-83

RADIO-FREQUENCY AMPLIFIERS 367

13-84
(*USB*) The limit of each sideband is governed by the highest audio frequency present in the modulation. For example, if the highest modulating frequency is 5,000 Hz, then the LSB ends at 1,000 − 5 or 995 kHz. Similarly, the USB ends at 1,000 + 5 or _____ kHz.

13-85
(*1,005*) For faithful reproduction, the RF amplifier in the receiver tuned to the modulated wave just described would have to have a passband extending from 995 to 1,005 kHz. The passband, therefore, would have to be _____ kHz wide.

13-86
(*10*) Some high fidelity stations use modulating frequencies as high as 10,000 Hz. To receive this signal without distortion or loss of audio frequencies, the RF amplifier would have to have a passband of _____ kHz.

13-87
(*20*) Thus, to determine the passband required of any receiving system for modulated waves, we take the highest anticipated modulating frequency (audio), expressed in kHz rather than Hz, then multiply it by _____.

13-88
(*2*) For example, a 6,600-kHz carrier is modulated by speech in which the highest audio frequency is 4,500 Hz. The passband of the system to receive this should be _____ kHz.

13-89
(*9*) Figure 13-89 illustrates 3 curves for 3 different RF amplifier systems. All 3 amplifiers are resonant to a carrier frequency of _____ kHz.

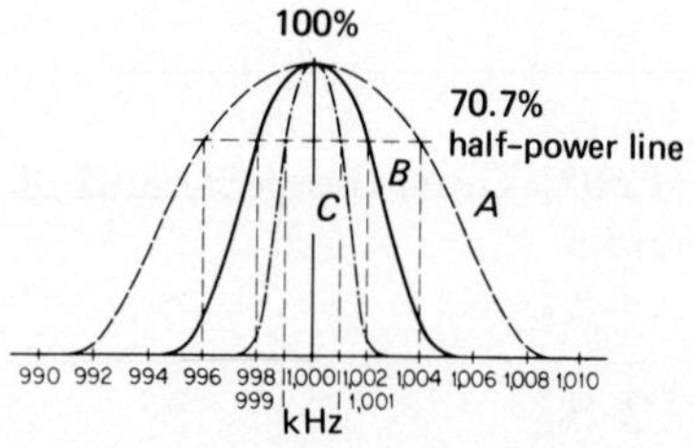

Fig. 13-89

13-90
(*1,000*) The amplifier having the highest selectivity of the 3 is responsible for curve _____.

13-91
(*C*) The passband of the amplifier that produces curve *C* is approximately _____ kHz.

13-92
(2) This amplifier probably consists of several high-Q tuned stages. The amplifier having the widest passband is the one responsible for curve _____.

13-93
(A) The passband of this amplifier is _____ kHz.

13-94
(8) We want the *highest possible selectivity* in an RF amplifier, yet its passband must be sufficiently wide to pass both the LSB and _____ of the received modulated wave.

13-95
(*USB*) Consider a certain marine radiotelephone station that has a carrier frequency of 6,000 kHz and is modulated with frequencies extending from 100 to 4,000 Hz. The bandwidth of the emission from this station is _____ kHz.

13-96
(8) The passband required of an RF amplifier in a radio to receive this station without distortion must be, therefore, _____ _____.

13-97
(*8 kHz*) Although the amplifier that produces curve C in Fig. 13-89 is the most selective of the 3, its passband is only 2 kHz wide. Hence, this amplifier would be _____ (satisfactory, unsatisfactory) for use in a radio intended to receive this station.

13-98
(*unsatisfactory*) It would be unsatisfactory, despite the fact that its selectivity is excellent, because its passband is too _____.

13-99
(*narrow*) Referring now to the amplifier of curve B, we find that its selectivity is not as good as that of the amplifier of curve C, but its passband is _____ kHz.

13-100
(4) This is a wider passband than that of curve C. This amplifier would be _____ (satisfactory, unsatisfactory) in a radio intended to receive the station in question.

13-101
(*unsatisfactory*) This leaves only amplifier A (curve A). Since the passband of this amplifier is 8 kHz, it would be _____ for use in a radio to receive the marine station described.

13-102
(*satisfactory*) Thus we see that as desirable as selectivity may be, we must also consider the _____ of the amplifier as related to the modulated wave it is to amplify.

13-103
(*passband*) Response curves are not always round-topped or even in appearance. In Fig. 13-103, curve B illustrates the kind of response with which we have been dealing. The passband of this amplifier is _____ kHz.

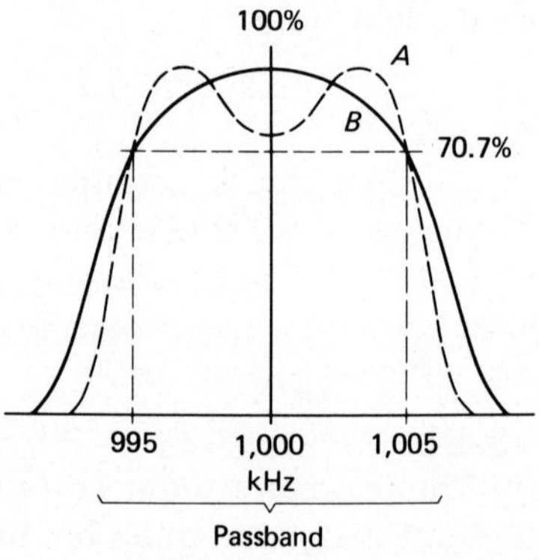

Fig. 13-103

13-104
(*10*) Curve A is a double-humped curve that is characteristic of certain types of double-tuned circuits. Since the dip in its center does not drop below the half-power line, the passband of this amplifier is _____ kHz.

13-105
(*10*) Note that curve A has steeper sides than curve B. This means that this amplifier will not respond as well to frequencies that lie just outside the upper and lower limits of its passband. Thus, the amplifier of curve A is _____ selective than that of curve B.

13-106
(*more*) Thus, despite the peculiar dip in the curve of A, it has better selectivity and a passband as good as that of amplifier B. In selecting an amplifier response for use in the RF section of a receiver, given the choice of A and B in Fig. 13-103, the better choice would be _____.

13-106
(*A*)

370 A PROGRAMMED COURSE IN BASIC ELECTRONICS

BANDWIDTH REGULATION

The bandwidth (or passband) of an amplifier may be controlled in several different ways. We shall discuss two of the more popular methods. The first of these involves the use of loading resistors which serve to increase the bandwidth of the amplifier. The second method makes use of the change in the resonance curve for double-tuned stages when the coupling between primary and secondary windings of the RF transformer is varied.

13-107
In any tuned stage, an increase in passband or bandwidth is inevitably accompanied by a loss of gain. It is necessary to compromise. First make certain that the amplifier passband is sufficiently wide, then attempt to raise the _____ to as high a figure as possible without causing the passband to narrow.

13-108
(*gain*) Several methods are in common use for increasing bandwidth. The first of these is the method shown in Fig. 13-108. Connected across each winding of a double-tuned coupling transformer we find a _____.

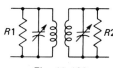

Fig. 13-108

13-109
(*resistor*) A resistor connected across a tuned circuit has the effect of lowering the Q of this circuit. When the Q is reduced, the output at resonance must _____ (increase, decrease).

13-110
(*decrease*) In Fig. 13-110, curve A represents a sharply resonant response curve. Curve B has a more gradual resonant response. One of these two curves results from the circuit in Fig. 13-108 without the loading resistors $R1$ and $R2$; the other curve is obtained with the resistors connected. Obviously, curve _____ is the one which shows the response without the resistors.

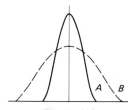

Fig. 13-110

13-111
(*A*) As shown by curve B, the use of loading resistors not only decreases the resonant output or gain, but also _____ the passband or bandwidth of the amplifier with which it is associated.

13-112
(*increases* or *widens*) As R1 and R2 in Fig. 13-108 are made smaller in resistance, the gain decreases and the bandwidth widens. Theoretically, we could increase the bandwidth almost indefinitely by reducing the resistances. The trouble with this procedure would be, however, that the _____ would become too low.

13-113
(*output* or *gain*) Thus, resistors R1 and R2 in Fig. 13-108 are selected so that they load the tuned circuits sufficiently to obtain a somewhat _____ passband but not so much that the gain is reduced to too low a value.

13-114
(*wider*) A *second* method of obtaining a greater bandwidth in an amplifier coupled by a double-tuned transformer is to change the *degree of coupling* between the primary and secondary coils. Coupling is discussed in terms of tight and loose coupling. If the 2 coils are very close to each other, the coupling is tight; if far apart, the coupling is _____.

13-115
(*loose*) In Fig. 13-115, 3 response curves—each a result of a different degree of coupling—are illustrated. Curve B is the kind of response obtained with loose coupling. As is evident from the figure, the gain is quite low and the bandwidth quite _____ for curve B.

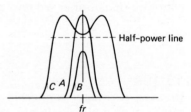

Fig. 13-115

13-116
(*narrow*) Curve C is the response obtained by tightly coupling the two coils. Although the curve has a dip in the center, the dip does not descend to the half-power line; hence both the gain and _____ are quite good.

13-117
(*bandwidth*) The coupling responsible for curve A is known as *critical coupling*. Critical coupling gives the highest possible gain for the circuit but produces a rather narrow _____.

13-118
(*bandwidth*) The coupling between two coils is increased if we bring them closer to one another; the coupling between the same two coils is _____ if we move them farther apart.

13-119
(*decreased*) In review, the degree of coupling between two coils is given by the coefficient of coupling, k, of the system. The equation for mutual inductance is $M = k\sqrt{L_p \times L_s}$, in which M is the mutual inductance, L_p and L_s are the primary and secondary inductances, respectively, and k is the _____ of _____.

13-120
(*coefficient, coupling*) Solving the equation $M = k\sqrt{L_p \times L_s}$ for k, we divide both sides by $\sqrt{L_p \times L_s}$. This yields $k = M/$_____.

13-121
($\sqrt{L_p \times L_s}$) The maximum value that M can assume under perfect coupling conditions is equal to $\sqrt{L_p L_s}$. Thus, if every line of force in the primary links with the secondary, the coupling will be perfect. For this condition (since $M = \sqrt{L_p L_s}$), the coefficient of coupling, as given in Frame 13-120, must be _____.

13-122
(*1*) Radio-frequency coupled coils are never coupled this perfectly. For example, k for curve A in Fig. 13-115 might be 0.02 and would therefore be the coefficient of critical coupling in this case. Thus, k for curve C would be _____ than 0.02.

13-123
(*greater*) As a matter of fact, the value of k for curve C would probably never exceed 0.5. Hence, when two coils have a coefficient of _____ of 0.5, they are said to be tightly coupled.

13-124
(*coupling*) By similar reasoning, two coils are said to be _____ coupled if the coefficient of coupling is 0.005.

13-125
(*loosely*) Consider another pair of coupled RF coils. If critical coupling is obtained when $k = 0.034$, then when $k = 0.1$, the coils will be _____ coupled.

13-126
(*tightly*) If the coefficient of coupling is 0.1 for these coils, a single sharp peak will no longer be obtained. Instead, the response curve will display _____ peaks.

13-127
(2) When k is reduced to a value of 0.005, a single sharp peak is obtained, but the gain of the circuit is greatly diminished. This condition indicates that the coils are _____ coupled.

13-128
(*loosely*) In Fig. 13-128, curve A is the curve obtained for a certain pair of coils when they are critically coupled. Judging by the width and gain illustrated by curves B and C, both of these curves represent a condition of _____ coupling.

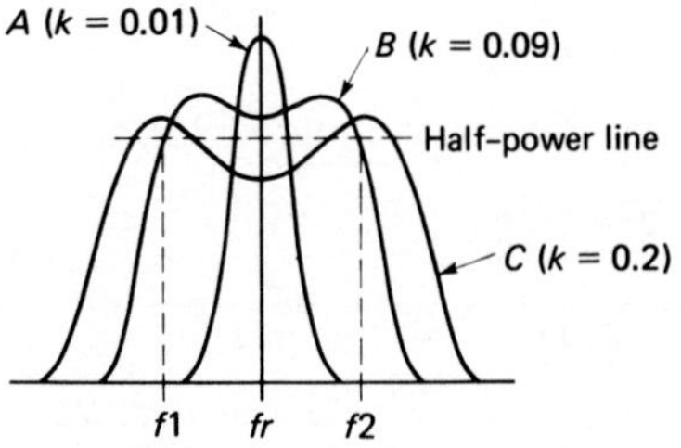

Fig. 13-128

13-129
(*tight*) Since curve B does not descend below the half-power line anywhere, it may be said to be flat in the range that extends from frequency $f1$ to frequency _____.

13-130
(*f2*) Thus, when the coefficient of coupling for these coils has a value of _____, they are coupled tightly enough for good bandwidth and gain.

13-131
(*0.09*) When the coefficient of coupling is _____, then the coils are said to be critically coupled.

13-132
(*0.01*) When k is increased beyond the value 0.09, the coils are said to be *overcoupled*. In this case, there are 2 distinct peaks as before, but this time the curve does dip below the _____-_____ line.

13-133
(*half-power*) The coefficient of coupling for the overcoupled condition in Fig. 13-128 is given as _____.

13-134
(*0.2*) Thus, when the coefficient of coupling is 0.2, the coupling is so tight as to render the amplifier unsatisfactory. Such an amplifier would show _____ distinct points of resonance with a low-gain gap in between them.

374 A PROGRAMMED COURSE IN BASIC ELECTRONICS

13-135
(2) In radio and TV circuits, the coupling of 2 coils is usually adjusted by moving an iron core in or out of the coils by means of an insulated screwdriver. If the core is moved into the coils, k is _____; if the core is backed out, k is _____.

13-135
(*increased, decreased*)

CRITERION CHECK TEST

_____13-1 A frequency of 300 GHz is equivalent to (*a*) 300,000 Hz, (*b*) 3,000,000 Hz, (*c*) 300,000,000 Hz, (*d*) 300,000,000,000 Hz.

_____13-2 FM broadcast stations transmit between 88 and 108 MHz. These frequencies fall into the (*a*) HF band, (*b*) VHF band, (*c*) UHF band, (*d*) SHF band.

_____13-3 The transformer in Fig. 13-17 is used to match the (*a*) high impedance tuned circuit to the low impedance of the base-emitter circuit, (*b*) high impedance of the tuned circuit to the high impedance of the base-emitter circuit, (*c*) low impedance of the tuned circuit to the low impedance of the base-emitter circuit, (*d*) low impedance of the tuned circuit to the high impedance of the base-emitter circuit.

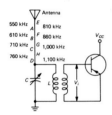

Fig. 13-17

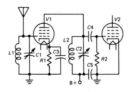

Fig. 13-34

Fig. 13-108

_____13-4 In Fig. 13-34, the RF grid signal voltage of V1 is developed by the components (*a*) L2C2, (*b*) L1C1, (*c*) R1C3, (*d*) R2C4.

_____13-5 RF amplifiers in receivers are usually biased in (*a*) class A, (*b*) class B, (*c*) class C, (*d*) class BC.

_____13-6 For *best selectivity* an RF amplifier tuned circuit should have maximum (*a*) bandpass, (*b*) resistance, (*c*) coupling, (*d*) Q.

_____13-7 In an AM wave (*a*) the magnitude of the carrier varies with the AF signal, (*b*) the RF and AF are sent out in alternate spurts, (*c*) RF is superimposed on the AF signal, (*d*) the RF signal must be an even multiple of the AF signal.

_____13-8 In Fig. 13-108, the bandwidth is increased at the expense of (*a*) coupling efficiency, (*b*) Q, (*c*) capacitance, (*d*) inductance.

_____13-9 The coefficient of coupling of 2 coils is determined by the (*a*) frequency of the circuit, (*b*) resistance of the circuit, (*c*) physical location of the coils relative to each other, (*d*) passband.

14 oscillator principles

OBJECTIVES

(1) Describe and graph the motion of a mass in an undamped spring-mass system. (2) Sketch and describe the motion of a damped spring-mass system. (3) Contrast the motion of an undamped pendulum with the motion of a damped pendulum. (4) Relate positive feedback to sustaining oscillation in a pendulum system. (5) Relate negative feedback to damped oscillation in a pendulum system. (6) Define *in-phase* and *out-of-phase* feedback. (7) Trace a complete cycle of oscillation in a simple *LC* circuit. (8) Sketch an *LC* circuit showing an initially charged capacitor. (9) Depict and discuss the voltage waveform when the capacitor discharges through the inductor. (10) Demonstrate that Lenz's law causes the coil to recharge the condenser. (11) Depict the magnitude and polarity of the voltage as the coil recharges the capacitor and as the capacitor again discharges through the coil. (12) Demonstrate that the oscillation is complete when the coil again recharges the condenser. (13) Compare the energy flow in the *LC* oscillator circuit with the energy flow in the spring-mass system. (14) Sketch a generalized transistor oscillator circuit. (15) Locate, label, and state the function of the tank circuit, amplifying device, and feedback circuit. (16) State the relative phase for the input and output voltages of a CE transistor amplifier. (17) State the requirement on the phase change of a feedback network for a CE oscillator. (18) Sketch and discuss a tickler coil oscillator. (19) Trace the inductive collector-to-base feedback of this oscillator. (20) Demonstrate that the feedback network of this oscillator can produce a 180° phase rotation. (21) List the 4 requirements which must be met by all electronic oscillators.

OSCILLATION IN MECHANICAL SYSTEMS

Mechanical oscillation is a vibratory process in which a mass moves with *periodic* motion. The swing of the pendulum is an example of mechanical oscillation, as is the vibratory motion of a mass on the end of a long spiral spring. An understanding of the terminology and concepts of mechanical oscillation is of unquestionable assistance when transferred to electric oscillation. We shall begin, therefore, with a study of mechanical oscillation.

14-1
Suppose that a weight were hung on a spiral spring as in Fig. 14-1. The weight would cause the spring to stretch as shown in the drawing. The amount of actual stretching that occurs would depend upon the stiffness of the spring and the size of the _____.

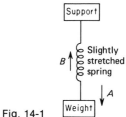

Fig. 14-1

14-2
(*weight*) Assume that the spring and weight are in equilibrium where there are no unbalanced forces. In this case, the downward pull of the weight would be exactly counteracted by the upward pull of the _____.

14-3
(*spring*) Now imagine that the weight is doubled in size. When this is done, we can properly expect that the weight would move in the direction of the arrow labeled _____ (Fig. 14-1).

14-4
(*A*) This situation is shown in Fig. 14-4. Here the spring has stretched so that the restoring force it now exerts again brings the system into equilibrium with the upward force of the spring being _____ to the downward force of the double weight.

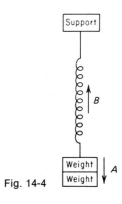

Fig. 14-4

14-5
(*equal*) Now let us remove 1 of the 2 weights suddenly. As this is done, the remaining weight would at once begin to move in a(n) _____ direction.

14-6
(*upward*) As the weight moves upward, it must accelerate since the unbalanced force of the spring is acting on the mass of the weight. By the time it reaches its initial position as in Fig. 14-1, it may be moving with considerable velocity. It does not stop here but continues to move in a(n) _____ direction.

OSCILLATOR PRINCIPLES 377

14-7
(*upward*) In reference to Fig. 14-7, the initial position of the weight was C and the stretched position was A. The spring was stretched a distance AC. When the spring was stretched to A, a certain *potential energy* was added to the system. According to the law of conservation of energy, when the weight returns to C, it will have kinetic _____ of motion.

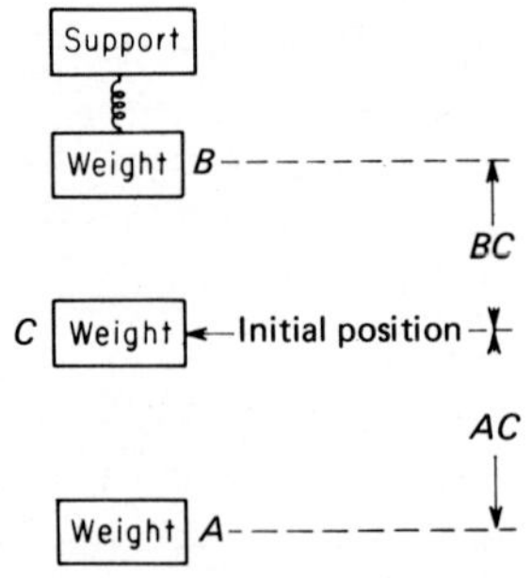

Fig. 14-7

14-8
(*energy*) Again according to the same law, the amount of kinetic energy will be exactly equal to the amount of _____ energy put into the system by stretching action. (We are assuming no friction or air resistance throughout this discussion until it is specifically introduced.)

14-9
(*potential*) Since the weight, on returning to C, has kinetic energy of motion, it continues to move upward until all this kinetic energy has been changed into _____ energy of its new raised position.

14-10
(*potential*) When all the kinetic energy possessed by the weight at C has been changed to potential energy at B, the weight comes to rest for an instant. A fraction of a second later, however, it begins to move in a(n) _____ direction again.

14-11
(*downward*) As it moves downward under the action of gravity, it again accelerates, picking up kinetic energy. When it reaches point _____, all the potential energy it had at B will have been converted back into kinetic energy.

14-12
(C) This kinetic energy keeps it moving in a(n) _____ direction past point C.

14-13
(*downward*) When the weight reaches point _____, all the kinetic energy at C will have been converted back into potential energy.

14-14
(*A*) We are now back at the beginning, where the weight will again move upward and execute the same motion as before. If friction and air resistance are both zero, the distance *AC* will always be equal to the distance _____ as the up-and-down motion repeats itself.

14-15
(*BC*) If no energy were lost to friction or air resistance, the vibration would continue forever and could be called *undamped oscillation*. Of these two words, _____ refers to the vibratory or back-and-forth motion.

14-16
(*oscillation*) And, of the same two words, _____ refers to the fact that the vibratory motion repeats itself without loss, each traversal upward and downward being as large as the previous one with respect to distance covered.

14-17
(*undamped*) If we graph the oscillation with respect to time, we obtain a curve such as that in Fig. 14-17. This is a sine curve and represents undamped oscillation because all the _____, both bottom and top, are the same size as the preceding ones.

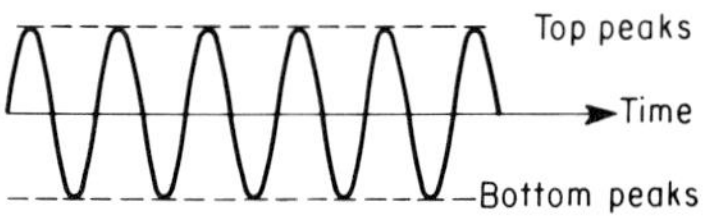

Fig. 14-17

14-18
(*peaks*) Now let us introduce some friction and air resistance. Starting again at the stretched position (*A* in Fig. 14-7), we release the weight. Again the weight will overshoot *C* and continue upward toward *B*, but because of the loss of energy to friction, the upward distance covered will be a little _____ than distance *AC*.

14-19
(*less* or *smaller*) This means that the potential energy at the new top position will be a little _____ than it was before friction entered the situation.

14-20
(*less*) Thus, the kinetic energy of return at point *C* will be slightly _____ than it was in the previous case where friction was zero.

14-21
(*smaller* or *less*) Energy continues to be lost during the downward motion. Because the weight started with less kinetic energy at position *C* and because more energy is lost to friction, the new bottom distance covered (formerly *AC*) will be measurably _____ than *AC*.

OSCILLATOR PRINCIPLES 379

14-22
(*smaller*) The new bottom potential energy of the stretched spring will, therefore, be _____ than it was when the position was actually *AC*.

14-23
(*smaller*) Thus, the weight again starts upward from a position of measurably less potential energy than position *A*. Energy continues to be lost in each sweep upward and downward. Therefore, each successive sweep becomes smaller until the motion of the weight _____ altogether.

14-24
(*stops*) Such a vibratory series constitutes a *damped oscillation*. The word *damped* refers to the fact that each successive vibration contains less _____ and potential energy than the one before it.

14-25
(*kinetic*) The waveform shown in Fig. 14-25 represents a damped vibration plotted against time. This waveform is characteristic of a damped vibration because each successive peak has a _____ amplitude than the one before it.

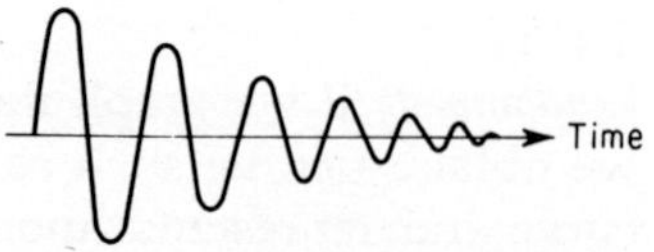

Fig. 14-25

14-26
(*smaller*) A pendulum swinging on a frictionless bearing in a perfect vacuum may be considered as an undamped *oscillatory system*. For example, in Fig. 14-26, the pendulum is hanging motionless and is in a condition of _____ because all forces acting on it are balanced.

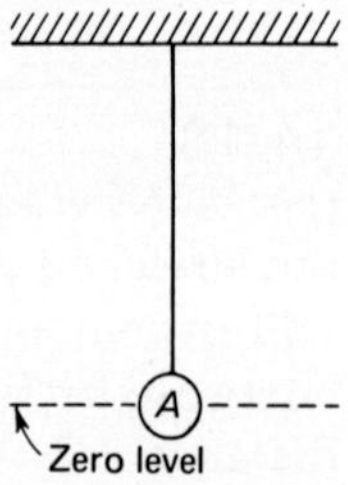

Fig. 14-26

14-27
(*equilibrium*) At the instant indicated in Fig. 14-26, the pendulum bob has neither _____ nor kinetic energy with reference to the zero level indicated.

14-28
(*potential*) In Fig. 14-28, the bob has been drawn to one side, thereby raising it slightly above the zero level. In this condition it has more _____ energy than it had before it was drawn to one side.

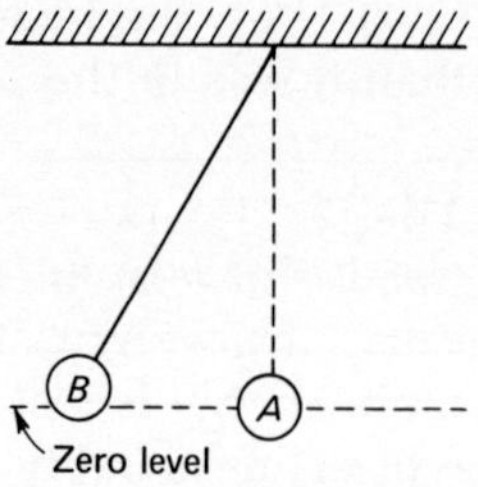

Fig. 14-28

14-29
(*potential*) The bob is now released from position B and is allowed to swing back toward A. It does not stop at A, however, because the potential energy given to it at B has changed to _____ energy. This energy explains why the bob continues to move.

14-30
(*kinetic*) The bob therefore overshoots A and goes on to C. At point C, all the kinetic energy the bob had at A has been converted to _____ energy of raised position (Fig. 14-30).

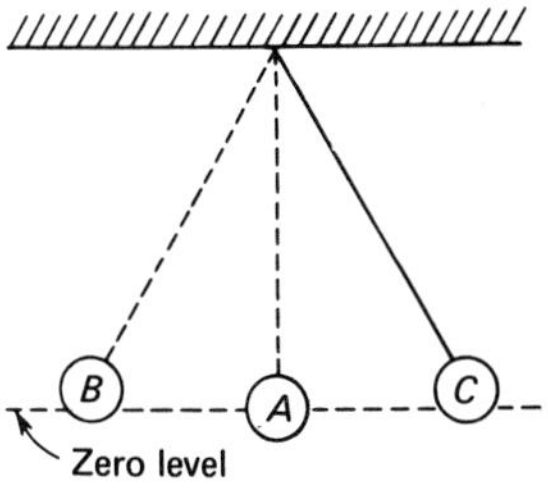

Fig. 14-30

14-31
(*potential*) In order for the conditions to be frictionless, the distance from A to C must be _____ to the distance from B to A.

14-32
(*equal*) If the bearing had friction and if air resistance were allowed to act on the bob, then the distance from A to C would be _____ than the distance from B to A.

14-33
(*smaller* or *less*) Under these conditions the bob would soon come to rest. For a system in which friction and air resistance exist, a pendulum would be used to illustrate _____ oscillation.

14-34
(*damped*) In all real situations, friction and air resistance are both present. Thus, on every swing in a real pendulum system, some _____ is lost to friction and air resistance.

14-34
(*energy*)

OSCILLATOR PRINCIPLES **381**

FEEDBACK IN AN OSCILLATING SYSTEM

14-35
Mechanical energy may be measured in *joules* (abbrev. J). Suppose we add 100 J to a system in equilibrium (as in Fig. 14-26), thus converting it to an energetic system (as in Fig. 14-28). Assume that 1 J is lost to friction and air resistance during each single swing (from *B* to *C* in Fig. 14-30). Then when the bob arrives at *C*, its potential energy would be _____ J.

14-36
(*99*) And when the bob returns from *C* to *B*, its potential energy would now have gone down to _____ J.

14-37
(*98*) If, however, when the bob gets to *C* in the first swing, we push it toward *B* with sufficient force to impart 1 J additional to it, it would arrive back at *B* with an energy content of _____ J.

14-38
(*99*) Then, if a joule is added to the bob by pushing it when it arrives at *B*, this time toward *C*, it would return to *C* with an energy content of _____ J.

14-39
(*99*) In other words, if we add sufficient energy to the bob at the end of each swing to make up for the loss due to friction during the previous swing, the oscillation will change from the damped type to the _____ type.

14-40
(*undamped*) This addition of energy by a small extra "push" is the function of the spring mechanism (or weight mechanism) of a pendulum clock. In such a clock, the pendulum's oscillation is the _____ variety.

14-41
(*undamped*) Undamped oscillations are often called *sustained oscillations*. In a sustained oscillatory system, each successive peak has the _____ magnitude as each previous peak.

14-42
(*same*) When just enough energy is added during each cycle of the oscillation to sustain it, we may refer to the added energy as *positive feedback*. When energy is fed back at the *right time* and of the *right magnitude* to sustain oscillation, the wave that we would get by plotting this oscillation against time would be an _____ sine wave.

14-43

(*undamped*) Imagine the bob of Fig. 14-30 to be swinging from *B* to *A* and to be just reaching *A* when energy is added to it *in the direction of B*. This energy would be added at the *wrong* time and would act to _____ the amplitude of the next swing.

14-44

(*reduce*) Addition of energy at the wrong time is often called *negative feedback*. In the case of an oscillatory system, negative feedback produces more heavily damped oscillation, thereby causing the system to come to _____ sooner than it would have if no energy had been added.

14-45

(*rest*) Instead of using the terms "right" and "wrong" *time*, it is conventional to talk of *in-phase* and *out-of-phase* feedback in oscillatory systems. Thus, when feedback energy is added to a pendulum bob at the right time to cause sustained oscillation, the feedback is said to be _____ phase.

14-46

(*in*) On the other hand, negative feedback is considered to be _____-_____-phase feedback.

14-47

(*out-of*) Thus, to sustain the oscillation of a pendulum the feedback must be positive in nature and of sufficient magnitude to make up for all _____ due to friction and air resistance.

14-47

(*losses*)

OSCILLATION IN AN *LC* ELECTRIC CIRCUIT

With the understanding obtained by analyzing mechanical oscillation, we are now prepared to transfer the terminology and concepts to *electric oscillation*. The similarity between the two kinds of oscillation is not accidental. Both actions follow closely related laws of nature and are quite comparable in all their aspects.

OSCILLATOR PRINCIPLES **383**

14-48
Electric oscillation is analogous to mechanical oscillation. A simple circuit is given in Fig. 14-48. In this circuit, we see a battery, a coil, a 2-position switch, a _____, and the connecting wires.

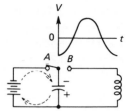

Fig. 14-48

14-49
(*capacitor*) In this circuit, Fig. 14-48, the switch is shown in position *A*. For this position, the battery is connected across the capacitor; the dotted arrow shows a surge of current, and the dotted "+" and "−" signs show the charge built up on the capacitor. Since position *B* is open, the _____ is not in the circuit.

14-50
(*coil*) The black waveshape above the circuit indicates the kind of voltage the circuit *will* produce shortly. We will show the growing voltage as a dotted line on the wave. Since no _____ line is yet visible on this waveshape, we can assume that no growing voltage has yet been generated.

14-51
(*dotted*) The capacitor is now charged to the potential of the battery. In Fig. 14-51, the switch position has been altered. The movable portion of the switch (switch arm) is now in contact with the terminal at position _____.

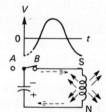

Fig. 14-51

14-52
(*B*) The voltage on the capacitor in Fig. 14-51 constitutes an emf that can drive a current through the coil. The direction of this current, from minus to plus, is shown by the _____ _____.

14-53
(*dotted arrows*) As the current flows through the coil, a _____ field must be built up around the coil.

14-54
(*magnetic*) The fact that this magnetic field is *building up* is shown by the _____ pointing outward from the coil.

14-55
(*arrows*) The dotted portion of the waveshape in Fig. 14-51 shows the voltage across the capacitor at that instant. This voltage started out as the full battery potential to which the capacitor was charged. Since the top plate is negative, we will show the dotted wave starting below the _____ axis as in the curve in Fig. 14-51.

14-56
(*zero* or *horizontal*) At the start of the capacitor discharge process—that is, *just as the switch moves to B*—we have the condition in which the capacitor still has its full charge. The full-charge voltage, as shown on the curve above, is represented as the point where the dotted wave intercepts the _____ axis.

14-57
(*vertical* or *voltage*) As soon as the switch is moved to position B in Fig. 14-51, the capacitor starts to discharge through the coil. Since the capacitor is no longer connected to the battery, the voltage across the capacitor must begin to _____.

14-58
(*decrease*) In Fig. 14-51, this decreasing voltage is shown by the dotted wave. It shows that the voltage is decreasing because the dotted wave is heading upward toward the _____ axis.

14-59
(*zero*) Now look at Fig. 14-59. In this drawing, the dotted wave has just reached the _____ axis.

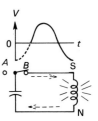

Fig. 14-59

14-60
(*zero*) The fact that the dotted wave has just reached the zero axis shows that, at this instant, the voltage across the _____ and coil is zero.

14-61
(*capacitor*) Note that in Fig. 14-59, the "+" and "−" signs on the capacitor have been omitted. This also shows that the system is at the point where the capacitor has fully discharged so that the voltage across it and the coil is _____.

OSCILLATOR PRINCIPLES 385

14-62

(*zero*) The lines around the coil have no arrows attached to them. This is intended to show that the _____ field *still exists* (having been built up before), but that it is neither building up nor collapsing *at this instant*.

14-63

(*magnetic*) Since the voltage across the capacitor is now zero, the current tends to stop flowing through the coil. Recall, however, from Lenz's law, that an inductance always tends to _____ the change of motion in a system.

14-64

(*oppose*) Thus, the coil opposes the *cessation* of current flow. Its magnetic field now begins to collapse (Fig. 14-64) as shown by the inward-pointing _____.

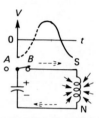

Fig. 14-64

14-65

(*arrows*) The collapsing field cuts through the coil turns, thereby inducing a new voltage across the coil which tends to _____ the cessation of the current.

14-66

(*oppose*) This causes the current to keep flowing in the same direction as it had been flowing all along. Electrons now enter the bottom plate of the capacitor, leaving the top plate and causing it to assume a _____ charge with respect to the bottom plate.

14-67

(*positive*) The reversal of charge polarity is shown by the "+" and "−" signs on the capacitor in Fig. 14-64. Since the top plate is becoming positive, we show the dotted wave as now appearing _____ the zero axis according to the convention adopted at the beginning of this section.

14-68

(*above*) The field continues to collapse until all the energy that had been stored in it has been converted to electric energy. The capacitor is now fully charged again, but with polarity opposite that of its initial charge. Current has stopped flowing. The absence of current is indicated in Fig. 14-68 by the absence of _____ on the circuit wires.

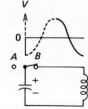

Fig. 14-68

14-69
(*arrows*) The fact that the capacitor has charged to its initial voltage but in reversed polarity is shown by the dotted waveshape in Fig. 14-68. The distance of the end of the dotted curve above the zero axis is _____ to the distance of the beginning of the dotted curve below the zero axis.

14-70
(*equal*) The system is not balanced, however, because the capacitor is fully charged in a closed circuit. Hence, the emf on the capacitor must now begin to drive a current back through the _____.

14-71
(*coil*) The direction of this current must be the _____ of its direction from the initial charge.

14-72
(*reverse*) As the capacitor discharges, the voltage across its terminals must again decrease. This is shown by the fact that the dotted wave (Fig. 14-72) is beginning to head back toward the _____ axis.

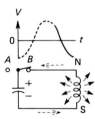

Fig. 14-72

14-73
(*zero*) The arrows around the coil indicate that a magnetic field is now again _____ in the coil.

14-74
(*growing* or *building up*) In Fig. 14-74, the capacitor voltage has again reached zero. This is indicated in the graph by the fact that the dotted curve has reached the _____ axis.

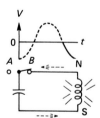

Fig. 14-74

14-75
(*zero*) Again the current tends to stop flowing. But the field around the coil now collapses and opposes the _____ of the current.

OSCILLATOR PRINCIPLES **387**

14-76

(*cessation*) The current, therefore, keeps flowing in the same direction, causing the capacitor to take on a charge (Fig. 14-76) with its top plate "−" and its bottom plate "+." This is the same polarity of charge that the capacitor had at the _____ of the cycle.

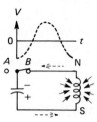

Fig. 14-76

14-77

(*start* or *beginning*) Since this polarity is the same as the polarity at the start of the cycle, the dotted curve is now shown _____ the zero axis and heading for a negative peak.

14-78

(*below*) The cycle is complete in Fig. 14-78. The capacitor is charged fully as it was initially, the magnetic field around the coil has completely collapsed, and the current has _____ _____.

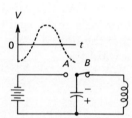

Fig. 14-78

14-79

(*stopped flowing*) In reference to damped and undamped oscillation (as discussed in mechanical oscillation), this is an example of a(n) _____ oscillation because the energy of the capacitor charge is just as large at the end of the cycle as it was at the beginning of the cycle.

14-80

(*undamped*) Comparing the electric oscillation to a mass vibrating on a spring, we see that the charged capacitor in Fig. 14-48 has stored up electric energy, just as energy was stored in the spring when it was first pulled down. Thus, the charged capacitor represents a condition of maximum _____ (potential, kinetic) energy.

14-81

(*potential*) Figure 14-59 shows that the potential energy of the capacitor is zero since the voltage on it is zero. This is analogous to the pendulum passing the midpoint of its full swing and is the condition where all the potential energy originally present has been converted to _____ energy.

14-82
(*kinetic*) Figure 14-68 indicates the condition where all the kinetic energy of the current flow has been changed back into potential energy, in the reverse direction. This is analogous to the pendulum having reached the _____ of its first half-vibration.

14-83
(*end* or *top*) In Fig. 14-74, the capacitor voltage is again zero, a condition of zero potential energy. All the potential energy now has taken the form of _____ energy of electron flow.

14-84
(*kinetic*) Thus, the situation in Fig. 14-74 is analogous to the pendulum having started back in its swing, reaching the _____ of its swing just at the instant when all its potential energy has been converted to kinetic energy.

14-85
(*midpoint*) Finally, initial conditions are restored in Fig. 14-78 where all the energy of the system has become _____ energy.

14-86
(*potential*) The fact that this electric oscillation has been assumed to restore full charge to the capacitor at the end of the first cycle is analogous to saying that the pendulum has reached the same _____ at the end of its first complete vibration as it had at the start.

14-87
(*amplitude* or *height*) To obtain undamped oscillation of a mass on a spring, or a pendulum on a string, it was necessary to assume that the magnitude of friction and air resistance was _____.

14-88
(*zero*) Similarly, to obtain undamped electric oscillations in a capacitor-coil combination, it is necessary to assume that the losses due to _____ are also zero.

14-88
(*resistance*)

OSCILLATOR PRINCIPLES

SUSTAINED ELECTRIC OSCILLATION

Sustained oscillations require energy feedback in the proper phase in electric oscillators. Now that we have analyzed the charge-discharge cycle in an *LC* circuit, we need to investigate how electric energy is added periodically, in phase, to overcome resistance losses and change the damped oscillation to undamped oscillation.

14-89
A mechanical system can never be freed entirely from friction. To cause a pendulum or mass on a spring to move with undamped oscillations, it is necessary to add _____ periodically to the system.

14-90
(*energy*) Not only must the added energy have the correct magnitude to sustain undamped oscillation, but it must also be added in the right phase. When energy is added in the right phase, it sets up a condition called _____ feedback.

14-91
(*positive*) In the pendulum, positive feedback was added through the use of a wound spring or unbalanced weights. In an electric oscillator, the in-phase feedback is produced by coupling back a portion of amplified power. The need for amplification at once suggests the use of a _____.

14-92
(*transistor*) The general feedback arrangement used for a transistor oscillator circuit is shown in Fig. 14-92. As this figure indicates, the coil-capacitor combination that produces the oscillatory waveform is generally known as a _____ circuit.

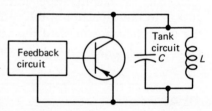

Fig. 14-92

14-93
(*tank*) Oscillators of the tank-circuit type almost invariably contain three distinct sections: the amplifier device, some feedback device or method, and the _____ circuit which usually appears as a part of the collector circuit of the transistor.

14-94
(*tank*) All electronic oscillations depend upon a random charge appearing on the capacitor through its connection with the transistor. Whatever voltage does appear on the capacitor is transmitted through the feedback device to appear as a voltage applied between the _____ and emitter of the transistor.

14-95
(*base*) The transistor then amplifies this feedback energy. A large part of the amplified energy is used to recharge the capacitor in the tank circuit to make up for resistance losses; a small part—whatever is needed—is fed back to the base of the transistor by means of the _____ device.

14-96
(*feedback*) Thus, the tank circuit produces sustained oscillation, provided enough energy is supplied to it by the amplified output and provided this energy reaches it in the right _____.

14-97
(*phase*) To have positive or in-phase feedback, the feedback voltage must have a phase relationship with the tank voltage that is either 0 or 360°. This is so because _____° represents exactly the same phase angle as 0°.

14-98
(*360*) You will recall that the output voltage of a transistor in the CE configuration is 180° out of phase with the input voltage. Thus, any attempt to feed the output back directly would result in _____ rather than positive feedback.

14-99
(*negative*) Negative feedback is out-of-phase feedback. Such feedback cannot produce _____ oscillation.

14-100
(*sustained* or *undamped*) Hence, the feedback device must be designed so that it will rotate the phase of the output by another 180°. When it does this, the feedback voltage will then have a phase relationship of _____° with the tank voltage.

14-101
(*360*) With the feedback voltage from the tank circuit reaching the base at an angle of 360°, the feedback is then positive. If the feedback voltage is sufficiently large to make up for all resistance losses, _____ oscillation will occur.

OSCILLATOR PRINCIPLES **391**

14-102

(*sustained*) Figure 14-102 illustrates the beginnings of a practical oscillator circuit. In this circuit, a second coil *L2* has replaced the _____ device in the diagram of Fig. 14-92.

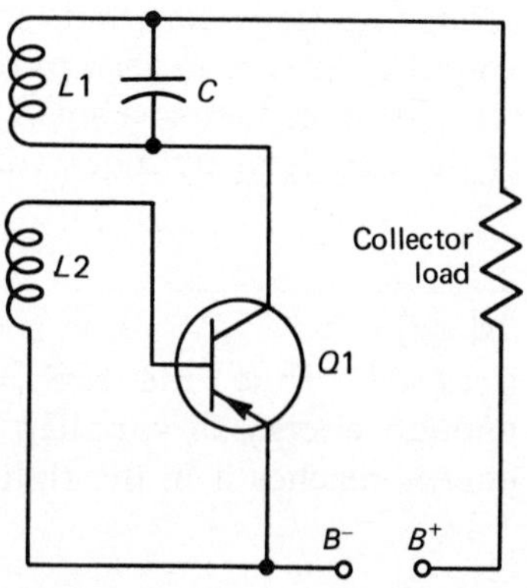

Fig. 14-102

14-103

(*feedback*) A source of dc power has also been inserted in the collector circuit to provide the energy required for amplification. Amplification will occur in the device symbolized by _____.

14-104

(*Q1*) When the dc power is switched on, transient voltages are induced in the circuit. These appear at the terminals of the capacitor, causing the capacitor to take on a _____.

14-105

(*charge*) The capacitor can discharge through *L1*. Thus, *C1* and _____ form a tank circuit in which oscillation can exist. The oscillating collector current passing through *L1* causes a magnetic field to build up around *L1*.

14-106

(*L1*) Because *L2* and *L1* are physically very close to each other, the magnetic lines of force from *L1* must cut through the turns of *L2*. This action causes a voltage to be _____ across the terminals of *L2*.

14-107

(*induced*) Since *L2* is in the base circuit of *Q1*, an oscillating feedback voltage appears between the base and _____ of transistor *Q1*. The transistor then amplifies the oscillating voltage.

14-108

(*emitter*) If the amplified oscillating voltage is in the correct phase to recharge the capacitor to the full voltage it had before the cycle began, _____ oscillation will be obtained.

14-109
(*sustained* or *undamped*) In order to obtain undamped or sustained oscillation, the feedback must be *positive* in nature. Positive feedback is possible only if the voltage that is fed to the tank circuit has the same _____ as the oscillatory voltage already there.

14-110
(*phase*) Since the transistor inverts the phase of the oscillatory voltage by 180°, the feedback coil or "tickler," as L2 is called, must add _____° more phase inversion.

14-111
(*180*) If the windings of L1 and L2 as given in Fig. 14-102 are incorrectly related to cause this phase inversion of 180°, the situation may be corrected by inverting coil _____ as shown in Fig. 14-111.

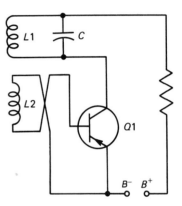

Fig. 14-111

14-112
(*L2*) Depending upon the directions of the windings of L1 and L2, *one* of the two arrangements (Fig. 14-102 or Fig. 14-111) will be correct for producing a phase inversion of _____°.

14-113
(*180*) In any case, once the proper phase for feedback has been established, sustained oscillation can occur if the amplitude of the power fed back is sufficient to make up for all the _____ losses in the oscillatory tank circuit.

14-114
(*resistance*) This "tickler-coil" oscillator demonstrates the requirements of all electronic oscillators. Such oscillators must have (1) a power supply, (2) an amplifier, (3) feedback of the proper amplitude and _____, and (4) an oscillatory or tank circuit.

14-114
(*phase*)

CRITERION CHECK TEST

____14-1 In Fig. 14-7, maximum kinetic energy occurs at point (a) A, (b) B, (c) C, (d) D.

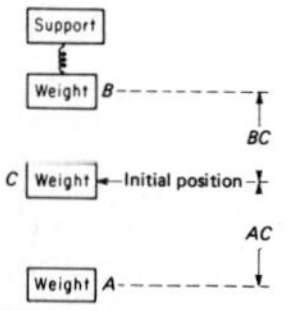

Fig. 14-7 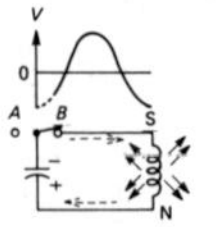   Fig. 14-51 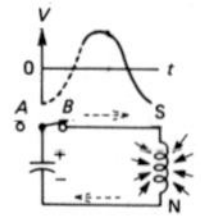   Fig. 14-64 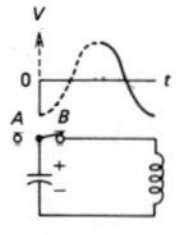   Fig. 14-68

____14-2 In Fig. 14-7, the speed at point B is (a) zero, (b) maximum, (c) the same as at point C, (d) twice that of point A.

____14-3 In Fig. 14-7, maximum potential energy occurs at (a) point C, (b) only point B, (c) only point A, (d) points A and B.

____14-4 A common synonym used for positive feedback is (a) in-phase feedback, (b) wrong feedback, (c) out-of-phase feedback, (d) extra feedback.

____14-5 In Fig. 14-51, the (a) capacitor is charging through the inductor, (b) capacitor is charging through the battery, (c) magnetic field of the inductor is collapsing, (d) capacitor is discharging through the inductor.

____14-6 In Fig. 14-64, the (a) magnetic field of the coil is zero, (b) capacitor is discharging, (c) magnetic field is collapsing, (d) current is zero.

____14-7 In Fig. 14-68, the (a) capacitor is charging to a negative voltage, (b) magnetic field is collapsing, (c) current is zero, (d) magnetic field is building.

____14-8 In Fig. 14-68, the (a) voltage is maximum, current is minimum, (b) voltage and current are maximum, (c) voltage is minimum, current is maximum, (d) voltage and current are minimum.

____14-9 For an LC tuned circuit, (a) when the capacitor energy is maximum, the inductor energy is maximum, (b) when the capacitor energy is maximum, the inductor energy is minimum, (c) when the capacitor voltage is maximum, the inductor voltage is minimum, (d) when the capacitor current is maximum, the inductor current is minimum.

____14-10 Tank circuit is the term always given to what part of an oscillator? (a) The amplifying device, (b) the feedback network, (c) the tuned circuit, (d) the bias network.

____14-11 In Fig. 14-102 a function of the L1L2 transformer is to (a) provide dc bias voltage, (b) couple Q1 to the next stage, (c) match the high impedance of the tank circuit to the low impedance of the base-emitter circuit, (d) block dc from the base circuit of Q1.

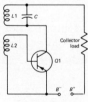

Fig. 14-102

____14-12 In Fig. 14-102 collector-to-base feedback is accomplished via (a) a feedback resistor, (b) a feedback capacitor, (c) magnetic induction, (d) a series RC filter.

15 LC oscillator circuits

OBJECTIVES

(1) List the design objectives for a transistor oscillator. (2) Define the alpha cutoff frequency of a transistor. (3) Relate the significance of frequency f_{max} for a transistor. (4) Describe a method for minimizing the effects of the collector-emitter interelement capacitance. (5) Sketch and describe the operation of a tuned-base oscillator, a tuned-collector oscillator, a Colpitts oscillator, and a Clapp oscillator. (6) For each of these oscillators, indicate the components which determine the dc operating point and the components which determine the frequency of oscillation, and describe the feedback network. (7) Contrast the tuned circuit of the Clapp oscillator with the tuned circuit of the Colpitts oscillator. (8) List at least one advantage of the Clapp oscillator over the Colpitts oscillator. (9) Demonstrate that the feedback networks of the Colpitts and Clapp oscillators provide the correct phase shift. (10) Describe the operation of the Hartley oscillator. (11) Sketch at least two Hartley oscillator circuits. (12) For each circuit, indicate the components which determine the frequency of oscillation, the ac bypass condensers, and the feedback elements. (13) Contrast the Hartley oscillator with the Colpitts oscillator.

INTRODUCTION
15-1
To generate ac power with a transistor amplifier (Fig. 15-1), a portion of the output power must be returned to the input in phase with the starting power. In Fig. 15-1, the output power from the amplifier is identified as ―――――.

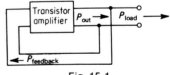

Fig. 15-1

15-2
(P_{out}) To produce sustained oscillations, the feedback power ($P_{feedback}$) must be *positive*. Regenerative feedback is another name for ―――― feedback.

15-3
(*positive*) The feedback power is lost insofar as the load is concerned. That is, the power delivered to the load is the output power minus the _____ power.

15-4
(*feedback*) Besides regenerative feedback, a transistor oscillator needs elements that determine the frequency of oscillation. The most common combination used as the frequency-determining element consists of a coil and a _____ constituting an *LC* network.

15-5
(*capacitor*) Transistor oscillators also employ *RC* networks, or _____-capacitance networks, for frequency control.

15-6
(*resistance*) Finally, good frequency control may be obtained by using a quartz _____ as the frequency-determining element.

15-7
(*crystal*) A transistor oscillator must also contain provisions for correct bias and for stabilization of the operating point. Instability of the operating point affects output amplitude, waveform, and _____ stability.

15-8
(*frequency*) Thus, for a practical transistor oscillator, it is necessary to provide regenerative _____, a frequency-determining system, and proper bias and control of operating point.

15-9
(*feedback*) Regenerative feedback requires that *power* be transferred from output to input. As always, for maximum power to be transferred it is important that the output _____ match the input _____ as closely as possible.

15-10
(*impedance, impedance*) A vacuum-tube amplifier has high input impedance and high output impedance. This provides a good impedance _____ between input and output, resulting in relatively little signal loss in the feedback process.

15-11
(*match*) In the common-base configuration, the input impedance is low and the _____ impedance is high.

396 A PROGRAMMED COURSE IN BASIC ELECTRONICS

15-12

(*output*) To transfer feedback power in a CB configuration, therefore, a *feedback network* may be needed to bring about a reasonably good impedance match between _____ and output circuits.

15-13

(*input*) In some cases, the loss due to mismatch may be compensated for by increasing the amount of feedback power. In these cases, the power delivered to the _____ will be reduced since much of the output power is being used as feedback to make up for mismatch loss.

15-14

(*load*) The problem of impedance matching is not as severe in a common-emitter oscillator since this configuration features moderate input and output _____.

15-15

(*impedances*) Also, the power gain of the _____ configuration is greater than the power gain of either the CB or CC configuration; hence this configuration is much used in the design of transistor oscillators.

15-16

(*CE*) Since a CC configuration has high input impedance and moderate output impedance, the problem of impedance _____ also exists in this configuration.

15-17

(*matching*) Thus, impedance-matching problems exist for all transistor configurations to a greater or lesser degree. These problems are overcome either by means of a feedback impedance-matching network or by using a large amount of _____ energy.

15-18

(*feedback*) The frequency characteristic of a transistor is stated in terms of its *cutoff frequency* in either the CB or CE configuration. To explain the meaning of cutoff frequency, look at Fig. 15-18. This is a graph showing how the relative _____ gain of a certain transistor varies with changing frequency.

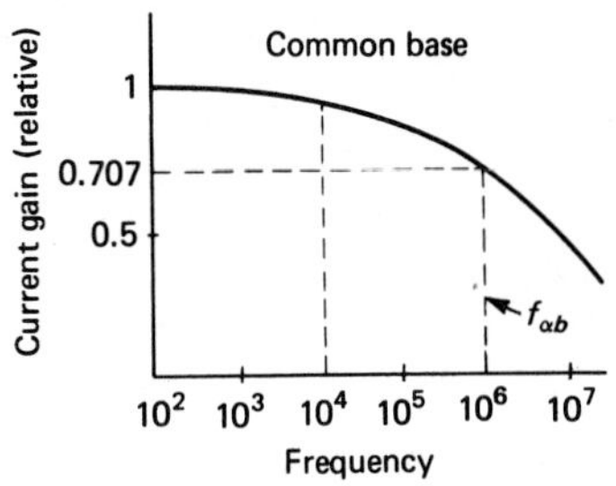

Fig. 15-18

LC OSCILLATOR CIRCUITS 397

15-19
(*current*) As shown in the graph, the current gain in the common-base configuration for this transistor remains relatively constant from 100 Hz to about _____ Hz where the gain begins to drop.

15-20
(*10,000*) When the current gain falls to 0.707 of its value at, say, 1,000 Hz we say that this frequency is the alpha cutoff frequency. This occurs at about _____ MHz for this particular transistor.

15-21
(*1*) The vertical line at the 1-MHz point indicates the alpha cutoff frequency for this transistor. Note that the symbol for alpha cutoff frequency in the CB configuration is given as _____.

15-22
($f_{\alpha b}$) Some manufacturers specify the alpha cutoff frequency for the CE rather than the CB configuration. When you see the symbol $f_{\alpha e}$, you will recognize this as the alpha cutoff frequency for the _____ configuration.

15-23
(*CE*) A falling off of 3 dB is the equivalent of 0.707 of the gain at the reference frequency (usually either 1,000 Hz or 270 Hz). That is, at the alpha cutoff frequency, the current gain has dropped _____ dB below its value at the reference frequency.

15-24
(*3*) For example, suppose $f_{\alpha b}$ for a certain transistor is 10 MHz. If the current gain at the reference frequency was 0.95, then at 10 MHz the current gain will be 0.95 × 0.707 or _____.

15-25
(*0.67*) Or, suppose that $f_{\alpha e}$ for a certain transistor is 100 kHz. If the current gain at 1,000 Hz is 20, then the current gain at 100 kHz will be _____.

15-26
(*14.14*) In other words, the alpha cutoff frequency provides information as to the frequency at which the current gain has become noticeably _____ than it was at the reference frequency.

15-27
(*less* or *smaller*) Oscillators *can* operate efficiently at frequencies well above the alpha cutoff frequency. When you see the specification f_{max} for a transistor, you will know that this symbol tells you the _____ frequency at which the transistor will oscillate reliably.

15-28
(*maximum*) The f_{max} of a transistor is defined as the frequency at which the *power gain* of the transistor as an amplifier is unity. That is, f_{max} is that frequency at which an input signal of a given level will appear in the output at exactly the same level with no gain or _____.

15-29
(*loss*) Power gain is required to overcome losses in the feedback circuit. At f_{max} there is no power gain. Hence, operation of the transistor as an oscillator at _____ is not possible.

15-30
(f_{max}) The operating frequency, therefore, is chosen at some value below _____ where sufficient power gain for feedback and output is obtained.

15-31
(f_{max}) A given transistor is rated as follows: $f_{ab} = 10$ MHz, $f_{max} = 18.3$ MHz. This transistor is capable of oscillation at 14 MHz but is not capable of _____ at 18.4 MHz.

15-32
(*oscillation*) We turn now to consideration of *frequency stability*. It is normally desirable to have an oscillator of good frequency stability; that is, it is desirable to have an oscillator whose _____ does not vary during operation.

15-33
(*frequency*) Of primary importance is the bias voltage source. It is especially true in high-frequency oscillators that variation of bias voltage will bring about a change in the transistor parameters. Such a change will always cause the oscillation frequency to _____.

15-34
(*change* or *vary, etc.*) Thus, a prime requirement for good frequency stability is that the voltage of the bias supply remain _____ during operation of the oscillator.

15-35
(*constant*) In a prior section we showed that the collector-to-emitter capacitance varies with changes of emitter current and collector voltage. This capacitance C_{ce} is an important reactive element in determining the _____ of the oscillating circuit.

15-36
(*frequency*) The effect of C_{ce} may be minimized by connecting a relatively large *swamping capacitor* across the collector-emitter terminals. The resulting large total capacitance makes the collector-to-emitter capacitance appear as a _____ percentage of the whole.

15-37
(*small*) Hence, the total capacitance of the two in parallel results in a circuit which is _____ sensitive to variations in supply voltage.

15-37
(*not* or *less*)

ANALYSIS OF BASIC OSCILLATOR CIRCUIT
15-38
The basic circuit of a transistor oscillator using a tickler coil is shown in Fig. 15-38. Feedback is accomplished by inductive coupling from one winding to the other of the component identified as _____ in this figure.

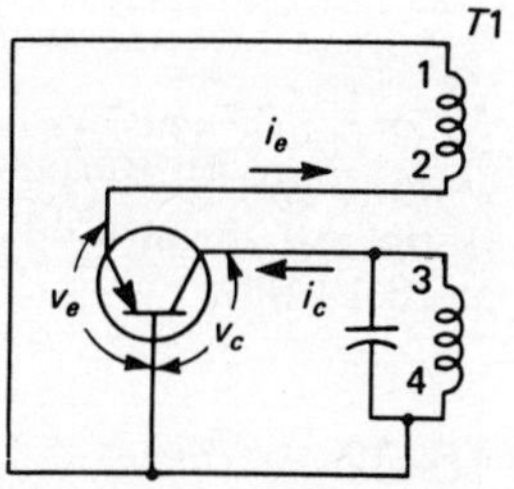

Fig. 15-38

15-39
(*T1*) Bias circuits have been omitted for simplicity. However, as in an amplifier, the base-emitter junction is forward-biased, while the base-collector junction is _____-biased.

15-40
(*reverse*) When power from the battery is first applied, a surge of collector current flows in the transistor. This current flows through the coil whose terminals are numbered _____ and _____.

15-41

(*3, 4*) This current does not reach full amplitude instantaneously but increases steadily between X and Y, as shown in Fig. 15-41b. The inductive coupling between coil 3-4 and coil _____ (Fig. 15-38) now causes a current to flow in the emitter circuit.

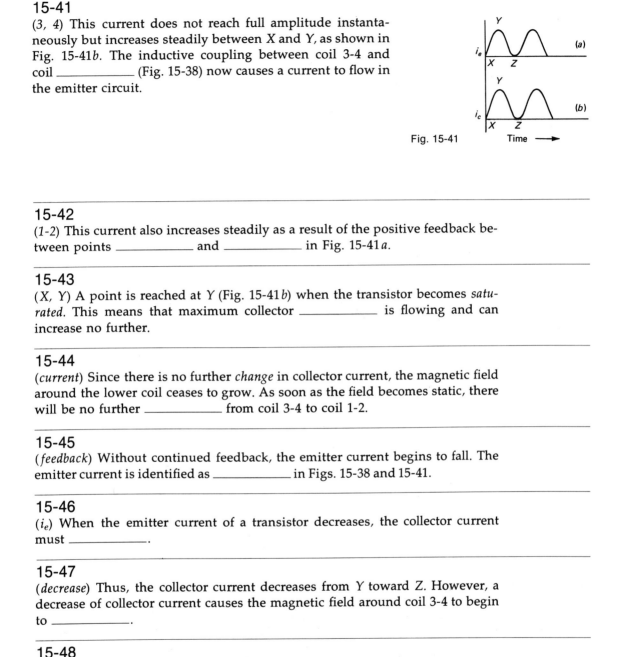

Fig. 15-41

15-42

(*1-2*) This current also increases steadily as a result of the positive feedback between points _____ and _____ in Fig. 15-41a.

15-43

(*X, Y*) A point is reached at Y (Fig. 15-41b) when the transistor becomes *saturated*. This means that maximum collector _____ is flowing and can increase no further.

15-44

(*current*) Since there is no further *change* in collector current, the magnetic field around the lower coil ceases to grow. As soon as the field becomes static, there will be no further _____ from coil 3-4 to coil 1-2.

15-45

(*feedback*) Without continued feedback, the emitter current begins to fall. The emitter current is identified as _____ in Figs. 15-38 and 15-41.

15-46

(i_e) When the emitter current of a transistor decreases, the collector current must _____.

15-47

(*decrease*) Thus, the collector current decreases from Y toward Z. However, a decrease of collector current causes the magnetic field around coil 3-4 to begin to _____.

15-48

(*collapse* or *decay*, *etc.*) Thus, the collapsing field cuts through coil 1-2 in a direction opposite from that caused by the growth of the field originally. This causes a further _____ of emitter current.

LC OSCILLATOR CIRCUITS

15-49

(*decrease*) This decrease continues to point Z where the transistor is "cut off." This means that both i_e and _____ cease to flow.

15-50

(i_c) With i_c at zero, the feedback current in coil 3-4 is also zero. Since it was this feedback current which initially drove the transistor into _____, when it ceases flowing, the bias conditions revert back to their original state.

15-51

(*cutoff*) The process now repeats. That is, the transistor is driven to saturation, then to cutoff, and then back to _____.

15-52

(*saturation*) The time for change from saturation to cutoff and back is determined by the constants of the tank circuit (coil and capacitor). Thus, in controlling the time between states, the tank circuit determines the _____ of oscillation.

15-53

(*frequency*) The oscillator shown in Fig. 15-53 is called a *tuned-base oscillator*. The grid of a vacuum tube and the base of a transistor are often compared functionally. The circuit in Fig. 15-53 is similar in many ways to the tuned-_____ vacuum tube oscillator.

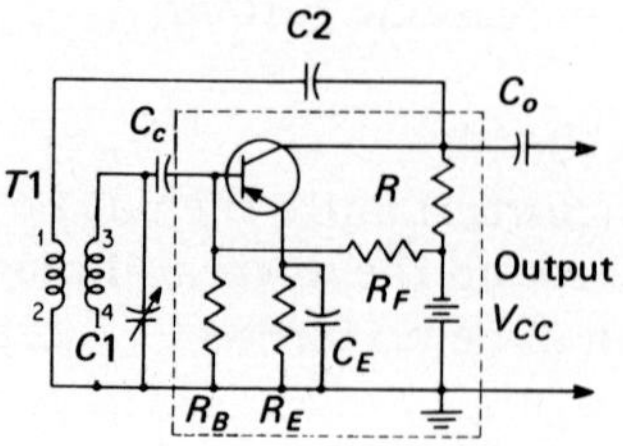

Fig. 15-53

15-54

(*grid*) First let us establish the type of configuration used in this circuit. The *input* from coil 3-4 of T1 is applied between base and emitter; output is taken across collector and emitter. Hence, this is a _____ configuration.

15-55

(*CE*) Base and collector bias are provided by a single source. This source is identified as _____.

15-56

(V_{CC}) The collector load resistor is R_C. Base bias is adjusted by the proper choice of values for the voltage divider consisting of R_F and _____.

15-57

(R_B) The emitter-swamping resistor is identified as _____.

15-58
(R_E) Collector dc flows through R_C but not through 1-2 of $T1$. Collector dc is prevented from flowing through 1-2 of $T1$ by the presence of the component identified as _____ .

15-59
($C2$) When the plate current flows through the tickler coil in a vacuum tube oscillator, we say that the tube is *series-fed;* if the plate is fed through a choke or resistor rather than the tickler, we say it is *shunt-fed.* The transistor oscillator in Fig. 15-53 is obviously of the _____-fed type.

15-60
(*shunt*) The frequency-determining elements in this circuit are coil 3-4 of $T1$ and _____ .

15-61
($C1$) Base bias dc cannot flow through the low-resistance winding of coil 3-4 because dc is blocked by capacitor _____ .

15-62
(C_C) Oscillatory currents can flow in the base-emitter circuit and the tuned circuit because these currents are coupled to the base through _____ .

15-63
(C_C) Degeneration in the emitter circuit is prevented by the bypassing element identified as _____ .

15-64
(C_E) When dc power is applied by connecting V_{CC} into the circuit, oscillations appear across the tuned circuit and are coupled into the base through _____ .

15-65
(C_C) The oscillations are amplified by the transistor, and a part of the oscillatory energy is coupled back to coil _____ of $T1$ through $C2$.

15-66
(*1-2*) If coil 1-2 is correctly polarized with respect to coil 3-4, the energy that is inductively coupled back into the base circuit through $T1$ will sustain the base circuit oscillation. That is, feedback will be in the same _____ as the oscillations in the base circuit.

15-67
(*phase*) When the feedback is in the same phase as the initial oscillatory energy, the feedback is said to be _____ or positive.

LC OSCILLATOR CIRCUITS

15-68
(*regenerative*) If the phase of the feedback energy is *degenerative,* either through error or circuit modifications, it may be made *regenerative* merely by reversing the connections to either 1-2 or to _____.

15-69
(*3-4*) Output oscillatory power is fed to the load through the component labeled _____.

15-70
(C_O) Now study the *tuned-collector* oscillator circuit in Fig. 15-70. First establish the configuration. This is a _____ configuration.

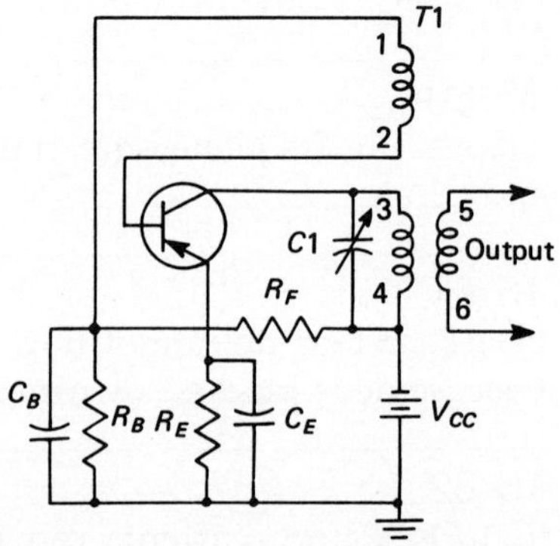

Fig. 15-70

15-71
(*CE*) Proper base bias is established by the values of _____ and _____.

15-72
(R_F, R_B) Degeneration in the emitter circuit is prevented by C_E, which bypasses the _____-_____ resistor R_E.

15-73
(*emitter-swamping*) The frequency-determining elements are _____ and coil _____.

15-74
(*C1, 3-4*) The tickler winding is identified as coil _____.

15-75
(*1-2*) The collector is fed the required dc through coil 3-4. Thus, this is a _____-fed oscillator.

15-76
(*series*) To keep the reactance to the oscillatory frequency low in the base return circuit, one of the base bias resistors is bypassed by a capacitor. This capacitor is identified as _____ in Fig. 15-70.

404 A PROGRAMMED COURSE IN BASIC ELECTRONICS

15-77
(C_B) In this circuit, the output is _____ coupled to the load rather than capacitively coupled as in the tuned-base oscillator.

15-78
(*inductively*) Output coupling is accomplished inductively by transferring energy from coil 3-4 to coil _____.

15-79
(*5-6*) This oscillator is a tuned-collector type. It is similar in circuit and function to the type of vacuum tube oscillator known as a "tuned-_____" type.

15-79
(*plate*)

STANDARD TRANSISTOR OSCILLATORS
15-80
Figure 15-80 illustrates a transistor *Colpitts* oscillator. A Colpitts oscillator, whether it uses tubes or transistors, is an oscillator in which feedback is obtained from a split-tank capacitor. In Fig. 15-80, the split-tank capacitor is made up of C1 and _____.

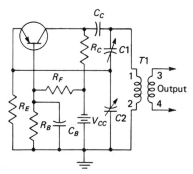

Fig. 15-80

15-81
(*C2*) Resonance is established by the natural frequency of the C1-C2 combination and the primary of _____.

15-82
(*T1*) The base-collector circuit is tuned while the input circuit contains the base and emitter. Thus, the common element of the transistor is the _____, making this a CB configuration.

15-83
(*base*) The oscillatory voltage appears across the ends of the C1-C2 series combination. A part of this oscillatory voltage is taken from the top of capacitor _____ and fed back to the emitter (input).

LC OSCILLATOR CIRCUITS

15-84
(C2) It can be shown that this feedback is positive in nature. This means that it has the right _____ to produce regeneration.

15-85
(*phase*) As a result of regeneration, the oscillatory signal is amplified so that the resistance losses that occur in the collector circuit are canceled. This makes undamped _____ possible.

15-86
(*oscillation*) Base bias, as usual, is provided by the voltage divider made up of R_B and _____.

15-87
(R_F) The collector load resistor in Fig. 15-80 is identified as _____.

15-88
(R_C) Resistor R_E develops the emitter input signal that is derived from the regenerative feedback. It also has a second function: it serves as the emitter- _____ resistor.

15-89
(*swamping*) Since C1 and C2 are both part of the resonant circuit, either one (or both) of these may be used to control the _____ frequency produced by the circuit.

15-90
(*oscillation* or *oscillatory*) The amount of feedback is governed by the *ratio* of C1 to C2. This ratio should be adjusted so that feedback losses are _____ to a minimum while oscillation is reliably maintained.

15-91
(*reduced*) It is found that if $X_{C1}/X_{C2} = Z_o/Z_i$, feedback losses are minimized. Thus, to minimize feedback losses the ratio of the capacitive reactance of C1 to C2 should be nearly equal to the ratio of the output impedance to the _____ impedance of the transistor.

15-92

(*input*) Next study the oscillator circuit in Fig. 15-92. Noting the tank circuit first, we see that an additional part, as compared with the Colpitts, has been added. This part is identified as _____.

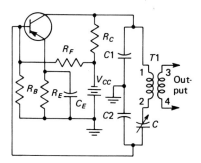

Fig. 15-92

15-93

(*C*) C1 and C2 are still present and are in series with each other but in *parallel* with the inductance. The tuning inductance is the _____ winding of T1.

15-94

(*primary* or *1-2*) Capacitor C, however, is in _____ connection with the tuning inductance.

15-95

(*series*) When the total capacitance of C1 and C2 is *large* as compared with the added capacitance C, then C1 and C2 no longer contribute significantly to the tuning of the circuit. That is, the oscillator frequency is effectively determined by winding 1-2 and capacitor _____.

15-96

(*C*) At resonance, the impedance of a *series* resonant circuit, such as that formed by C and winding 1-2, is _____.

15-97

(*low* or *small*) Thus, at resonance the shunting impedance of C and winding 1-2 is quite low. This makes the oscillation frequency relatively independent of transistor parameter variations. This would tend to improve the _____ stability of the oscillator.

15-98

(*frequency*) This modification of the Colpitts circuit is known as a *Clapp* oscillator. In the circuit of Fig. 15-92, the emitter is kept at signal ground potential by _____, which shunts R_E.

15-99
(C_E) Input or feedback voltage is applied between base and ground while the output is taken between collector and ground. Since the emitter is at ground potential, this is a _____ configuration.

15-100
(CE) R_E is the emitter-swamping resistor, R_F and R_B make up the voltage divider needed to produce base bias, and R_C is the _____ load resistor.

15-101
(*collector*) The simplified feedback network in Fig. 15-101 shows why the phase of the feedback is correct for developing regenerative action in the Colpitts oscillator. The Colpitts uses a _____ configuration.

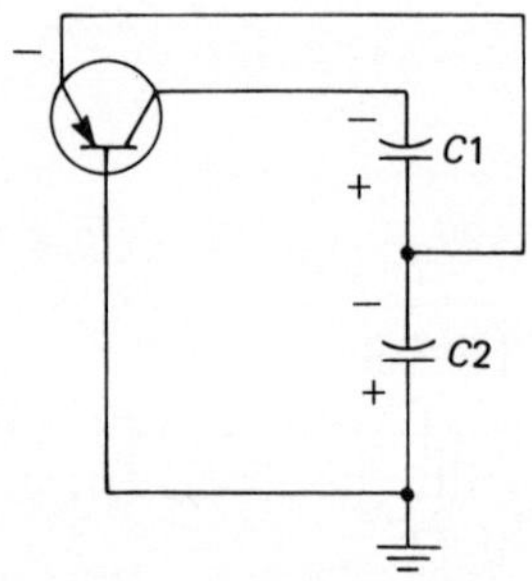

Fig. 15-101

15-102
(*CB*) In a common-base configuration, such as the Colpitts, there is no phase shift between the signal on the collector and the signal on the _____ of the transistor.

15-103
(*emitter*) As shown in Fig. 15-101, the feedback voltage that develops across C2 is applied between the ground of the system and the _____ of the transistor.

15-104
(*emitter*) Assume that the emitter is negative-going as indicated in Fig. 15-101. Since there is no phase shift between emitter and collector, then when the emitter is negative-going, the collector is _____-going.

15-105
(*negative*) Thus, C1 and C2 develop a voltage whose polarity is shown in Fig. 15-101. The top of C2 is negative-going, and since this is the voltage fed back to the emitter, the emitter in turn is made more _____-going.

15-106
(*negative*) Hence, the phase of the feedback is regenerative. Compare this with the Clapp oscillator. The Clapp oscillator is a common-_____ type.

15-107
(*emitter*) In a CE configuration, there is a phase shift of _____° between the signal on the collector and the signal on the base.

15-108
(*180*) Referring to Fig. 15-108, a simplified feedback network applicable to the CE configuration, it is seen that the feedback voltage developed across C2 is fed back between ground and the _____ of the transistor.

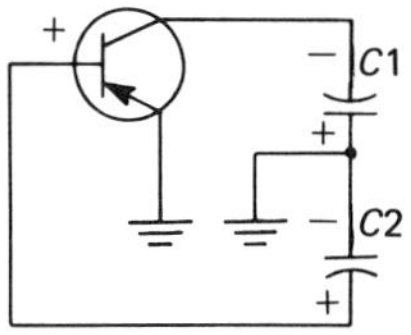

Fig. 15-108

15-109
(*base*) As the base is made positive-going, the collector must be made _____-going since there is a phase shift of 180° between the two elements.

15-110
(*negative*) As the collector goes negative, the voltages shown on C1 and C2 are developed. The bottom of C2 is positive-going and is connected to the base. Hence, the feedback voltage is positive-going while the base is positive-going, making the feedback of the proper _____ to sustain oscillation.

15-111
(*phase*) The *Hartley* oscillator shown in Fig. 15-111 is operationally similar to the Colpitts type except that feedback in the Hartley is obtained from a split inductance rather than a split _____ as in the Colpitts.

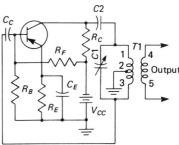

Fig. 15-111

15-112
(*capacitance*) In the Hartley of Fig. 15-111, the collector dc flows through R_C rather than through the tuning inductance. Thus, this circuit represents a _____-fed type of oscillator.

15-113
(*shunt*) The dc is blocked from the tank circuit by the action of the component identified as _____.

15-114
(*C2*) C1 tunes the entire inductance 1-3. Feedback voltage, however, appears across the terminals numbered 2-_____.

15-115
(3) The feedback voltage is coupled to the base of the transistor through the component identified as _____.

15-116
(C_C) The presence of a split-tank capacitor in the Colpitts oscillator makes series feed impossible. However, in the Hartley oscillator, series feed is possible. In Fig. 15-116 series feed is shown. This is series feed because the collector dc flows through that part of the tuning inductance terminated by 2 and _____.

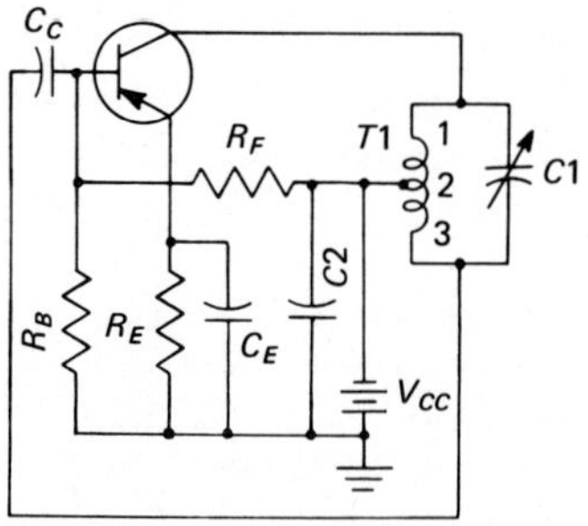

Fig. 15-116

15-117
(1) In the circuit of Fig. 15-116, base bias is provided by the action of resistors R_F and _____.

15-118
(R_B) Resistor R_E is the emitter-swamping resistor, and _____ provides an ac bypass that permits the emitter to maintain itself at ground potential for the signal.

15-119
(C_E) Terminal 2 of the tuning inductance is placed at ac ground potential through the action of the component labeled _____.

15-119
(C2)

CRITERION CHECK TEST

___ 15-1 The transistor oscillator with the closest existing impedance match would be in which configuration? (a) CB, (b) CC, (c) CE, (d) any configuration with proper bias.

___ 15-2 To sustain oscillations, the power gain of an amplifier (a) may be between 0.1 and 0.5, (b) may be any value from 0.5 upward, (c) must be greater than 1.0, (d) may be equal to or greater than 1.0.

___ 15-3 The greatest amount of frequency instability is caused by variations in (a) the emitter-swamping resistor, (b) coil flux linkages, (c) the swamping-resistor by-pass capacitor, (d) the bias.

____15-4 Referring to Fig. 15-41, we see there is no feedback voltage developed (a) between point Y and point Z, (b) at point Y only, (c) at point Z only, (d) at points Y and Z.

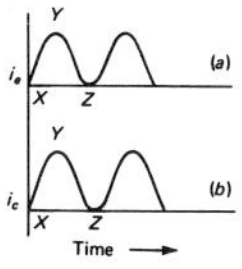

Fig. 15-41 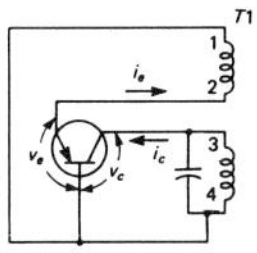   Fig. 15-38 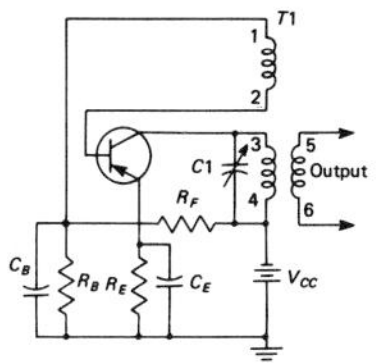   Fig. 15-70 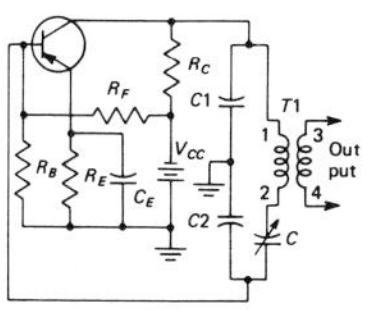   Fig. 15-92

____15-5 In Fig. 15-38, the feedback is from (a) emitter to collector, (b) collector to emitter, (c) base to collector, (d) collector to base.

____15-6 In the circuit of Fig. 15-70, the emitter is tied to ac ground by (a) R_E, (b) C_E, (c) R_F, (d) C_B.

____15-7 In the circuit of Fig. 15-70, base bias is provided by (a) $R_F R_E$, (b) $R_E R_B$, (c) $R_F R_B$, (d) $R_F C_B$.

____15-8 The transistor in Fig. 15-70 is in the (a) common-base configuration, (b) common-emitter configuration, (c) common-collector configuration, (d) common-drain configuration.

____15-9 In the Colpitts oscillator, feedback is (a) taken from the center tap of a coil, (b) obtained by magnetic induction, (c) obtained by a tickler coil, (d) taken from the center of split capacitors.

____15-10 When the CE configuration is used for an oscillator, the voltage fed back (a) must be inverted by 180°, (b) must have a 0° phase shift, (c) must have either a 0 or 180° phase shift, depending on the type of oscillator, (d) must be taken from a capacitor.

____15-11 The operation of a Clapp oscillator is usually arranged so that the oscillation frequency depends on the (a) magnitudes of the split capacitors, (b) ratio of the split capacitors, (c) magnitude of the series capacitor, (d) ratio of the resistance of the split capacitors.

____15-12 In Fig. 15-92, when the ac base voltage is positive, (a) the ac collector voltage is positive and the ac emitter voltage is negative, (b) the ac collector voltage is positive and the ac emitter voltage is zero, (c) the ac collector voltage is negative and the ac emitter voltage is zero, (d) the ac collector voltage is negative and the ac emitter voltage is positive.

16 crystal oscillators

OBJECTIVES

(1) Define the *piezoelectric effect*. **(2)** Depict the amplitude of oscillation of a piezoelectric crystal versus the applied frequency. **(3)** State the advantage of using a crystal in an oscillating circuit. **(4)** Sketch a quartz mother stone, labeling the optical axis. **(5)** Depict and discuss the relationship among the optical, electrical, and mechanical axes of a quartz crystal. **(6)** Depict an X-cut, Y-cut, and AT-cut crystal. **(7)** Discuss the temperature coefficient of a quartz crystal and compute the frequency shift of a crystal using the temperature coefficient. **(8)** Sketch and discuss frequency change versus temperature curves for several modern crystal cuts. **(9)** Draw an electrical equivalent circuit for a crystal. **(10)** Demonstrate that a crystal has both a series and a parallel resonant mode. **(11)** Compute the resonant frequencies of a crystal and demonstrate that the parallel and series resonant frequencies are nearly the same. **(12)** Compute the Q of a typical crystal. **(13)** Describe the operation of an oscillator using a crystal in the parallel resonant mode. **(14)** Sketch a tuned-grid, tuned-plate vacuum tube oscillator using a crystal as a tuned circuit. **(15)** Trace the development of the oscillatory signal and state the necessary conditions for sustained oscillation to occur. **(16)** Describe the operation of an oscillator using a crystal in the series resonant mode. **(17)** Sketch a transistor Hartley oscillator using a crystal as the feedback element. **(18)** Demonstrate that the crystal produces a precise frequency of oscillation. **(19)** Contrast the frequency stability of LC and crystal oscillators with respect to transistor gain changes.

PIEZOELECTRIC EFFECT

The crystal oscillator was developed to improve frequency stability of oscillator output. In the years that have followed its invention, dozens of new circuits have been devised in which a quartz crystal is employed to maintain a constant oscillation frequency. Other materials such as tourmaline have also been tried as frequency-controlling elements but have been largely abandoned in favor of quartz. For a thorough study of crystal oscillators, we must first learn something of the *piezoelectric effect*.

16-1

Certain crystals, notably quartz, exhibit an effect called *piezo-electricity* which is used in the design of oscillators having excellent frequency stability. Fundamentally, the piezoelectric effect, as shown in Fig. 16-1, is an effect in which pressure on the faces of a crystal gives rise to a(n) _____ charge on these faces.

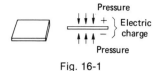

Fig. 16-1

16-2

(*electric*) Carried a bit further, if a piece of quartz is squeezed, it develops a potential between its faces. If the potential is positive on the top face and negative on the bottom face when pressure is applied (Fig. 16-2a), then when the pressure is suddenly released (Fig. 16-2b), the polarity of the potential _____.

Fig. 16-2

16-3

(*reverses*) The converse also occurs: When a potential is applied to a wafer of quartz, it may contract by reducing the distance between the two faces. Also, when the potential is reversed, the wafer will expand by _____ the distance between faces.

16-4

(*increasing*) Quartz has relatively large mechanical strength. If an alternating voltage is applied to a crystal wafer, the wafer will vibrate at the frequency of the applied voltage. Since this vibration may be violent, greater mechanical _____ is advantageous.

16-5

(*strength*) For one particular thickness of crystal cut from the mother crystal in a specific way, there is one frequency at which the amplitude of vibration is tremendously greater than for all other frequencies. In Fig. 16-5, this frequency is labeled _____.

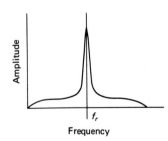

Fig. 16-5

16-6

(f_r) This is called the resonant frequency of the crystal. For all other frequencies, the vibration amplitude is comparatively _____ (Fig. 16-5).

16-7
(*small*) Note the sharp resonance peak at the resonant frequency in Fig. 16-5. If this kind of peak were obtained from a tuned *LC* circuit, we would agree at once that the circuit had a very _____ Q.

16-8
(*high*) If the vibration of the crystal could be made self-sustaining by some form of feedback as used in *LC* oscillators, it would have very little tendency to wander from its resonant frequency because the instant that it did wander, its vibration amplitude would _____ seriously and stop the oscillation.

16-9
(*decrease*) An oscillator that tends to oscillate at a very constant frequency is said to have good frequency _____.

16-10
(*stability*) Since a crystal of quartz behaves like a very high-*Q* resonant device, an oscillator that uses a quartz crystal as a frequency-controlling element would have excellent frequency _____.

16-11
(*stability*) On the other hand, certain factors, particularly temperature, have a marked effect on the resonant frequency of quartz crystals that are cut in a certain way from the mother crystal. For such crystals, high-frequency stability can be maintained only by keeping the temperature _____.

16-12
(*constant*) There are ways to cut a crystal wafer from the mother stone that result in crystals which are not very sensitive to temperature changes. Such crystals will have good frequency stability even when few, if any, precautions are taken to maintain _____ temperature.

16-12
(*constant*)

TEMPERATURE STABILITY OF CRYSTALS

Quartz crystals found in nature have a typical shape and form. These crystals have been scientifically explored from both optical and electrical points of view. They have clearly determinable axes, all of which have been identified by nomenclature used in the next section of our work. Many of the crystals now used in electronics have been cut from synthetic mother stones grown in the laboratory.

16-13
The quartz mother stone is a crystal with a hexagonal cross section (Fig. 16-13) and a pointed end. The axis of this crystal that connects to the pointed end and runs parallel to the crystal corners is known as the optical or _____ axis.

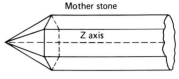

Fig. 16-13

16-14
(Z) Looking at a cross section of the mother stone, as in Fig. 16-14, we note that an axis is drawn from one corner of the hexagon to the opposite corner of the hexagon. This is labeled the _____ axis.

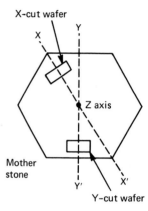

Fig. 16-14

16-15
(X) The X axis is called the *electrical* axis of the crystal. Only one X axis is shown, but it is possible to draw in _____ X axes by connecting appropriate corners.

16-16
(3) In Fig. 16-14, another axis called the mechanical axis of the crystal is shown. This axis is perpendicular to the opposite faces of the crystal and is labeled the _____ axis.

16-17
(Y) If lines are now drawn perpendicular to the remaining faces, we would have _____ Y axes.

16-18
(3) A crystal wafer cut from the mother stone so that one of its major or large faces is parallel to the Z axis but at right angles to the X axis will oscillate freely in the proper piezoelectric circuit. Such a crystal, according to Fig. 16-14, is called a(n) _____-cut crystal.

CRYSTAL OSCILLATORS **415**

16-19
(X) Another type of cut involves slicing a wafer again parallel to the Z axis but at right angles to one of the Y axes. This cut is called a(n) _____ cut.

16-20
(Y) Another very popular crystal cut, called the AT cut, is illustrated in Fig. 16-20. This cut is not parallel to the Z axis but is tipped toward one of the faces at an angle of _____°.

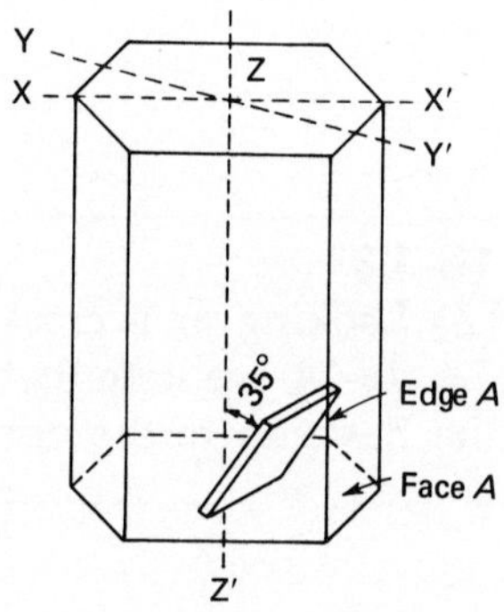

Fig. 16-20

16-21
(35) The tipped outline as shown in Fig. 16-20 is one way of making an AT cut. You will notice that one edge (edge A) is kept parallel to face _____.

16-22
(A) Different cuts have different temperature-frequency characteristics. This means that different cuts show different amounts of variation of _____ when the temperature of the crystal is allowed to vary.

16-23
(*frequency*) The frequency of a typical X-cut crystal *decreases* as the temperature is *increased*. This is called a *negative temperature coefficient*. The frequency of a Y cut *increases* with *increasing* temperature. The Y cut is said to have a positive _____ coefficient.

16-24
(*temperature*) A typical X-cut crystal ground for 1,000 kHz (1,000,000 Hz) might decrease 20 Hz for a single degree rise in Celsius temperature (formerly centigrade temperature). Its temperature coefficient (abbrev. TC) is −20 hertz per million hertz per _____ Celsius.

16-25
(*degree*) The minus sign before the 20 in the above statement refers to the fact that the X-cut crystal has a negative temperature _____.

16-26
(*coefficient*) Instead of saying −20 hertz per million hertz per degree Celsius, the usual statement is −20 *parts* per *million* parts per degree _____.

16-27
(*Celsius*) In the literature, this expression is abbreviated as follows: temperature coefficient = −20 ppm/°C. In this shortened statement, the abbreviation ppm stands for parts per _____.

16-28
(*million*) If a given crystal has a temperature coefficient of −20 ppm/°C, then this crystal will decrease in frequency to the extent of 20 Hz for each million hertz per second for each degree Celsius. Thus, if the crystal has a fundamental frequency of 1,000,000 Hz and its temperature rises 1°C, its new frequency will be _____ Hz.

16-29
(*999,980*) If the fundamental frequency of a crystal having the same temperature coefficient is 2,000,000 Hz, there will be a decrease of 20 Hz for each of the 2 MHz per degree Celsius. Hence, this crystal will go down to the extent of _____ Hz for each degree Celsius rise.

16-30
(*40*) To find the frequency change for any crystal, therefore, we multiply the temperature coefficient by the fundamental frequency expressed as a power of 10. That is, if the fundamental frequency is 2,000,000 Hz, expressed as a power of 10, the fundamental frequency is _____ Hz.

16-31
(2×10^6) For example, if the coefficient is −20 ppm/°C and the fundamental is 2,000,000 Hz, first express the frequency as a power of ten thus: 2×10^6. Then multiply by −20 to obtain −40. This tells you that the frequency will decrease by _____ Hz.

16-32
(*40*) Thus, as long as the frequency is expressed as some number times 10^6, you need multiply only the number by the coefficient to obtain the frequency change. If a crystal has a frequency of 3,000 kHz, you would consider this as 3,000,000 Hz and express it as _____ $\times 10^6$.

16-33
(*3*) If a crystal has a frequency of 1.6 MHz, you would consider this as 1,600,000 Hz and express it as _____ $\times 10^6$.

16-34
(*1.6*) Another example would be a crystal with a frequency of 4.7 MHz. This would be expressed as _____ $\times 10^6$.

16-35
(*4.7*) Returning to our X-cut example, suppose its fundamental frequency is 2.66 MHz. Since its temperature coefficient is −20 ppm/°C, then it would drop in frequency to the extent of _____ Hz for a 1°C temperature rise.

16-36
(*53.2*) Suppose this same crystal now increased 2°C in temperature. For each degree rise, the frequency drop is 53.2 Hz. Thus, for a 2°C rise in temperature, the frequency would drop _____ Hz from its original frequency.

16-37
(*106.4*) Thus, to find the frequency change for any crystal that changes in temperature by any amount, simply multiply the fundamental frequency expressed as a number times 10^6 by the temperature coefficient in parts per million per degree Celsius and then multiply this product by the change in _____ in degrees Celsius.

16-38
(*temperature*) In equation form, we can write this statement as freq. change = $f \times TC \times$ temp. change. Consider a rise in temperature as a "+" change and a drop in temperature as a _____ change.

16-39
(*minus* or "−") Then handle the algebraic signs as you would for ordinary algebraic multiplications. That is, a "+" × "+" = "+"; a "−" × "−" = "+"; and a "+" × "−" = _____.

16-40
(*minus* or "−") Example: The temperature coefficient of a certain crystal is −15 ppm/°C. If its temperature *falls* 5°C and its frequency is given as 2.8 MHz, then its frequency will change: 2.8 × (−15) × (−5) = _____ _____.

16-41
(*+210 Hz*) The "+" before the 210 in the above answer indicates that the frequency will _____.

16-42
(*rise* or *increase*) Example: A crystal has a fundamental frequency of 3,500 kHz and a TC of −25 ppm/°C. For a 3°C rise in temperature, the frequency change will be 3.5 × (−25) × (+3) = _____ _____.

16-43
(*−262.5 Hz*) The "−" before the 262.5 Hz in the above answer indicates that the frequency will _____.

16-44
(*fall* or *decrease*) Example: An X-cut crystal whose frequency is 7.1 MHz has a TC of −20 ppm/°C. If the temperature now falls 15°C, the frequency change of the crystal will be _____ Hz.

16-45
(*+2,130*) This is better stated in kilohertz. Thus, in kilohertz the frequency change would be _____ _____.

16-46
(*+2.13 kHz*) Y-cut crystals have very large temperature coefficients. For example, a value of +70 ppm/°C is not unusual. This means that an oscillator controlled by a Y-cut crystal of these characteristics will not have good _____-frequency characteristics.

16-47
(*temperature*) For this reason, Y-cut crystals are seldom if ever used in any equipment where reasonable temperature stability is expected. The same is true of ordinary X-cut crystals since the _____ coefficients of these are larger than one would want for excellent frequency stability.

16-48
(*temperature*) The temperature-frequency characteristics of four of the more modern cuts are shown in Fig. 16-48. Consider the *CT-cut* curve. Going from 5 to 10°C, the frequency rises (goes in a positive direction toward less negative numbers) from −35 to − _____ ppm.

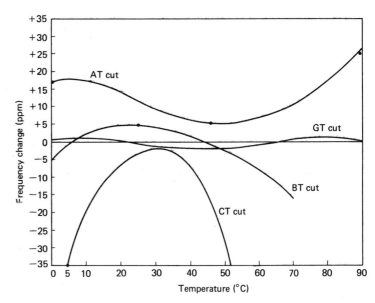

Fig. 16-48

CRYSTAL OSCILLATORS **419**

16-49
(*19*) For a temperature change of 5° (from 5 to 10°), there is a frequency change from −35 to −19 ppm. Thus the frequency change for 5° is _____ ppm.

16-50
(*16*) To find the TC over this range in parts per million per degree Celsius, we must divide the frequency change by the number of degrees. Thus, the TC of the CT-cut crystal in this range is +_____ _____.

16-51
(*3.2 ppm/°C*) The "+" sign before 3.2 indicates a positive temperature coefficient. This is true because the frequency is _____ as the temperature rises.

16-52
(*rising*) For the CT cut between the temperatures of 10 and 20°C, the frequency rises from −19 to −6 ppm, for a total frequency change of +_____ _____.

16-53
(*13 ppm*) This is the frequency change for 10°C. Thus, the TC for the CT cut in this range in parts per million per degree Celsius is _____ _____.

16-54
(*+1.3 ppm/°C*) Thus, the frequency stability with respect to temperature changes is better for the CT cut in the range from 10 to 20° since the TC is _____ in this range than it is in the 5 to 10° range.

16-55
(*smaller*) Still studying the CT-cut curve, we see that the frequency change from 20 to 30° goes from −6 to −2 ppm. This is for a 10° change. Hence, the TC over this range is _____ ppm/°C.

16-56
(*+0.4*) This again indicates that the performance characteristics of the CT cut improved since the TC is even _____ over the range from 20 to 30° than it was over previous ranges.

16-57
(*smaller*) The TC over the range from 5 to 30°C for the CT cut was consistently positive since the frequency rose as the temperature rose. Beyond 30°, however, the frequency begins to fall with rising temperature. This shows that the TC must become _____ in the range from 30 to 50°.

16-58
(*negative*) The slope of the CT-cut curve is very steep between 5 and 10° and also between 45 and 50°. Where the slope is steep, the TC is _____, indicating poor frequency-temperature characteristics.

16-59
(*large*) The slope of the CT-cut curve is very gentle between 25 and 35°. Over this range, the TC is quite small as we have seen. Thus, the best frequency-_____ characteristics are obtained around 30° for the CT cut, suggesting that this crystal should be used at this temperature.

16-60
(*temperature*) Generalizing this concept so that it can be used for quick analysis of any crystal curve, we may say the TC is smallest, hence the frequency-temperature characteristics are best, where the _____ of the curve is most gentle.

16-61
(*slope*) Now we can observe that the BT cut has a considerably gentler overall curve than the CT cut, thus informing us that the frequency-temperature characteristics of the BT cut, on the average, are _____ than those of the CT cut.

16-62
(*better*) The TC of the BT cut is positive from 0°C up to about 25°C since the curve rises over this range. Beyond 25°C, however, the TC becomes _____.

16-63
(*negative*) The curve of the BT cut is virtually flat over the range from about 20°C to about _____°C.

16-64
(*30*) The best temperature-frequency stability for the BT cut is obtained, therefore, between the temperatures of _____ and 30°C.

16-65
(*20*) For the AT-cut crystal, the TC reverses sign at 5°C and at about _____°C.

16-66
(*46*) The AT-cut curve is very flat between 35 and 60°C, giving us a 25° range over which the AT cut has excellent temperature-frequency _____.

16-67
(*stability*) The most excellent stability curve, however, belongs to the _____-cut crystal.

16-68
(*GT*) Note that, although the sign of the TC for the GT cut reverses 3 times over the range from 0 to 90°C, the slope of the curve is extremely _____ over the entire range.

16-69
(*gentle*) The TC of the GT cut is virtually zero over the entire range of temperatures shown on the graph. Thus, of all the modern cuts shown in the figure, the crystal having the most desirable temperature-frequency stability properties is the _____ _____.

16-69
(*GT cut*)

CRYSTAL RESONANT FREQUENCY

16-70
As we have described, when a crystal is made to vibrate by connecting it in the proper electronic circuit, it generates an alternating electric potential across its faces. Thus the crystal becomes an oscillator. Because the crystal oscillator is extremely sharp in frequency response, it is said to behave like a very high-_____ resonant circuit.

16-71
(*Q*) To be connected in a circuit, a crystal must be secured in a special holder. Such a holder is illustrated in Fig. 16-71. The crystal is placed into the square hollow of the holder body; the upper electrode has four "feet" that rest on the crystal. This provides an air gap between the upper electrode and the upper face of the _____.

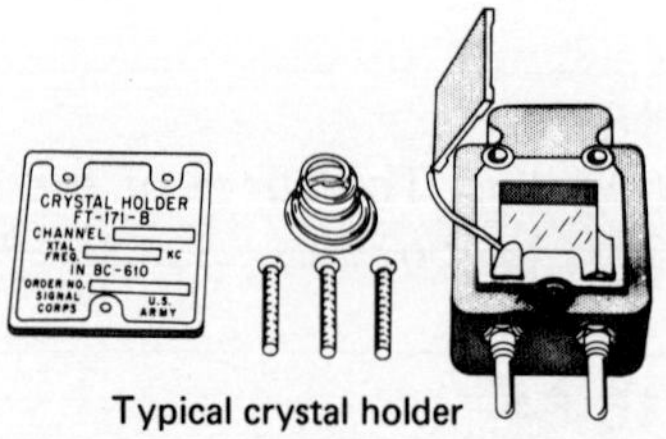

Typical crystal holder
Fig. 16-71

16-72
(*crystal*) The spring (Fig. 16-71) holds the upper electrode firmly against the edges of the crystal. The air gap permits the major portion of the crystal to _____ freely.

16-73
(*vibrate* or *oscillate*) When the holder is assembled, the crystal is sandwiched between two flat metal electrodes that act as a capacitor. This capacitor is an integral part of the crystal assembly and has a significant effect upon the _____ of oscillation (Fig. 16-73).

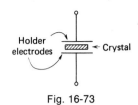

Fig. 16-73

16-74
(*frequency*) When a crystal in a holder is connected in an amplifier circuit, the assembly has the equivalent action of the parts shown in Fig. 16-74. The crystal alone acts like a very high-Q series _____ circuit in which the frequency is controlled by the equivalent C and L of the crystal.

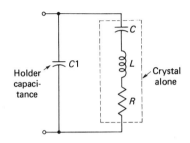

Fig. 16-74

16-75
(*resonant*) The capacitance of the holder behaves as though it were connected in _____ with the equivalent crystal circuit.

16-76
(*parallel*) This parallel holder capacitance produces a second resonant frequency. However, this second resonant frequency represents a parallel resonant circuit, rather than a _____ resonant circuit.

16-77
(*series*) Whereas the series resonance was determined by the L and C of the crystal, the parallel resonance is determined by the inductance L and the *two* capacitances, C and _____.

16-78
(*C1*) In fact, the effective capacitance of the series combination of C and $C1$ is given by the formula.

$$C_{\text{eff}} = \frac{C\,C1}{C + ?}$$

In some oscillators, the crystal is used as a parallel resonant circuit; in other applications, the crystal is used as a series resonant circuit.

CRYSTAL OSCILLATORS **423**

16-79
(C1) A typical low-frequency crystal has dimensions and characteristics as shown in Fig. 16-79. For this low frequency, the crystal is quite thick as compared with units ground for use at the higher frequencies. The thickness of a 430-kHz crystal of this type is about _____ in.

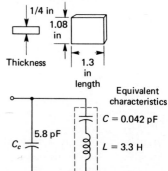

Fig. 16-79

16-80
(¼) The crystal measurements are also given as 1.08 in in width and _____ in in length.

16-81
(1.3) By separate measurements of the crystal characteristics, it is found that the equivalent capacitance of the crystal is 0.042 pF and the equivalent inductance of the crystal is _____ H.

16-82
(3.3) The crystal frequency is given as 430 kHz. Let us check this value against the series resonant frequency obtained by using the equivalent L and C given in Fig. 16-74. You will recall that the frequency of oscillation is given by the equation:

$$f = \frac{1}{?}$$

16-83
$(2\pi \sqrt{LC})$ In this equation, L must be in henrys and C in farads. L is already stated as 3.3 H, but C is given in picofarads. Thus, in farads, the value of C is _____ .

424 A PROGRAMMED COURSE IN BASIC ELECTRONICS

16-84
(0.042×10^{-12} F) Since 2π is 6.28, we may substitute in the equation. In substituted form, the equation appears thus:

$$f = \frac{1}{?}$$

16-85
($6.28 \times \sqrt{3.3 \times 0.042 \times 10^{-12}}$) Multiplying out the factors under the radical, we obtain

$$f = \frac{1}{6.28 \times \sqrt{?}}$$

16-86
($\sqrt{0.1386 \times 10^{-12}}$) The easiest way to handle this is to convert the 0.1386 to 1,386 by multiplying by 10^4. When we do this, we must then divide 10^{-12} by _____.

16-87
(10^4) When 10^{-12} is divided by 10^4 we obtain 10^{-16}. Inserting this into the substituted form of the equation and remembering the change from 0.1386 to 1,386, we can write

$$f = \frac{1}{6.28 \times \sqrt{?}}$$

16-88
($\sqrt{1,386 \times 10^{-16}}$) We can obtain the square root of 1,386 by arithmetic methods — by looking it up in a table of square roots, or by using the slide rule or calculator. The square root of 1,386 to 3 significant figures is _____.

16-89
(37.2) The square root of 10^{-16} is found by dividing the exponent by 2 and writing the new number in exponential form. Thus, the square root of 10^{-16} is _____.

16-90
(10^{-8}) The equation, without the radical, thus may be written

$$f = \frac{1}{6.28 \times 37.2 \times ?}$$

CRYSTAL OSCILLATORS

16-91
(10^{-8}) The factor 10^{-8} may be moved up to the numerator merely by changing the sign of the exponent. The product of 6.28 and 37.2 to 3 significant figures is 234. Thus, dividing 10^8 by 234 gives us the frequency in hertz. Stated in kilohertz to 3 significant figures, this frequency comes out _____ _____.

16-92
(*428 kHz*) Considering the presence of other factors such as the series resistance of 390 Ω and the capacitance of the holder, this result agrees substantially well with the manufacturer's statement of the actual frequency at which the crystal will operate. The figure given by the manufacturer is _____ kHz.

16-93
(*430*) The parallel resonant frequency could be calculated by the same method. The only change would be to replace the crystal capacitance C by the effective capacitance, _____.

16-94
(C_{eff}) C_{eff} can be computed from the formula:

$$C_{eff} = \frac{C\,C1}{C + C1} = \frac{0.042 \times 10^{-12} \times 5.8 \times 10^{-12}}{0.042 \times 10^{-12} + 5.8 \times 10^{-12}} = \underline{} \text{ pF}$$

16-95
(*0.0416*) Thus, for this crystal the effective capacitance C_{eff} is very close in value to the crystal capacitance _____. It then follows that the parallel resonant frequency will be *very* close to the series resonant frequency.

16-96
(*C*) Summarizing: In general crystals have two resonant frequencies, one a series mode, the other a parallel mode. Further, the values of these two frequencies are generally quite _____ (different, close).

16-97
(*close*) Let us now check the Q of the crystal. The figure given by the manufacturer is that $Q = 23{,}000$ correct to 2 significant figures. To check this, we must first write the formula: $Q = 2\pi fL/R$. In this equation, $f = 430 \times 10^3$ Hz, $L = 3.3$ H, and $R = $ _____ _____.

16-98
(*390 Ω*) Substituting all known figures in the equation, we may write it as $Q = 6.28 \times 430 \times 10^3 \times $ _____ /390.

16-99
(3.3) Multiplying through, we find that the Q (to 2 significant figures) is _____.

16-100
(23,000) This is in agreement with the manufacturer's statement of the value of Q. A resonant system with a Q of this magnitude will have a very sharp _____ curve.

16-100
(*resonance*)

CRYSTAL OSCILLATOR CIRCUITS

Two oscillator circuits will now be discussed. The first will use a crystal in the parallel resonant mode, the second in the series resonant mode. Remember that a parallel resonant circuit has a high impedance; a series resonant circuit has a low impedance.

16-101
Since a crystal in its holder behaves like a high-Q resonant circuit, it may be used in place of an LC circuit for frequency control. In Fig. 16-101, the crystal is used to replace the tuned circuit that would normally be found in the _____ circuit of a tuned-grid, tuned-plate oscillator.

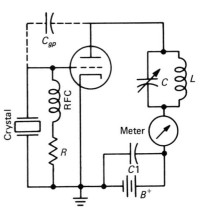

Fig. 16-101   Crystal-controlled oscillator

16-102
(*grid*) In this circuit, the feedback required to produce sustained oscillation occurs through the _____ capacitance of the tube itself.

16-103
(*interelectrode, grid-plate, etc.*) When the system is turned on, a surge of plate current flows through L, feeding energy back through the grid-plate _____ of the tube to the grid circuit.

16-104

(*capacitance*) The oscillatory energy produced by oscillations in the *LC* plate circuit is thus fed back to the crystal. On the first voltage pulse, the crystal contracts or bends because of the _____ effect.

16-105

(*piezoelectric*) When the first surge of voltage is past, the crystal springs back to its normal shape. In doing this it produces a _____ of its own. This is applied to the grid.

16-106

(*voltage*) Assume that this first voltage pulse due to crystal flexure and restoration is positive. Thus, the grid draws current from the cathode via the RF choke (RFC) and _____.

16-107

(*R*) The voltage drop across *R* appears as a pulse of negative voltage. Hence, the grid which is connected through RFC to the top of *R* becomes _____ in polarity.

16-108

(*negative*) This places an opposite charge on the crystal and causes it to flex or bend in the direction opposite the first mechanical change. When it springs back, it generates a new pulse of voltage, this time a _____ one.

16-109

(*negative*) Thus, the grid cycle is completed. It is clear, however, that sustained crystal vibration and, therefore, sustained oscillation can continue only if energy is fed back through the grid-plate capacitance in sufficient amplitude and in the correct _____.

16-110

(*phase*) Next, we shall analyze a transistor oscillator which uses a crystal in the series resonant mode. The circuit shown in Fig. 16-110 is basically a Hartley oscillator. The collector tank circuit consists of the elements *L* and _____.

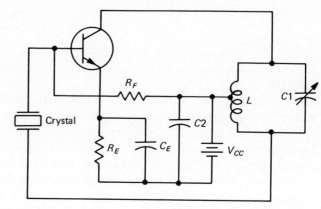

Fig. 16-110

16-111
($C1$) If the circuit of Fig. 16-110 is compared with the Hartley circuit of Fig. 15-116, it is seen that the crystal has replaced the feedback capacitor, _____.

16-112
(C_C) Recall that a series LC resonant circuit has a very low impedance at the resonant frequency, and a _____ impedance at frequencies away from resonance. Since a crystal has an extremely sharp resonance (Fig. 16-5), the impedance becomes high at frequencies off resonance.

16-113
(*high*) In Fig. 16-110, the magnitude of the voltage fed back to the base is a function of the voltage divider formed by the crystal and the internal transistor base resistance. The larger the crystal impedance, the _____ the voltage developed at the base of the transistor.

16-114
(*smaller*) Thus, because of the sharp resonance peak of the crystal, only a signal at the _____ frequency will feed back sufficient voltage to cause oscillation.

16-115
(*resonant*) The _____-type oscillator just discussed would produce good frequency stability even if the gain of the transistor varied considerably. Two factors which can affect the gain of a transistor are temperature variations and switching from one transistor to another.

16-116
(*Hartley*) For example, changing transistors in a crystal-controlled oscillator may produce frequency variations of less than 0.001 percent, whereas similar changes in an LC oscillator may produce frequency changes of about 1 percent. Thus, _____ oscillators are commonly found in receivers and transmitters where the oscillation frequency must remain constant.

16-116
(*crystal*)

CRITERION CHECK TEST

____16-1 The piezoelectric effect in a crystal is (*a*) a change in frequency with temperature, (*b*) an ultrasonic wave caused by pressure, (*c*) a change in resistance because of temperature, (*d*) a voltage developed because of mechanical stress.

____16-2 The crystal oscillator frequency is very stable due to the (*a*) rigidity of the crystal, (*b*) vibration of the crystal, (*c*) high Q of the crystal, (*d*) crystal structure.

_____16-3 The axis connecting the corners of a crystal is the (a) X axis, (b) Y axis, (c) Z axis, (d) mechanical axis.

_____16-4 Most crystals are affected by changes in temperature. If the crystal frequency increases with temperature, we say the crystal has (a) a positive temperature coefficient, (b) a negative temperature coefficient, (c) a zero temperature coefficient, (d) poor stability.

_____16-5 A crystal is rated 10 ppm/°C. This means (a) for every degree of temperature increase, the crystal frequency increases 10 Hz, (b) for every degree of temperature increase, the crystal frequency increases 10 Hz per MHz, (c) the crystal has a negative temperature coefficient, (d) the crystal frequency may increase or decrease 10 Hz per degree temperature change.

_____16-6 A crystal holder must provide an air gap to (a) keep the stray capacitance at a minimum, (b) keep contact resistance at a minimum, (c) allow the crystal to vibrate, (d) provide a point-contact rectifying junction.

_____16-7 Compute the Q of a quartz crystal whose parameters are: resonant frequency = 430 kHz, series resistance = 780 Ω, inductance = 6.6 H: (a) 11,500, (b) 23,000, (c) 46,000, (d) 92,000.

_____16-8 In Fig. 16-101, there is a tuned circuit (a) in the cathode lead, (b) only in the grid, (c) only in the plate, (d) in both the plate and grid.

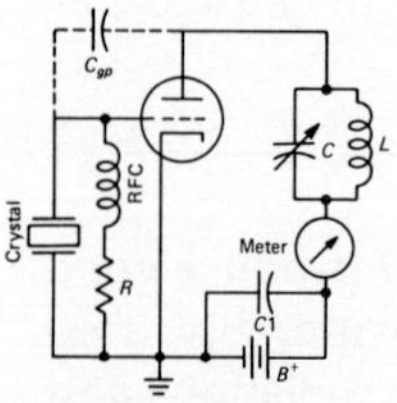

Fig. 16-101

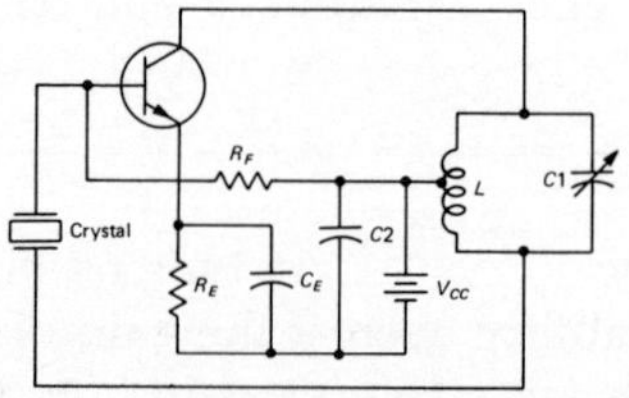

Fig. 16-110

_____16-9 In Fig. 16-101, energy is fed back from the plate by (a) magnetic induction, (b) a tickler coil, (c) the interelectrode capacitance, (d) a coupling capacitor.

_____16-10 The frequency of oscillation in Fig. 16-101 is controlled by the (a) plate-grid capacitance, (b) quartz crystal, (c) LC tuned circuit, (d) RF choke (RFC).

_____16-11 The crystal in Fig. 16-101 is used (a) in its series resonant mode, (b) in its parallel resonant mode, (c) as the grid bias, (d) as a choke.

_____16-12 In Fig. 16-110, dc base bias is provided by (a) R_F, (b) R_E, (c) the crystal, (d) C2.

_____16-13 The application where one would most likely find a quartz crystal oscillator would be (a) a commercial radio transmitter, (b) a kitchen-type AM receiver, (c) a hi-fi test AF sweep generator, (d) an electronic organ for the home.

17 modulation fundamentals

OBJECTIVES

(1) List the basic elements and give several examples of communication systems. **(2)** Sketch unmodulated sine waves of various frequencies and a pulsed sine wave, indicating the information carried by the pulsed wave. **(3)** Discuss the operation of a single-stage pulsed RF transmitter, including a sketch of a crystal-controlled oscillator circuit. **(4)** Define *modulation, carrier wave,* and *amplitude modulation.* **(5)** Describe the operation of an AM transmitter using a single RF oscillator. **(6)** Sketch the block diagram of an AM transmitter using an AF amplifier and RF oscillator. **(7)** Describe the operation of the AF modulator and sketch a carrier wave modulated by an AF wave. **(8)** Draw the block diagram of a complete transmitter using RF amplifiers. **(9)** Demonstrate the use of a feedback capacitor to prevent unwanted oscillations. **(10)** Describe the operation of an *RC*-coupled oscillator-amplifier circuit. **(11)** Sketch an *RC*-coupled 2-stage transmitter. **(12)** Indicate the function of the RF chokes, the component used for neutralization, and the tank circuit capacitor. **(13)** Sketch the schematic of a typical RF transformer. **(14)** Relate the load on a tuned circuit to its Q. **(15)** Give the equation for the resonant frequency of a tuned circuit. **(16)** State the requirement on the L/C ratio to give a high-Q tuned circuit. **(17)** Sketch a link-coupled 2-stage RF system. **(18)** State the function of the transmission line and the physical and electrical properties of the link coils. **(19)** Indicate the relative magnitudes of currents and voltages in the links and transmission line. **(20)** Demonstrate that the link coils behave as RF transformers. **(21)** Draw the schematic of a 3-transistor transmitter. **(22)** Indicate the tank circuit of the RF oscillator, the frequency-stabilizing element, the coupling used for the oscillator-amplifier stages, the function of the RF choke, and the transistor used as the AF amplifier, the bias resistors, emitter resistors, and bypass capacitors. **(23)** Demonstrate that the AF amplifier varies the RF output.

PULSED MODULATION

Communication of intelligence from one point in space to another demands that there be variations of some form in the signal used for such communication. A steady tone or a steady radio wave cannot convey intelligence. The variations produced in a radio wave for the purpose of including a message on this wave are produced by a process called *modulation*. Modulation may take many forms. At this point we shall introduce a form of modulation which uses a dot-dash code for communication.

17-1
When one person speaks to another in the same room, all the elements of a *communication system* are present. The voice box generates recognizable symbols (words), the air serves to carry the sound, and the _____ of the listener acts to receive the symbols, passing them to the brain for interpretation.

17-2
(*ear*) The elements of a communication system are (1) the *transmitter,* or device that produces symbols in suitable form to be carried by (2) the *medium,* or element which transfers the symbols from the transmitter to (3) the *receiver* which accepts the symbols for interpretation. In the example of Frame 17-1, the _____ was the medium between voice and ear.

17-3
(*air*) In the case of an ordinary telephone, the medium is a length of _____ which connects the transmitter to the receiver.

17-4
(*wire*) In the telephone, the *transmitter* must convert the sound of the voice into symbols suitable for transmission by wire. Thus, in this case, the symbols take the form of variations in _____ potentials and currents.

17-5
(*electric*) Continuing with the telephone example, it is clear that a dc voltage applied to the transmitter cannot communicate, or convey intelligence. In the same sense, a steady sound such as "ah-h-h-h" cannot communicate. Thus, communication requires that the symbol _____ from moment to moment in order to convey intelligence.

17-6
(*change* or *vary*) In radio communication, the *transmitter* generates electromagnetic waves that move through *space* to the *receiver*. This is the radio communication system. In this case, the medium is _____.

17-7
(*space*) Three waves are shown in Fig. 17-7. All 3 are waves of the same *steady* amplitude. The first is a 60-Hz power wave, the second is an audio wave, and the third is a radio wave. They are all electromagnetic in nature, differing only in their wavelength or _____.

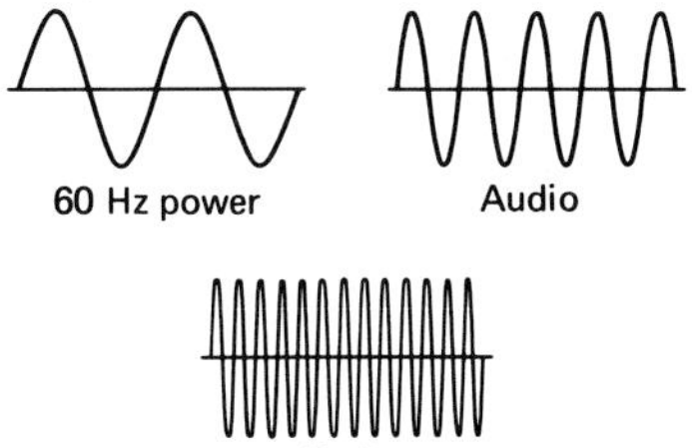

Fig. 17-7

17-8
(*frequency*) None of these 3 waves as shown in Fig. 17-7 conveys intelligence because none of these waves contains _____ that might be interpreted as communication symbols.

17-9
(*variations*) The simplest way to add suitable symbolic variations to a radio wave is to break the wave up into bursts of emission of different lengths. For example, in Fig. 17-9, we have a short burst, a long burst, and finally a second _____ burst of energy.

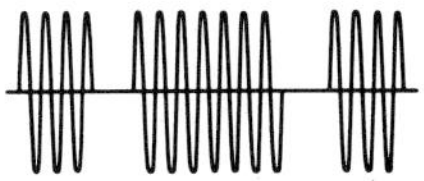

Fig. 17-9

17-10
(*short*) In the standard International Radio Code, this would be interpreted as the letter R; hence, intelligence has been conveyed. Similarly, as in Fig. 17-10, the letter K is symbolized by a long burst, a _____ burst, and a final long burst of energy.

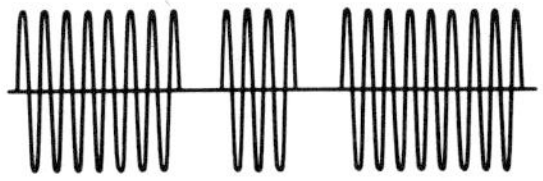

Fig. 17-10

17-11
(*short*) We call these long and short bursts "dots and dashes," a different combination used for every letter and number in common use. Thus, the use of the International Code permits complete communication by introducing _____ into the otherwise steady radio wave.

17-12
(*variations*) When a radio wave is varied in *any* way, we say it is *modulated*. The process of introducing the variations is called *modulation*. It is only by modulation that a radio wave may serve as the means of communication between a transmitter and _____

MODULATION FUNDAMENTALS 433

17-13
(*receiver*) Figure 17-13 illustrates a radio-frequency oscillator with an additional coupled circuit on the right side. The type of oscillator is a _____ oscillator.

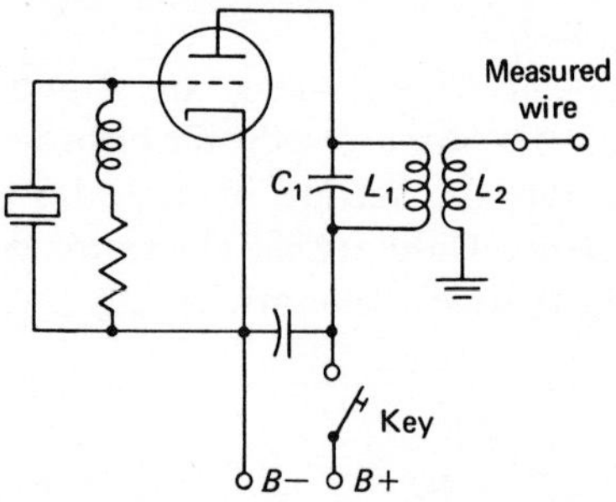

Fig. 17-13

17-14
(*crystal*) The radio-frequency energy present in $L1$ due to sustained oscillation is transferred by electromagnetic induction to $L2$. Thus, $L1$ and $L2$ together may be considered as a radio-frequency _____.

17-15
(*transformer*) The measured wire connected to the top of $L2$ is an *antenna*. The function of an antenna is to transfer the radio-frequency variations in $L2$ into space which will act as a _____ to carry the wave to the receiver.

17-16
(*medium*) When the radio-frequency variations in the antenna radiate into space, the waves thus produced are electromagnetic in nature. As the system continues to oscillate, a *carrier* wave is produced. The carrier wave cannot convey intelligence in itself since it contains no _____ that could be symbolically interpreted.

17-17
(*variations*) By opening and closing the power circuit, however, the carrier wave can be broken up into dots and dashes according to an established code. The device shown in Fig. 17-13 which accomplishes the addition of intelligence to the carrier wave is the _____.

17-18
(*key*) When the dots and dashes are intercepted at the receiver, they may be made audible in the form of short and long whistling sounds. These sounds spell out words which may then be interpreted as a message. Hence, the process of keying an oscillator is a form of _____.

17-19
(*modulation*) Thus, an oscillator becomes a *transmitter*, the antenna "matches" the transmitter to space, which serves as a *medium*, and the electromagnetic waves carrying their dot-and-dash symbols move to the _____ where the symbols may be interpreted into a message.

17-19
(*receiver*)

AM TRANSMITTER—BLOCK DIAGRAM

When the intelligence present in an audio waveform is combined with a radio-frequency waveform in a manner such that the *amplitude* of the RF varies in step with the AF, the radio waveform is said to be *amplitude-modulated.* In this section we shall work with a block diagram of an amplitude-modulated RF oscillator with a view toward interpreting the resulting waveforms.

17-20
Another important form of modulation is known as *amplitude modulation.* Modulation means to vary or change. Amplitude modulation (abbrev. AM) is therefore a process in which the amplitude of the radio frequency wave is _____.

17-21
(*varied* or *changed*) For communication, speech is used to vary the amplitude of the carrier wave. The carrier wave is a radio-frequency wave. The wave produced by speech that is changed into electric impulses is a(n) _____-frequency wave.

17-22
(*audio*) Figure 17-22 illustrates in block form the system that might be used to produce speech AM. In this system, the carrier wave is produced by the _____ oscillator.

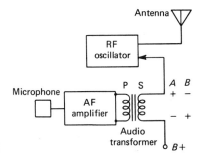

Fig. 17-22

17-23
(*RF*) As long as no one speaks into the microphone, the output of the RF oscillator does not vary. The carrier wave fed to the _____ under these conditions goes out into space without variations.

17-24
(*antenna*) When the microphone is activated by speech, it converts the sound waves into _____ impulses.

MODULATION FUNDAMENTALS

17-25
(*electric*) These electric impulses have the same frequency and waveform as the original _____ waves.

17-26
(*sound*) The electric impulses are then amplified in the AF amplifier. The output of the AF amplifier is fed to the primary winding of an _____ transformer.

17-27
(*audio*) By induction, these impulses then appear as voltage drops across the secondary winding of the audio transformer. The secondary voltage of *any* transformer is always ac. At a given instant, the top of the secondary may be positive with respect to the bottom as for condition *A*. At the next instant, a polarity reversal may occur (condition _____).

17-28
(*B*) You will note that the secondary winding of the transformer is in series with the *B+* feed to the RF oscillator. When condition *A* prevails, the voltage across the secondary is *series aiding*. This means that at this instant the voltage drop across the secondary is _____ to the magnitude of the dc voltage fed to the oscillator.

17-29
(*adding*) At the next instant when condition *B* prevails, the action is *series opposing*. When this occurs, the voltage drop across the transformer secondary _____ from the dc voltage fed to the oscillator.

17-30
(*subtracts*) The amplitude of the RF wave generated by the oscillator depends upon the power it receives from the power supply. When the audio adds to the power, the amplitude of the RF wave must _____.

17-31
(*increase*) This happens for condition *A*. But for condition *B*, where the audio voltage subtracts from the dc voltage fed to the oscillator, the amplitude of the carrier must _____.

17-32
(*decrease*) Thus, the amplitude of the carrier rises and falls in synchronism with the variations in the _____ voltage applied through the secondary of the transformer.

17-33
(*audio*) Figure 17-33 illustrates this effect in diagrammatic form. In this diagram, the high-frequency wave labeled wave *A* is the _____ wave.

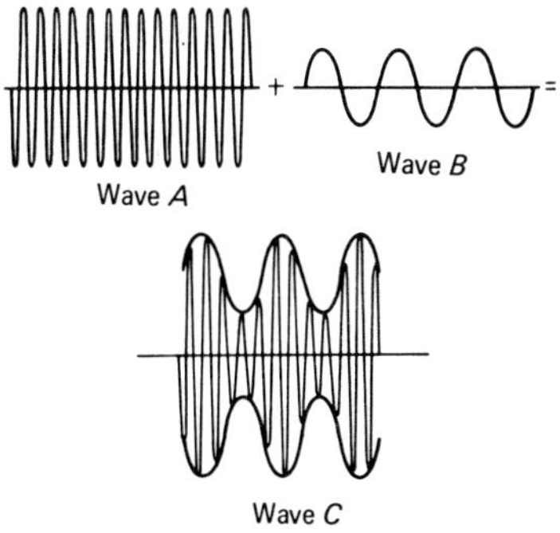

Fig. 17-33

Wave *A*

Wave *B*

Wave *C*

17-34
(*carrier* or *RF*) The low-frequency wave that is added to the carrier wave, indicated as wave *B*, is the _____-frequency wave used to modulate the carrier.

17-35
(*audio*) When these two waves are added, the resulting composite wave contains both the original RF variations and an audio variation that acts to increase and decrease the _____ of the carrier.

17-36
(*amplitude*) An AM wave may be used to communicate intelligence because the carrier contains variations which can be interpreted as a message when the wave reaches a suitable _____ that can extract the audio variations and form them into speech.

17-37
(*receiver*) Modulation of an oscillator has many limitations. Among these is the problem of poor frequency stability. When an oscillator is coupled to an antenna directly, changes in antenna capacitance due to moisture, vibration, swinging motions, etc., cause the _____ of the oscillator to vary as well.

17-38
(*frequency*) The frequency of an oscillator such as a Hartley or Colpitts is determined by the *L* and *C* values of its tank, but also to some extent by the voltages applied. When modulation occurs, plate voltage changes in accordance with the modulation. This must also cause the _____ to vary with the modulation.

MODULATION FUNDAMENTALS

17-39

(*frequency*) If we attempt to achieve good frequency stability by using a crystal oscillator and modulating this oscillator, we find that the maximum power output we can obtain is poor because we cannot feed too much power to the oscillator without endangering the _____.

17-40

(*crystal*) Thus, practical modern transmitters do not use an oscillator alone as the RF output for modulation. As we have seen, when this is attempted, the transmitter either has poor frequency _____ or low output.

17-41

(*stability*) Instead, modern transmitters are designed along lines shown in the block diagram of Fig. 17-41. According to the waveform shown above the block labeled *oscillator*, this section of the transmitter is the source of the _____ wave.

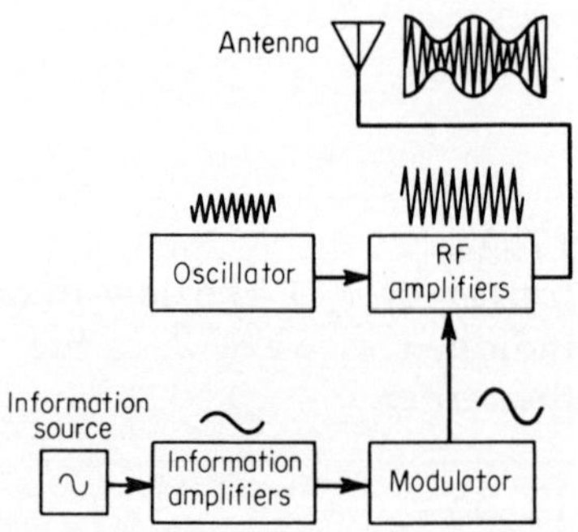

Fig. 17-41

17-42

(*RF* or *carrier*) The oscillator does not feed the antenna directly. Its output is fed to the block labeled *RF amplifiers* where the power of the RF or carrier is greatly _____.

17-43

(*amplified* or *increased*) Along the lower line, we first encounter an *information source*. This is the source of the audio-frequency power that is to be used to modulate the _____ wave.

17-44

(*carrier* or *RF*) The information source may be a microphone, tape recorder, record player, or any other device that produces the _____ required to endow the carrier with a waveform capable of transmitting a meaningful message.

17-45

(*variations*) The power output of the information source, however, is usually quite _____.

17-46

(*small* or *low*) For this reason, the information source must be followed by an _____ that is capable of increasing this power.

17-47

(*amplifier*) The information amplifiers are followed by a circuit or device that is capable of *superimposing* the information on the carrier. This circuit or device is shown labeled as a _____ in Fig. 17-41.

17-48

(*modulator*) After the carrier and information voltages are mixed, the final output into the antenna is in the form of a(n) _____-modulated wave.

17-48

(*amplitude*)

NEUTRALIZATION

We must pause in our development of modulation principles in order to introduce an important aspect of transmitting RF amplifiers: unwanted oscillation. The technique we shall discuss to prevent unwanted oscillation is equally applicable to RF receiving circuits.

17-49

Study Fig. 17-49. This shows one form of oscillator amplifier in skeleton circuit form. This arrangement could be used as a code transmitter in which the oscillator circuit is started and stopped by the operation of the _____.

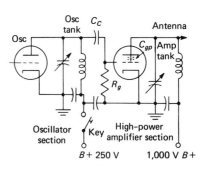

Fig. 17-49

17-50

(*key*) Note that the plate voltage used for the oscillator is 250 V, while the plate voltage used for the amplifier is _____ V.

17-51

(*1,000*) The RF amplifier is coupled to the oscillator by capacitive coupling. The coupling capacitor is the one labeled _____ in Fig. 17-49.

17-52
(C_C) The oscillator tank belongs properly to the oscillator circuit, of course. Yet, by viewing the oscillator tank as part of the amplifier circuit, we note that the amplifier now has a tuned-grid circuit and a tuned-_____ circuit.

17-53
(*plate*) Since the amplifier is connected to the oscillator tank, this tank circuit may form a part of the amplifier grid circuit. Noting this, we can see at once all the components required to form a tuned-_____, tuned-plate *oscillator* are present in the *amplifier* circuit.

17-54
(*grid*) We do not want the amplifier to oscillate; for good frequency stability, the oscillator should serve as the stable frequency-controlling device while the amplifier should merely amplify, not _____.

17-55
(*oscillate*) If the amplifier breaks into oscillation, all the desirable features of stable oscillator-frequency control are lost. The feedback "component" in a tuned-grid, tuned-plate oscillator, shown in dotted lines in Fig. 17-49, is the _____-_____ capacitance.

17-56
(*grid-plate*) But the grid-plate capacitance of triodes is large; hence, oscillation of the amplifier is likely to occur in transmitting RF amplifiers where _____-type tubes are used.

17-57
(*triode*) Since C_{gp} is the cause of oscillation, it is possible to prevent oscillation from occurring by *neutralizing* the effects of this capacitor. Remember that this capacitance feeds back an oscillatory voltage in the proper _____ to make up for losses in the grid tank circuit.

17-58
(*phase*) Figure 17-58 shows one type of *neutralization* circuit. A neutralizing capacitor C_N is connected from the bottom of the oscillator tank to the _____ of the tube.

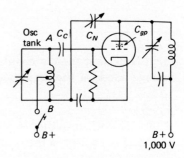

Fig. 17-58

17-59
(*plate*) The B+ to the oscillator plate in this circuit is fed through only a *part* of the tank coil. Note the tap on the oscillator tank coil. Note also that C_{gp} feeds back its voltage through C_C to point A on the coil, while C_N feeds back an equivalent voltage to point _____ on the oscillator tank coil.

17-60
(*B*) By adjusting the capacitance of C_N, it is possible to feed back exactly as much voltage through this capacitor as is fed back through C_{gp}. Since the voltages are fed to *opposite* ends of the tank coil, they cancel each other in the tank coil. With the effect of C_{gp} canceled, _____ cannot occur since there is no positive feedback.

17-60
(*oscillation*)

RC OSCILLATOR-AMPLIFIER COUPLING

The efficiency of power transfer from an RF oscillator to an RF amplifier is governed by the character of the coupling system. The selection of coupling method depends upon factors such as separation of oscillator from amplifier, power being transferred, economy of components, and economy of tuning. We start our discussion of coupling with a system already familiar to you.

17-61
If the oscillator and amplifier can be placed physically close to each other, they may be coupled with an *RC* circuit just as audio amplifiers are often coupled (Fig. 17-61). In this circuit, the component labeled _____ is the coupling capacitor.

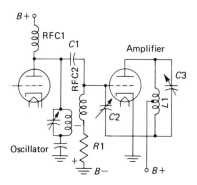

Fig. 17-61

17-62
(*C1*) RFC1 is present in the oscillator plate circuit to prevent _____-frequency energy generated by the oscillator being lost in the power supply.

17-63
(*radio*) The capacitance of C1 must be selected so as to provide a _____-impedance path for RF.

MODULATION FUNDAMENTALS 441

17-64
(*low*) The voltage drop across the low impedance of C1 will be small; hence most of the RF voltage developed at the plate of the oscillator will appear at the _____ of the amplifier tube.

17-65
(*grid*) To keep the voltage at the grid of the amplifier tube high, the RF voltage must find a high-impedance path to ground at this point. The high impedance to ground at the grid of the amplifier is presented by the component labeled _____.

17-66
(*RFC2*) In operation, the oscillator voltage appearing at the grid of the amplifier (via C1) causes the grid circuit to draw current. This current flows down through RFC2 and R1 to ground. The voltage drop serves as negative _____ bias for the amplifier.

17-67
(*grid*) A somewhat different neutralizing circuit is used in Fig. 17-61. The neutralizing capacitor is _____ since this capacitor allows energy from the bottom of the amplifier tank to be fed back into the grid circuit.

17-68
(*C2*) This fed-back energy comes from the *bottom* of the tapped tank coil; the positive feedback that would tend to cause oscillation comes from the _____ of the tank coil and passes back into the grid circuit through the C_{gp} of the amplifier.

17-69
(*top*) When C2 is properly adjusted, the magnitudes of the two fed-back voltages are equal. They cancel each other, thereby preventing oscillation. Thus, the phase relationship between these voltages must be _____°.

17-70
(*180*) The capacitor which resonates the plate circuit of the amplifier to the same frequency as that of the oscillator is _____.

17-71
(*C3*) The grid of the amplifier is driven positive by the oscillator voltage; hence the amplifier grid circuit draws current. Thus, power must be transferred from the _____ plate circuit to the amplifier grid circuit.

17-72
(*oscillator*) For maximum transfer of power, the impedance of the source must equal the impedance of the load. The source in this case is the oscillator plate circuit, and the load is the amplifier _____ circuit.

17-73
(*grid*) C1 is merely a coupling device, not an impedance-matching device. Thus, if the output impedance of the oscillator is not equal to the input impedance of the amplifier, maximum _____ transfer will not occur.

17-74
(*power*) This is one defect of capacitive coupling. Here is another: The impedances of both the oscillator tank circuit and the grid circuit of the amplifier are quite high. This means that large RF voltages exist in the connecting wires to and from C1. Power can *radiate* into space from these leads. Radiated _____ represents a *loss*.

17-75
(*power*) Power radiated from a *high-impedance* line such as that connecting the plate of the oscillator to the grid of the amplifier is power that never reaches the _____ of the amplifier tube. Thus the power is not efficiently transferred from oscillator to amplifier.

17-76
(*grid*) Radiation losses are not severe if the connecting wires between C1 and its terminations are very short. The power loss may be considerable, however, if these leads are _____ .

17-77
(*long*) Therefore, capacitive coupling between RF transmitting stages is never used when the stages must be located some _____ from each other.

17-77
(*distance*)

INDUCTIVE OSCILLATOR-AMPLIFIER COUPLING

An improvement in impedance matching is achieved by utilizing inductive rather than capacitive coupling. Inductive coupling is more efficient, but it also requires that the oscillator plate tank and the amplifier grid tank be closely adjacent to each other.

MODULATION FUNDAMENTALS 443

17-78

A second method of coupling RF transmitting amplifiers is shown in Fig. 17-78. This method is called *inductive coupling*. In this system, $L1$ and $L2$ are placed physically adjacent to each other so that currents flowing in $L1$ will _____ voltages across $L2$.

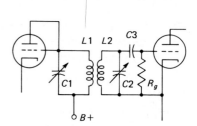

Fig. 17-78

17-79

(*induce*) When $L1C1$ is tuned to resonance at the transmitting frequency, this combination forms a parallel resonant circuit. In common with all parallel resonant circuits, $L1C1$ has a _____ impedance.

17-80

(*high*) As previously explained, the connecting wires that lead to a high-impedance termination must be kept reasonably _____ in order to avoid radiation losses.

17-81

(*short*) The combination of $L2$ and $C2$ also forms a _____ resonant circuit having a high impedance.

17-82

(*parallel*) Hence, the connecting wires that go to the following amplifier must also be _____ in length.

17-83

(*short*) This automatically means that the amplifiers must be _____ to each other if efficient transfer of energy from one to the other is to occur.

17-84

(*close*) In this respect, then, inductive coupling and _____ coupling have the same limitation; the amplifiers to be coupled must be located physically close to each other.

17-85

(*capacitive*) $L1$ and $L2$ form an *air-core* transformer. That is, no iron is present in the coupling circuit as there is when audio amplifiers are coupled. Without iron, the number of lines from the primary that cut the secondary must be _____ than in the iron-core transformer.

17-86
(*fewer, less, etc.*) Since the lines of force from the primary that cut through the secondary are *fewer* in the air-core transformer, the coefficient of coupling between these coils must be _____ than it is for an iron-core transformer.

17-87
(*smaller*) We have shown that, in transmitting amplifiers, the grid of the second amplifier will always draw current under excitation from the first amplifier. This means that _____ must be transferred from the first to the second amplifier, not merely voltage.

17-88
(*power*) When the secondary of a coupled system requires power, we say that the primary circuit is *loaded* by the secondary circuit. Thus, in Fig. 17-78, the second amplifier grid circuit causes the primary resonant circuit to be _____.

17-89
(*loaded*) Loading a tuned circuit always reduces its Q. The combination L1C1 may have a high Q when not coupled to L2C2, but the moment L2C2 begins to load L1C1, the Q of L1C1 must _____.

17-90
(*decrease*) When the Q of one of the coupled circuits is low, power transfer by inductive coupling becomes difficult to achieve, especially when the coefficient of coupling is also _____.

17-91
(*low*) Thus, to ensure adequate power transfer, it is necessary to make the Q of both tuned circuits as _____ as possible.

17-92
(*high*) To improve Q, we must review the resonance equation. The resonant frequency of a circuit may be found from the equation $f = 1/$_____.

17-93
($2\pi \sqrt{LC}$) This equation shows that we can bring about resonance at a given frequency by using a medium-sized L and a medium-sized C, a large L and a small C, or a small L and a _____ C.

17-94
(*large*) If the L and C are both medium in value, we say we have a normal L/C ratio; if L is large and C is small, then the L/C ratio is large; finally, if L is small compared with C, the L/C ratio is _____.

17-95

(*small*) When a circuit is loaded and its Q tends to decrease, the Q of the circuit may be improved by using a *smaller L* and a *larger C* to establish the desired resonant frequency. This means that we must decrease the _____ ratio to raise the Q of a loaded circuit.

17-96

(L/C) Thus, in an inductively coupled circuit such as that of Fig. 17-78, where loading is heavy, coefficient of coupling is small, and Q is relatively low, adequate power can be coupled only by ensuring a higher Q. This is best done by using a _____ L/C ratio.

17-96
(*small*)

LINK COUPLING

For good impedance matching between amplifiers located at some distance from each other, *link coupling* is to be preferred. In addition, the radiation losses in a link-coupled cascade are substantially smaller than with the arrangements previously discussed.

17-97

Figure 17-97 illustrates a third method of coupling RF transmitting amplifiers. In this system, there are, altogether, _____ coils.

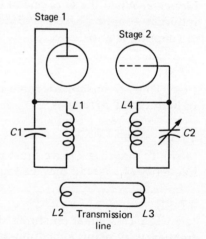

Fig. 17-97

17-98

(4) Two of these 4 coils are tuned by tuning capacitors ($C1$ and $C2$). The other 2 coils have very few turns and are nonresonant because there are no tuning _____ associated with them.

17-99
(*capacitors*) The wires connecting the two coils, L2 and L3, compose what is known as a _____ _____. This line may be several feet long, but is so designed that it produces almost no radiation losses. A coaxial cable is an example of a transmission line.

17-100
(*transmission line*) Thus, this system of coupling has the advantage over the other two in that the stages need not be very _____ to each other physically.

17-101
(*close*) This system of coupling is called *link coupling*. Coils L2 and L3 are known as the *links*. Each of these are closely coupled to their respective tank coils, establishing as large a _____ of coupling as possible.

17-102
(*coefficient*) Both L2 and L3 have very few turns. At high frequencies, either of the coils may consist of 2 or even a single turn. This means that the inductances of the individual coils are quite _____.

17-103
(*low* or *small*) The transmission line starts and ends with a low inductance in the form of the links. The *impedance* at either end is also quite _____.

17-104
(*low* or *small*) Such a line is said to terminate in low impedances. Nowhere in the line is there any impedance that can be considered high. When the impedance of any inductive system is low, the voltage that can develop across any of its components must also be _____.

17-105
(*low* or *small*) This follows from the consideration that $V = IZ$. If Z is small, the V must be _____.

17-106
(*small*) The power required by the grid circuit of stage 2 can be supplied by either voltage or _____ since power is determined by the product of V and I.

17-107
(*current*) Hence, if power is transmitted to the grid of stage 2, there must be a relatively large _____ flowing in the transmission line.

MODULATION FUNDAMENTALS

17-108
(*current*) The RF voltage, however, as we have shown is nowhere high. When the RF voltage is low, the amount of radiation is _____.

17-109
(*low* or *small*) Thus, a link-coupled circuit radiates very little energy. This accounts for the fact that the lines may be several feet in length. Even with a line length of this size, the radiation losses will be _____.

17-110
(*small*) Realizing that the current in the lines may be large, however, it is necessary to take certain precautions to avoid heat losses in the lines. All we need do is be sure that the _____ of the coils and lines are kept small to avoid large I^2R losses.

17-111
(*resistance*) The link-coupled system may be viewed in a somewhat different way. The combination of $L1$ and $L2$ may be thought of as a transformer. Since $L1$ has more turns than $L2$, this would be a step _____ transformer.

17-112
(*down*) The RF voltage developed in $L1$, then, would be reduced by stepdown action in $L2$. Hence, the lines are seen to carry power in the form of low _____ and high current.

17-113
(*voltage*) At the remote end of the line, the transformer is of the _____ type.

17-114
(*stepup*) Thus, the low voltage appearing across $L3$ is restored to nearly the original value by stepup induction from $L3$ to _____.

17-115
(*L4*) The link-coupled system may also be viewed as an impedance-transformation arrangement. In the $L1L2$ combination, the impedance is transformed from a high value in $L1$ to a low value in _____.

17-116
(*L2*) Power is transferred from one end of the line to the other through the low impedance of the system. At the remote end, the power is transferred from the low impedance of $L3$ to the high impedance of _____.

17-117
(*L4*) Thus, the impedance requirements are satisfied: The plate circuit of stage 1 must be a high-impedance circuit at the resonant frequency; the transmission system must be a _____-impedance system to avoid radiation losses; and finally, the impedance must be transformed upward to meet the demands of the grid circuit.

17-117
(*low*)

TRANSISTORIZED AM TRANSMITTER
17-118
Figure 17-118 depicts an AM transmitter. This transmitter uses 3 transistors, denoted by _____, _____, and _____.

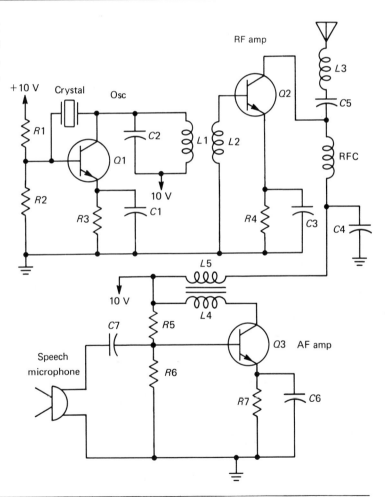

Fig. 17-118

17-119
(*Q1, Q2, Q3*) The first transistor, *Q1*, is used in the RF oscillator stage. The collector tank circuit of *Q1* consists of the two elements _____ and _____. This tank is tuned to the carrier frequency.

17-120
(*C2, L1*) The element providing frequency stability is a quartz crystal. Since the quartz crystal is the feedback element, it is operating in the _____ (series, parallel) mode.

17-121
(*series*) Transistor *Q2* is used in the RF amplifier stage. The oscillator and RF stages are indirectly coupled via the coils _____ and _____.

17-122
(*L1, L2*) The antenna is connected to *Q2* by the two components _____ and _____.

17-123
(*L3, C5*) The load for the collector of *Q2* is a coil denoted by _____.

17-124
(*RFC* or *radio frequency choke*) The 2 stages described so far would alone produce a single-frequency carrier wave. Thus, the transistor _____ must be providing the modulation.

17-125
(*Q3*) The function of *Q3* is to amplify the audio frequencies (speech) which are picked up by the _____. The amplified speech appears as an alternating voltage at the collector load of *Q3*.

17-126
(*microphone*) The collector load of *Q3* is the coil denoted as *L4*. The alternating voltage in *L4* in turn induces an alternating voltage in coil _____.

17-127
(*L5*) Moreover, the induced voltage in *L5* is in series with the collector supply (10 V) of *Q2*. Thus the collector of *Q2* sees a net supply voltage consisting of _____ V dc plus an alternating voltage.

17-128
(*10*) Summarizing: The collector of the RF amplifier sees a dc voltage plus an alternating voltage in the _____-frequency range.

17-129
(*audio*) In Fig. 17-129, we have depicted the effective supply voltage of Q2. It is seen that we have chosen the audio to be a single-frequency sine wave of amplitude _____ V.

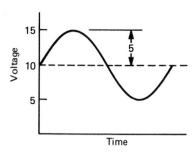

Fig. 17-129

17-130
(5) Thus the effective supply voltage will have a maximum voltage of 15 V and a minimum voltage of _____ V.

17-131
(5) As the effective supply voltage varies, the RF carrier output will also vary in magnitude. In fact, the RF carrier amplitude is directly proportional to the _____ _____ voltage.

17-132
(*effective supply*) When the effective supply voltage is maximum, the carrier voltage is maximum; when the effective supply voltage is minimum, the carrier voltage is minimum. Thus the output of the RF amplifier is an amplitude-_____ carrier wave.

17-133
(*modulated*) Summarizing: The RF carrier is generated in the _____ stage.

17-134
(*oscillator*) Summarizing: The carrier is amplified in the _____ _____ stage.

17-135
(*RF amplifier*) Summarizing: The voice energy from the microphone is raised to the levels needed for modulation by the _____ _____ stage.

17-136
(*audio amplifier*) Summarizing: The audio frequencies are inductively coupled into the RF amplifier stage to modulate the _____ wave.

17-137
(*carrier*) The remainder of the circuit is analyzed as follows: The emitter-swamping resistor for Q1 is _____.

17-138
(*R3*) C1 is the bypass capacitor for resistor _____.

17-139
(*R3*) Forward bias for transistor Q1 is provided by resistors _____ and _____.

17-140
(*R1, R2*) The emitter-swamping resistor and bypass capacitor for Q2 are _____ and _____.

17-141
(*R4, C3*) Q2 has no forward bias; hence it is operating in class B. The output stage of a transmitter often uses class B or C since these offer much better efficiency than class _____.

17-142
(*A*) Forward bias for Q3 is provided by resistors _____ and _____.

17-143
(*R5, R6*) The emitter-swamping resistor and bypass capacitor for Q3 are _____ and _____.

17-143
(*R7, C6*)

CRITERION CHECK TEST

___17-1 Modulation is the process whereby (*a*) a carrier is generated, (*b*) a carrier is transmitted, (*c*) a signal is transmitted, (*d*) intelligence is impressed on the carrier.

___17-2 In radio the medium of transmission is (*a*) wire, (*b*) an antenna, (*c*) space, (*d*) coaxial cable.

___17-3 In a pulsed modulation system, the carrier (*a*) is frequency-modulated, (*b*) is either full on or full off, (*c*) has a frequency equal to the pulse width, (*d*) has a frequency less than 60 Hz.

___17-4 In amplitude modulation, the (*a*) carrier wave has a frequency about equal to the frequency of the modulating signal, (*b*) carrier wave has a frequency much higher than the frequency of the modulating signal, (*c*) modulating signal must have a single fixed frequency, (*d*) amplitude of the modulating signal must be much smaller than the amplitude of the carrier wave.

___17-5 In high-power AM transmission the stage being modulated is usually the (*a*) oscillator, (*b*) buffer, (*c*) RF power amplifier, (*d*) microphone.

___17-6 In Fig. 17-49, the amplifier will oscillate if C_{gp} provides (*a*) negative feedback, (*b*) out-of-phase feedback, (*c*) positive feedback, (*d*) degenerative feedback.

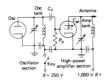
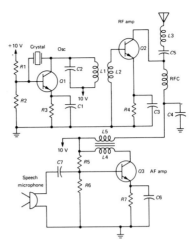

Fig. 17-49   Fig. 17-118

____17-7 The term *neutralization* refers to (*a*) stabilizing the frequency of an oscillator, (*b*) modulating at very low levels, (*c*) inhibiting spurious oscillations, (*d*) tuning an antenna for maximum gain.

____17-8 The function of an RF choke is to provide a (*a*) high RF impedance, high dc resistance; (*b*) high RF impedance, low dc resistance; (*c*) low RF impedance, low dc resistance; (*d*) low RF impedance, high dc resistance.

____17-9 An advantage of inductive coupling is that (*a*) there is good impedance match, (*b*) there is low radiation loss, (*c*) oscillator and amplifier need not be close together, (*d*) air-core transformers are very efficient.

____17-10 A typical line used in link coupling is a (*a*) microwave horn, (*b*) coaxial cable, (*c*) stub antenna, (*d*) parabolic reflector.

____17-11 In Fig. 17-118, the feedback element of the oscillator is (*a*) a crystal, (*b*) R1, (*c*) R2, (*d*) C2.

____17-12 In Fig. 17-118, the oscillator tank circuit consists of (*a*) L1C2, (*b*) L2C3, (*c*) R3C1, (*d*) L3C5.

____17-13 In Fig. 17-118, the oscillator is coupled to the RF amplifier by (*a*) RC coupling, (*b*) inductive coupling, (*c*) link coupling, (*d*) cathode coupling.

____17-14 In Fig. 17-118, the function of Q3 is to act as an (*a*) RF amplifier, (*b*) RF oscillator, (*c*) AF amplifier, (*d*) AF oscillator.

____17-15 In Fig. 17-118, the collector supply of Q2 is modulated by (*a*) C5, (*b*) L3, (*c*) L5, (*d*) C4.

MODULATION FUNDAMENTALS **453**

18 detection and detectors

OBJECTIVES

(1) Define the terms *frequency coverage, sensitivity, selectivity,* and *fidelity*. **(2)** Sketch a typical AM receiver antenna system. **(3)** List the desirable characteristics of an RF tuned circuit. **(4)** Diagram an RF transformer and modify it to obtain frequency selectivity. **(5)** Relate the load on a tuned circuit to the Q of the circuit. **(6)** Define and describe the process of *detecting* an AM wave. **(7)** Draw a modulated RF carrier wave and state the reason for using an RF carrier wave. **(8)** Specify the response of ordinary headphones to an RF-modulated signal. **(9)** Define and describe the process of rectification applied to an AM wave. **(10)** Draw an AM wave, labeling the positive and negative envelopes. **(11)** Draw a rectified AM wave. **(12)** Draw an AM wave which has gone through the process of detection and rectification. **(13)** Draw a simple diode AM radio receiver and describe its operation. **(14)** State the function of the diode and the headphones, and indicate the component which performs the detection process. **(15)** State the relative sensitivity, relative fidelity, and selectivity characteristic of the diode receiver. **(16)** Describe the operation of a single-transistor AM receiver. **(17)** Draw an AM receiver circuit using a single FET. **(18)** Indicate the component used to provide gate bias. **(19)** State the operating point of the drain circuit. **(20)** Demonstrate that the FET provides rectification of an AM wave. **(21)** Contrast the performance of the diode receiver and the transistor receiver.

RADIO RECEIVER CRITERIA

As an introduction to our study of receiver detectors, we first define the four criteria by which a receiver is judged. These are frequency coverage, sensitivity, selectivity, and fidelity.

18-1
The radio receiver has the job of completing the communications cycle that starts at the transmitter. At the transmitter, some form of modulation is added to the carrier. At the receiver, this same _____ must be removed from the carrier.

18-2
(*modulation*) The excellence of a receiver is judged by four criteria. The first of these is called its *frequency coverage*. The frequency coverage of a receiver is defined by the range of _____ frequencies it is capable of tuning to.

18-3
(*radio*) Frequencies may be radiated anywhere from 10 kHz to above 30,000 MHz. No one receiver has yet been designed that will cover this entire range successfully. To cover this entire range, it is necessary to have more than _____ receiver.

18-4
(*1*) Thus, a standard broadcast receiver covers the band from 550 to 1,600 kHz. We might therefore say that a broadcast receiver has a frequency _____ that extends from 550 to 1,600 kHz.

18-5
(*coverage* or *range*) Other receivers are designed for use at the higher radio frequencies. For example, a receiver designed to cover from 1.6 to 28 MHz has a _____ _____ over a wide portion of the medium-frequency spectrum.

18-6
(*frequency coverage*) The second criterion for judging a receiver has to do with its *sensitivity*. A sensitive receiver, or a receiver with *high* sensitivity, can extract the intelligence from weak signals. If the receiver requires a strong signal for extraction of intelligence, it is said to have _____ sensitivity.

18-7
(*low*) A receiver may be called upon to receive and interpret signals of widely varying strengths. If its sensitivity is high, it will extract information from any type of signal. Thus, a good high-sensitivity receiver is capable of operating on either weak or _____ signals.

18-8
(*strong*) A third criterion for judging receiver performance is *selectivity*. A selective receiver is one which can separate radio frequencies that are closely adjacent in frequency. As it selects one, it must reject others. Thus high selectivity means the ability to _____ one frequency while rejecting all others.

18-9
(*select*) When in the course of ordinary use, a receiver makes audible several signals of different frequencies, it is said to have poor _____.

18-10
(*selectivity*) If a receiver can differentiate between two signals that lie very close to each other in the frequency spectrum, this receiver is said to have _____ selectivity.

18-11
(*high*) The fourth criterion of receiver performance is *fidelity*. A receiver intended for the reproduction of symphonic music must have high _____.

18-12
(*fidelity*) Thus, fidelity may be defined as the accuracy with which a receiver is capable of reproducing the original modulation used for transmission. If a receiver renders a transmission unintelligible, it must have low _____.

18-13
(*fidelity*) A receiver which produces *distortion* of the signal to the point where the reproduced signal is very different from the original modulation has very _____ _____.

18-14
(*low fidelity*) The start of the receiver is its antenna and input tuning system. The antenna is the link between space and the receiver; the input tuning system is the first of several stages which will work with each other in selecting one of the many RF signals intercepted by the antenna for further processing.

18-15
The receiver system starts with a receiving *antenna*. The function of the antenna is to intercept the passing wave. If the antenna is outdoors, it may be a length of _____ stretched between two insulated masts (Fig. 18-15).

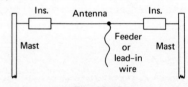

Fig. 18-15

18-16
(*wire*) As the antenna comes into the electromagnetic field of the passing wave, induction occurs. This means that a radio-frequency _____ is induced in the wire, causing a radio-frequency current to flow in it.

18-17
(*voltage*) As shown in Fig. 18-17, the symbol for the receiving _____ is a triangle with its apex pointing downward.

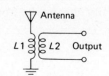

Fig. 18-17

456 A PROGRAMMED COURSE IN BASIC ELECTRONICS

18-18
(*antenna*) The RF voltage induced in the antenna causes a current to flow from the antenna to ground. In the circuit of Fig. 18-17, this RF current must flow through the coil labeled _____ on its way from the antenna to ground.

18-19
(*L1*) Since *L1* and *L2* are wound in such a manner as to establish a mutual inductance between them, the two coils may be considered to be an air-core transformer. In this transformer, coil _____ is the primary winding.

18-20
(*L1*) Coil *L2* is the _____ winding of this air-core transformer.

18-21
(*secondary*) This means that a radio-frequency voltage must be induced across the ends of *L2* whenever a radio-frequency _____ flows in the primary winding *L1*.

18-22
(*current*) An antenna is not sufficiently selective by itself. That is, it cannot choose one particular RF signal and ignore others. Since many different electromagnetic waves are intercepted by any antenna, then _____ of many different frequencies must flow in *L1*.

18-23
(*currents*) In the circuit of Fig. 18-17, *L2* is not selective in itself. That is, it cannot discriminate between closely adjacent frequencies. This means that voltages of various _____ would appear across *L2* since currents of various frequencies flow in *L1*.

18-24
(*frequencies*) A new component has been added to the circuit in Fig. 18-24. The new component is a capacitor. It is labeled _____.

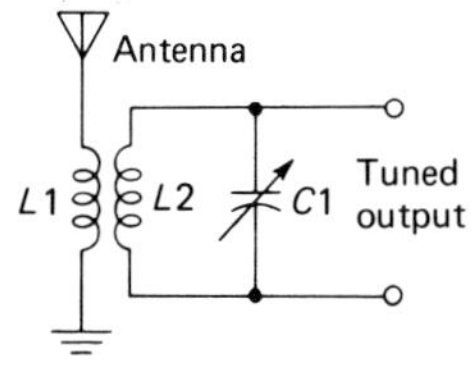

Fig. 18-24

18-25
(*C1*) A capacitor connected in parallel with a coil may set up a condition of parallel resonance. A resonant circuit *is* frequency selective. Thus, the addition of *C1* must be for the purpose of improving the _____ of the circuit.

18-26
(*selectivity*) The degree of success with which a resonant circuit is able to separate two adjacent frequencies depends upon the Q of the tuned circuit. Good selectivity is obtained only when the circuit has a high _____.

18-27
(Q) Assume that the Q of the $L2C1$ circuit has been made as high as possible by choosing the right wire for $L2$, the right L/C ratio, etc. We have seen that *loading down* a tuned circuit causes its _____ to decrease.

18-28
(Q) The circuit of Fig. 18-24 has no load whatsoever shown connected to the tuned output terminals. For this condition, that is, no load at all, the selectivity or sharpness of resonance of this circuit will be good because the Q will be _____.

18-29
(*high*) There is an open circuit between the tuned output circuit looking to the right. That is, the impedance between these two terminals is very _____.

18-30
(*large, high, etc.*) Reasoning in reverse, we may say that when the output terminals "look into" a high-impedance load, the Q of the circuit will be _____.

18-31
(*high*) When the terminals connect to a high-impedance load, this means that the tuned circuit will not have to deliver much current to the load. Thus, if the load draws very little current from the tuned circuit, the Q of the tuned circuit will be _____.

18-32
(*high*) If the Q of a tuned circuit is high, its selectivity is good. If the load draws very little current from the tuned circuit, the Q is high. Thus, when a load that draws little current is connected to a tuned circuit, we may expect the selectivity to be _____.

18-33
(*good*) Thus, the tuned circuit connected to a receiving antenna will be able to select one frequency out of many if the Q of the coil itself is high, if the L/C ratio is correctly chosen, and if the load draws very _____ current from the tuned circuit.

18-33
(*little*)

DEMODULATION

18-34
There are two components that go into making an amplitude-modulated wave. As shown in Fig. 18-34, an _____ component must be added to the RF component in order to produce an AM wave.

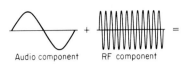
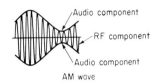

Fig. 18-34

18-35
(*audio*) Thus, an AM wave is the *composite* of an audio component that has been superimposed on an _____ component.

18-36
(*RF*) The carrier wave is the wave that can be radiated out into space to form an electromagnetic wave. The audio component cannot be radiated over long distances. Hence, it is the _____ wave that is the carrier.

18-37
(*RF*) The carrier wave serves merely as a vehicle to carry the intelligence originated at the transmitter. Hence, the intelligence is present in the form of the _____ component that rides on the carrier.

18-38
(*audio*) In the process of modulation, the audio component shown as a single sine wave in Fig. 18-34 is impressed on the carrier. The AM wave, however, carries two audio components. One of these forms the top envelope of the AM wave, and the other forms the _____ envelope.

18-39
(*lower* or *bottom*) The AM wave is radiated out into space. When it intercepts a receiving antenna, it induces a voltage that has exactly the same waveform as the AM wave in the antenna. In Fig. 18-39, the voltage waveform that closely resembles the AM wave is labeled _____.

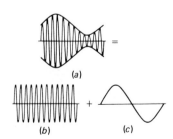

Fig. 18-39

18-40

(*a*) This is the voltage waveform that is induced in the receiving antenna. This voltage causes a _____ having exactly the same waveform to course down through the antenna coil to ground.

18-41

(*current*) Now assume that the tuned circuit (*L2C1* in Fig. 18-41) is resonated to the frequency of this AM wave. Since the AM wave is essentially an RF wave with varying amplitude, the tuned circuit would be made to resonate with the _____ component of the wave, not the audio component.

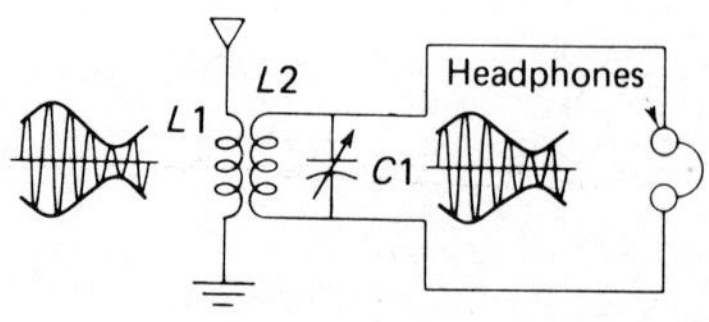

Fig. 18-41

18-42

(*RF*) If a pair of headphones were now connected across the terminals of the tuned circuit as in Fig. 18-41, the current through the headphones would be a _____-frequency current, not an audio-frequency current.

18-43

(*radio*) The highest frequency to which normal headphones will respond is about 10,000 Hz. The RF component of the AM wave being received might have a frequency of 10,000,000 Hz (10 MHz) or higher. Since the headphones will not respond to this frequency, no _____ will be heard in them.

18-44

(*sound, signal, etc.*) Returning to Fig. 18-39, we now recognize that the RF component, or the wave labeled *b*, cannot be heard in the headphones because its frequency is too _____.

18-45

(*high*) Even if the RF component could be heard, it would mean nothing because this component does not carry the intelligence. The intelligence is carried in the _____ component.

18-46

(*audio*) In Fig. 18-39, the audio component is labeled _____.

18-47

(*c*) Thus, in order to hear anything at all in the headphones, it will be necessary to eliminate the RF component altogether and make available to the headphones the _____ component of the AM wave.

18-48
(*audio*) The process by which this separation is accomplished is called *detection*. A signal is said to be *detected* when the _____ component of the signal is eliminated and the audio component made available to the headphones.

18-49
(*RF*) Thus, the detection process involves some method of selecting waveform *a* (Fig. 18-39) from all other frequencies, getting rid of waveform *b* or the carrier, and allowing waveform _____ to act upon the sound-reproducing equipment.

18-50
(*c*) Figure 18-50 illustrates the waveform of the voltage developed across the tuned circuit $L2C1$ of Fig. 18-41. In this drawing, the portion of the voltage labeled A is the RF carrier, while the portions labeled B and C are the _____ components carried by the RF.

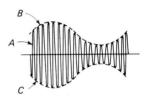

Fig. 18-50

18-51
(*audio*) One function of the detector circuit is to eliminate the carrier from the receiving equipment. If the carrier were eliminated, as Fig. 18-51 shows, there would be _____ audio components left, 1 above and 1 below the axis.

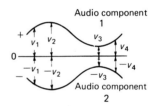

Fig. 18-51

18-52
(2) Since one of these components is always above the zero axis, it represents a voltage that is pulsating and always positive. Since the other component is always below the zero axis, it represents a voltage that is pulsating and always _____.

18-53
(*negative*) Choosing any instantaneous voltage such as v_1 at random, we find that there is an equal voltage $-v_1$ at the same instant in time. These voltages cancel each other since they are _____ in magnitude and opposite in phase.

18-54
(*equal*) For instantaneous voltage v_2 there is a corresponding voltage $-v_2$; similarly, for instantaneous voltage v_3, there is a corresponding voltage _____.

18-55
($-v_3$) Since there is an equal and opposite instantaneous voltage for any voltage we select on the curves, and since the net voltage at any instant is zero due to cancellation, the net output of both audio components taken together must also be _____ .

18-56
(*zero*) Since the net audio output is zero due to cancellation of the two components, it is clear one of these two components must be eliminated. If either one is eliminated, there will then be a definite net _____ output voltage which may then be reproduced in headphones or loudspeaker.

18-57
(*audio*) Let us return to the RF carrier and its audio envelope. Figure 18-57 shows a portion of the waveform enlarged to make the individual RF variation clearer. In this figure, the waveform labeled _____ is the RF carrier.

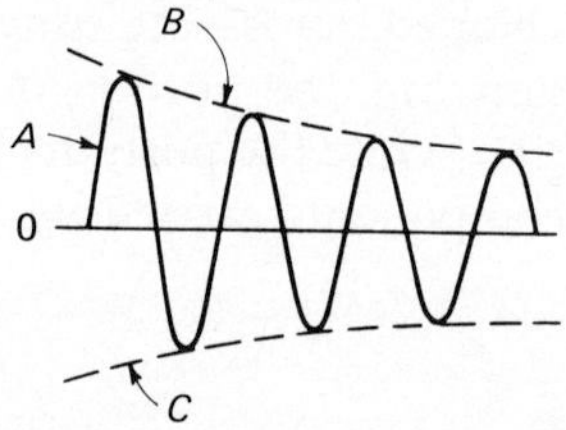

Fig. 18-57

18-58
(*A*) The waveform labeled *B* is one audio component, while the waveform labeled _____ is the second component.

18-59
(*C*) This enlarged view shows clearly that we are looking at a radio-frequency voltage in which each successive cycle is smaller in amplitude than the next, and that the audio component is merely a line we draw connecting each RF peak to the RF _____ next in line.

18-60
(*peak*) Since this voltage varies above and below the zero axis, it must be an _____ voltage.

18-61
(*alternating*) If this alternating voltage is *rectified*, it will have the appearance of the waveform in Fig. 18-61. This waveform is not alternating. It is a _____ direct voltage.

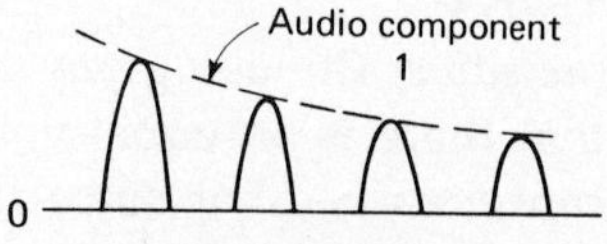

Fig. 18-61

18-62
(*pulsating*) Since one-half of the carrier *together with audio component 2* has been wiped out, this leaves only _____ audio component (Fig. 18-61).

18-63
(1) We show one complete audio component cycle riding on the carrier in Fig. 18-63. Since the lower halves of all the RF cycles are missing as well as audio component 2, we know that this wave has been _____ to change the original ac to pulsating dc.

Fig. 18-63

18-64
(rectified) The audio component alone is shown in Fig. 18-64. This is the waveform we require since it represents the original audio added to the carrier in the _____ process at the transmitter.

Fig. 18-64

18-65
(modulation) To obtain this component alone, it is now evident that we must first rectify the carrier, thus eliminating 1 audio component. Then we must separate the audio component that is left from the remaining RF so that we may feed the _____ component alone to the reproducing device.

18-65
(audio)

BASIC DIODE DETECTOR

The simplest detector uses a diode. The diode may be either a semiconductor type or an electron tube suitable for operation at radio frequencies. We shall start by tracing the action which ultimately results in recovery of one of the two audio components present in the AM wave.

18-66
Figure 18-66 presents the elements of a simple detector circuit. As we have seen, the passing electromagnetic wave induces a voltage in the _____ of the receiver.

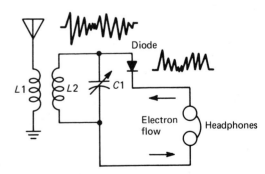

Fig. 18-66

18-67
(antenna) The antenna coil transfers the RF current thus produced to the tuned circuit consisting of C1 and _____.

DETECTION AND DETECTORS 463

18-68
(*L2*) Here frequencies other than those for which the tuned circuit is in resonance are rejected, while the signal for which *L2C1* resonates develops a _____ across the terminals of the tuned circuit.

18-69
(*voltage*) This voltage is due to the modulated carrier and has an ac waveform. As illustrated in Fig. 18-66, a *semiconductor diode* is connected in series with this resonant voltage and the _____ that are intended to reproduce the audio component.

18-70
(*headphones*) The diode _____ the ac waveform of the voltage due to the signal and transforms it thereby into pulsating dc.

18-71
(*rectifies*) The current flowing through the headphones consists of very short pulses of RF energy. The audio component is the line connecting the _____ of the RF pulses.

18-72
(*peaks*) This waveform is not yet suitable for reproduction of the original modulation. The RF component and the audio component must be separated from each other. This is done with the aid of an additional part, as in Fig. 18-72. The added part is the one labeled _____.

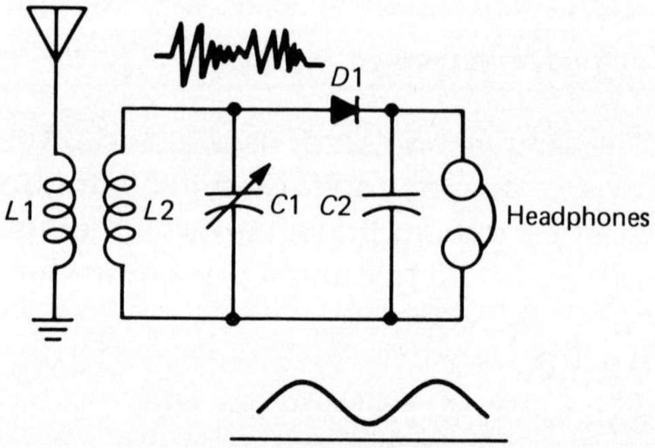

Fig. 18-72

18-73
(*C2*) This part is a bypass capacitor. For normal broadcast and shortwaves, this capacitor might have a value of 0.001 µF. Its reactance is low for radio frequencies; hence these frequencies cannot develop a large _____ drop across it.

18-74
(*voltage*) On the other hand, the audio component for which *C2* has a high reactance is *low* frequency. Thus, the audio-frequency component can develop a _____ voltage drop across *C2*.

18-75
(*large*) C2, with a large audio voltage drop across it, may then act as a *source* of voltage that can drive an audio-frequency current through the _____ which can then reproduce it as sound.

18-76
(*headphones*) Thus, the detection process is complete. Summarizing: In this process, *interception* of the radio waves occurs in the _____ circuit.

18-77
(*antenna*) The current flowing in the antenna circuit is then transferred by transformer action to the _____ circuit consisting of $L2$ and $C1$, where *selection* occurs.

18-78
(*tuned, resonant, etc.*) The current due to the voltage developed across the tuned circuit is then _____ by the diode. This is where detection takes place.

18-79
(*rectified*) *Separation* of the pulsating RF component from the audio component is accomplished by the component labeled _____.

18-80
(*C2*) Reproduction of the audio component is accomplished by means of the _____.

18-81
(*headphones*) Let us appraise the diode detector in terms of the criteria discussed in a previous section. First, its sensitivity: A diode does not amplify; without amplification, only very strong signals have a sufficiently large audio component to be reproduced satisfactorily. Thus, the _____ of a diode detector is not very high.

18-82
(*sensitivity*) For this reason, the _____ detector is not well adapted for receiving very weak signals.

18-83
(*diode*) A radio-frequency amplifier can amplify the modulated wave that intercepts a receiving antenna. If such an amplifier is connected to a receiving antenna, it can _____ the amplitude of the received signal.

18-84
(*increase*) This signal, having increased amplitude, may then be satisfactorily detected by a diode. Thus, if the signal is initially weak, the diode detector should be preceded by a(n) _____ amplifier.

18-85
(*RF*) A diode detector causes virtually no distortion of the incoming signal. It merely rectifies without disturbing the waveform. Since the audio component is altered very little, the diode may be said to have _____ fidelity as a detector.

18-86
(*good, high, etc.*) A diode is capable of handling a wide range of signal amplitudes without *overloading*. That is, neither weak nor strong signals cause positive or negative peak clipping. This is further evidence that a diode has high _____.

18-87
(*fidelity*) We are now concerned with selectivity. You will recall that selectivity depends upon circuit Q. You will also recall that a tuned circuit that feeds a high-impedance load may have a high Q, but that its Q may be _____ if it feeds a low-impedance circuit.

18-88
(*low*) Therefore, to determine how good the selectivity of a diode detector is, we must examine the kind of _____ into which the tuned circuit has to work.

18-89
(*impedance* or *load*) If the load draws current from the tuned circuit, it tends to "load down" the tuned circuit. For this condition, the tuned circuit sees the load as a _____ impedance.

18-90
(*low*) On the other hand, if the load draws very little current from the tuned circuit, it does not load it down, and hence behaves as a _____ impedance.

466 A PROGRAMMED COURSE IN BASIC ELECTRONICS

18-91

(*high*) Referring to Fig. 18-91, we see that the load on the tuned circuit consists of the diode $D1$ in series with the _____.

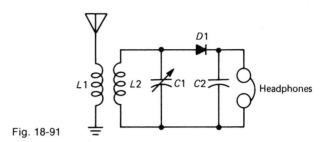

Fig. 18-91

18-92

(*headphones*) During the conduction half of each cycle of the incoming wave, the diode conducts. A good conductor represents a *low* impedance. Hence, the diode as a circuit element in the detector is a _____-impedance device.

18-93

(*low*) The headphones may have an impedance that is generally lower than 5,000 Ω. In terms of circuit loading, this is a low impedance too. Thus, the headphones as a circuit element in the detector load represent a low _____.

18-94

(*impedance*) The load on the tuned circuit in the diode detector thus consists of 2 low impedances in series. These impedances, even when added, are still low. Hence, the tuned circuit in the diode detector is seen to work into a _____-_____ load.

18-95

(*low-impedance*) Since the impedance of the load is low, current will flow in the load. This causes loading down of the tuned circuit. We may therefore conclude that the selectivity of a diode detector is _____.

18-95
(*fair*)

SINGLE-TRANSISTOR RECEIVER

The diode detector has two weaknesses: (1) it causes loading of the tuned circuit, thereby broadening the response and lowering the selectivity and (2) it does not amplify and therefore has relatively poor sensitivity. Both of these weaknesses are partially corrected in the detector to be studied next.

18-96
Another type of detector is shown in Fig. 18-96. This detector uses a _____-_____ _____ rather than a diode.

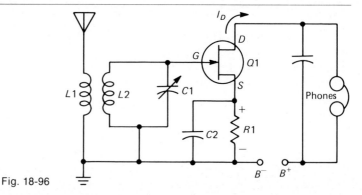

Fig. 18-96

18-97
(*field-effect transistor*) The output from the tuned circuit (*L2C1*) is applied between the _____ of the field-effect transistor (FET) and ground.

18-98
(*gate*) Resistor *R1* has a relatively high value, say 10,000 Ω. As the drain current flows through this resistor, there will be a relatively _____ voltage drop across it.

18-99
(*large*) This voltage drop occurs between the source of the FET and ground. This drop, therefore, places the source at a relatively high _____ potential above ground.

18-100
(*positive*) The gate of the FET is connected directly to ground through *L2*. With respect to dc voltages, then, the gate of the FET is _____ with respect to the source.

18-101
(*negative*) Thus, the FET is biased negatively by *R1*. If *R1* is sufficiently large, the bias will be close to _____ bias as shown in Fig. 18-101.

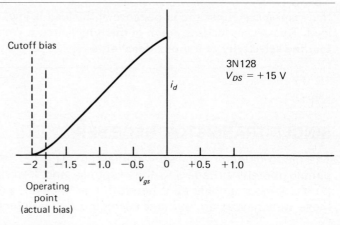

Fig. 18-101

468 A PROGRAMMED COURSE IN BASIC ELECTRONICS

18-102
(*cutoff*) If the FET is a 3N128, for example, operating at a drain-source voltage of 15 V, then cutoff bias is _____ V.

18-103
(−2) As shown in Fig. 18-101, the operating bias established by the voltage drop across $R1$ is _____ V.

18-104
(−1.8) Under these conditions of bias, the drain current of the FET will be small. This is true because the operating bias is almost at the drain current _____ point for the FET.

18-105
(*cutoff*) Since the gate is virtually at cutoff, any additional *negative* voltage applied to the gate will have very _____ effect upon the drain current.

18-106
(*little*) Figure 18-106 shows the voltage from the tuned circuit being applied to the gate of the FET. As may be seen from this drawing, the negative half-cycles of the modulated AM wave have a negligible effect upon the _____ current of the FET.

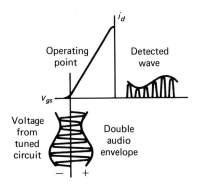

Fig. 18-106

18-107
(*drain*) The positive half-cycles of the voltage from the tuned circuit, however, cause a varying drain _____ to flow.

18-108
(*current*) If the v_{gs}-versus-i_d curve is linear, the drain current variations will be exact duplicates of the variations in the positive half-cycles coming from the _____ circuit.

18-109
(*tuned*) The voltage from the tuned circuit contains 2 out-of-phase components as shown. The rectifier has the job of removing _____ of the 2 components.

DETECTION AND DETECTORS

18-110
(*1*) It is evident that this circuit does cause rectification because the drain-current variations contain only 1 _____ component.

18-111
(*audio*) The detector just described uses an amplifying FET. Hence, some amplification of the signal occurs during the detection process. This suggests that the sensitivity of the FET detector is _____ than that of the diode detector.

18-112
(*better*) This is indeed true. Thus, the FET detector does have this advantage to start over the diode detector: its _____ is better.

18-113
(*sensitivity*) If the v_{gs}-versus-i_d curve of the FET used for detection is linear, there will be little distortion of the signal during detection. This suggests that the _____ of the FET detector is good.

18-114
(*fidelity*) Recall that the gate current of a FET is practically zero. A zero gate current means no power is taken from the tuned circuit. Hence, the loading of the tuned circuit by the gate circuit of the FET is close to zero. When the loading is small, the Q of the tuned circuit is _____.

18-115
(*high*) A high-Q tuned circuit is capable of _____ selectivity.

18-116
(*good*) Thus, since the FET detector does not load down the tuned circuit connected to the antenna, we may expect good _____ from this detector.

18-117
(*selectivity*) Summarizing: A FET detector has better sensitivity than the diode detector, almost as good fidelity, and considerably better _____ than the diode detector.

18-117
(*selectivity*)

CRITERION CHECK TEST

_____18-1 A high-Q tuned circuit will permit an amplifier to have high (*a*) fidelity, (*b*) selectivity, (*c*) frequency range, (*d*) bandpass.

____18-2 If a receiver produces an accurate reproduction of the modulating signal, then we say it has good (*a*) selectivity, (*b*) sensitivity, (*c*) fidelity, (*d*) frequency coverage.

____18-3 In Fig. 18-17, L_1L_2 form a complete (*a*) tuned circuit, (*b*) tank circuit, (*c*) antenna, (*d*) air-core transformer.

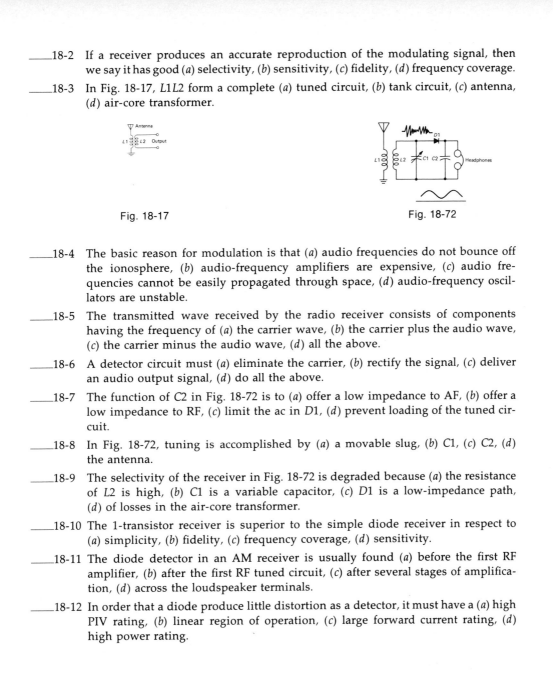

Fig. 18-17   Fig. 18-72

____18-4 The basic reason for modulation is that (*a*) audio frequencies do not bounce off the ionosphere, (*b*) audio-frequency amplifiers are expensive, (*c*) audio frequencies cannot be easily propagated through space, (*d*) audio-frequency oscillators are unstable.

____18-5 The transmitted wave received by the radio receiver consists of components having the frequency of (*a*) the carrier wave, (*b*) the carrier plus the audio wave, (*c*) the carrier minus the audio wave, (*d*) all the above.

____18-6 A detector circuit must (*a*) eliminate the carrier, (*b*) rectify the signal, (*c*) deliver an audio output signal, (*d*) do all the above.

____18-7 The function of C_2 in Fig. 18-72 is to (*a*) offer a low impedance to AF, (*b*) offer a low impedance to RF, (*c*) limit the ac in D_1, (*d*) prevent loading of the tuned circuit.

____18-8 In Fig. 18-72, tuning is accomplished by (*a*) a movable slug, (*b*) C_1, (*c*) C_2, (*d*) the antenna.

____18-9 The selectivity of the receiver in Fig. 18-72 is degraded because (*a*) the resistance of L_2 is high, (*b*) C_1 is a variable capacitor, (*c*) D_1 is a low-impedance path, (*d*) of losses in the air-core transformer.

____18-10 The 1-transistor receiver is superior to the simple diode receiver in respect to (*a*) simplicity, (*b*) fidelity, (*c*) frequency coverage, (*d*) sensitivity.

____18-11 The diode detector in an AM receiver is usually found (*a*) before the first RF amplifier, (*b*) after the first RF tuned circuit, (*c*) after several stages of amplification, (*d*) across the loudspeaker terminals.

____18-12 In order that a diode produce little distortion as a detector, it must have a (*a*) high PIV rating, (*b*) linear region of operation, (*c*) large forward current rating, (*d*) high power rating.

19 superheterodyne principles

OBJECTIVES

(1) Draw a block diagram of a TRF receiver. (2) State the performance of the TRF receiver with respect to sensitivity and selectivity. (3) Indicate how the superheterodyne receiver overcomes the limitations of the TRF receiver. (4) Draw a block diagram of the front end of a superheterodyne receiver. (5) Give the characteristics of a mixer. (6) State the signals which are fed into the mixer and the frequencies which leave the mixer. (7) State the effect of heterodyning on a modulated signal. (8) Define and compute the intermediate frequency. (9) Discuss the advantage of the heterodyning process for amplification purposes. (10) Describe the method used to keep the intermediate frequency constant. (11) Draw a block diagram of a superheterodyne receiver up to and including the IF stage. (12) Define, indicate on the block diagram, and discuss the use of a ganged capacitor. (13) Give a numerical example to demonstrate the tracking of the RF and local oscillator stages. (14) Compare unwanted signal rejection of an IF stage and an RF stage. (15) Draw a block diagram of a complete superheterodyne receiver. (16) Trace the signal flow through the block diagram. (17) Review the function of the RF amplifier, local oscillator, IF stage, detector stage, and audio amplifier stage.

TRF RECEIVER

Radio reception today—and this includes all forms of reception such as television, long- and shortwave communications, and radar—is based almost universally on a multistage arrangement which forms a *superheterodyne* receiver. The superb performance of a well designed and well adjusted superheterodyne has never been equaled by any other form of receiver. Most of the circuit concepts of this system are already part of your electronic vocabulary, but some of them still remain to be discussed. As a beginning, we will run through a block diagram of a simpler form of receiver, known as a *tuned RF* type, for the express purpose of pointing out its defects, so that the manner in which these defects are cured by the superheterodyne may be more fully appreciated.

19-1
A successful radio receiver can be built according to the plan shown in Fig. 19-1. This receiver consists of 3 stages: an RF amplifier, a _____, and an audio amplifier system.

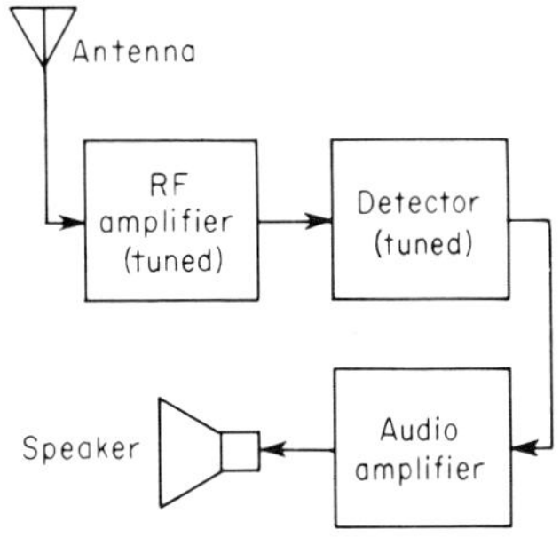

Fig. 19-1

19-2
(*detector*) Its selectivity is provided by 2 tuned circuits. One of these is in the input to the RF amplifier, and the other is in the input to the _____.

19-3
(*detector*) The RF amplifier tuned circuit is resonant to the carrier frequency, and hence is a *high-frequency* tuned circuit. The detector tuned circuit is also resonant to the carrier frequency. Thus, the detector tuned circuit is also a _____-_____ tuned circuit.

19-4
(*high-frequency*) The higher the frequency of a signal, the more difficult it is to amplify this signal. Thus, it is difficult to amplify the carrier efficiently in either the RF stage or the detector because the carrier is a _____-frequency signal.

19-5
(*high*) The receiver in Fig. 19-1 is called a tuned radio-frequency (abbrev. TRF) receiver. Since all amplification in a TRF receiver is carried on at high _____, the amplification is not as effective as it might be.

19-6
(*frequencies*) Thus, a significant defect of the TRF receiver is that its _____ is not particularly good because its circuits work at high frequencies.

19-7
(*amplification*) To recognize the cause of another defect, let us look back at the influence of Q. Recall that the Q of a tuned circuit determines how sharp the _____ of the circuit will be.

19-8
(*selectivity*) If the Q of a tuned circuit is high, the selectivity will be _____.

SUPERHETERODYNE PRINCIPLES

19-9
(*sharp* or *high*) You will also recall that Q depends upon the L to C ratio. This means that the Q of a tuned circuit depends upon having an optimum ratio of tuning inductance to tuning _____.

19-10
(*capacitance*) The Q of a tuned circuit improves as the L to C ratio goes down. Thus, to obtain a high Q, we should have a small L and a _____ C.

19-11
(*large*) When a tuned circuit is used to select one carrier in preference to another, it must be made to resonate with one of the _____ but not with the other.

19-12
(*carriers*) This will require a certain value of L and a certain value of C, chosen so as to _____ with the desired carrier.

19-13
(*resonate*) Assume that this particular carrier is selected by a tuned circuit having a low L/C ratio. For this carrier, the _____ of the tuned circuit will be sharp.

19-14
(*selectivity*) Now we wish to tune to another carrier. To do this, either the _____ or the C value, or possibly both, will have to be changed.

19-15
(*L*) The moment we change either the L or the C, the L/C _____ must also change.

19-16
(*ratio*) But if the L/C ratio changes, then the Q of the circuit must also change. This, in turn, causes the _____ of the circuit to change as well.

19-17
(*selectivity*) Thus, in a TRF receiver, as we tune from station to station we do not get uniform _____.

19-18
(*selectivity*) This lack of uniformity in selectivity and sensitivity of a TRF receiver is a definite defect. The lack of sensitivity is due to the fact that amplification is not effective at _____ frequencies.

19-19
(*high*) The selectivity is nonuniform because, in a *variable* tuned stage, the L/C ratio is always changing, resulting in a changing _____ which causes the selectivity characteristics to change.

19-20
(*Q*) To cure the defect of poor sensitivity, it would be necessary to arrange things so that amplification could take place at lower _____.

19-21
(*frequencies*) And to cure the defect of nonuniform selectivity, it would be necessary to use principles that permit the use of a fixed _____ circuit rather than a variable tuned circuit.

19-22
(*tuned*) These are the underlying advantages of the *superheterodyne* receiver. The superheterodyne receiver has fixed tuned circuits, and its amplifiers work at _____ frequencies than the TRF type.

19-22
(*lower*)

FREQUENCY HETERODYNING

We are now ready to investigate how the superheterodyne succeeds in minimizing the two principal defects of the TRF receiver: (1) ineffective amplification due to high-frequency tuning and (2) variable selectivity. Again we will approach the theory by means of a block diagram, reserving the actual circuits until later.

19-23
The fundamental principle of the superheterodyne receiver is best studied by referring to Fig. 19-23. The signal to be received is picked up by the _____ connected to the stage called an *RF tuned amplifier*.

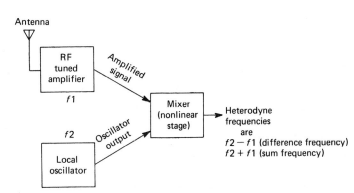

Fig. 19-23

SUPERHETERODYNE PRINCIPLES 475

19-24
(*antenna*) The amplified RF signal from the tuned RF circuit is then fed to a stage called a _____.

19-25
(*mixer*) Simultaneously, an oscillator produces a second RF signal which is also fed to the _____ stage.

19-26
(*mixer*) As indicated in the diagram, a mixer is a *nonlinear stage*. This merely means that the amplifying device in the mixer (a transistor) has a curved rather than a _____ base-collector transfer characteristic (Fig. 19-26).

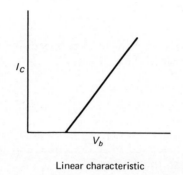
Linear characteristic

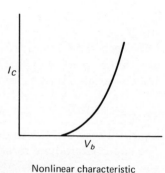
Nonlinear characteristic

Fig. 19-26

19-27
(*straight*) It may be shown that when 2 signals are mixed in a nonlinear amplifier, two additional frequencies appear in the output of the mixer (Fig. 19-23). One of these is the sum of the 2 original frequencies, while the other is the _____ between the 2 original frequencies.

19-28
(*difference*) If the two original frequencies are denoted by $f1$ (the lower of the two) and $f2$ (the higher of the two), then the *sum frequency* emerging from the mixer is denoted by _____.

19-29
($f2 + f1$) Similarly, the difference frequency resulting from the mixing of $f2$ and $f1$ where $f2$ is the higher of the two frequencies may be denoted by the symbols _____.

19-30
($f2 - f1$) The two original frequencies are also present in the output of the mixer. That is, there are altogether *four* frequencies coming from the mixer: (1) $f1$, (2) $f2$, (3) $f2 - f1$, and (4) _____.

19-31
($f2 + f1$) The process of producing sum and difference frequencies by mixing 2 signals is called *heterodyning*. The sum and difference frequencies, as indicated in Fig. 19-23, are therefore called the _____ frequencies.

19-32
(*heterodyne*) Let us use some actual numbers. Suppose that an incoming radio-frequency signal of 1,000 kHz is fed to the RF tuned amplifier. We will designate this frequency as $f1$. Thus, the frequency of $f1$ is _____ _____.

19-33
(*1,000 kHz*) Assume that a local oscillator stage feeds a frequency of 1,100 kHz to the second input point of the mixer. Calling this local oscillator frequency $f2$, we recognize that the mixer now has fed to it a frequency $f1$ of 1,000 kHz and a frequency $f2$ of _____ kHz.

19-34
(*1,100*) Both of these original frequencies will appear in the output of the mixer. Thus, if nothing else happened, the mixer output would contain a frequency of 1,000 kHz and a frequency of _____ _____.

19-35
(*1,100 kHz*) If the mixer is nonlinear in operation, the heterodyne process will occur in it. This means that in addition to $f1$ and $f2$, the mixer output will also contain the sum heterodyne frequency $f1 + f2$ which, for this case, will be _____ _____.

19-36
(*2,100 kHz*) Also, the mixer output will contain the difference heterodyne frequency, $f2 - f1$, which, for this case, will be _____ _____.

19-37
(*100 kHz*) Thus, in the output of the mixer we will find a total of _____ different frequencies.

19-38
(*4*) The 4 frequencies we will find are 1,000 kHz; 1,100 kHz; _____ kHz; and 100 kHz.

19-39
(*2,100*) If one of the original signals $f2$ or $f1$ is modulated, it is found that the modulation is carried along unchanged through the heterodyne process. Thus, if $f1$ as induced in the antenna is amplitude-modulated, then $f2 + f1$ is amplitude-modulated and also $f2 - f1$ is _____-_____.

19-40
(*amplitude-modulated*) We are particularly interested in the difference frequency, $f2 - f1$. As an example, suppose that the 1,000-kHz incoming signal is amplitude-modulated by a pure tone of 2,000 Hz. Then the difference frequency $f2 - f1$ will be a 100-kHz signal modulated by a pure tone of _____ Hz.

19-41
(*2,000*) The heterodyning process is often called the process of *frequency conversion*. In our example, an incoming frequency of 1,000 kHz has been converted to a difference frequency of _____ kHz by heterodyning it with a local oscillator frequency of 1,100 kHz.

19-42
(*100*) We shall be interested only in the difference frequency, $f2 - f1$, from this point on. We shall also remember that the modulation on $f1$ is not affected by the conversion process. We shall call the difference frequency by the name *intermediate frequency* (abbrev. IF). Thus, the IF in the example just given is _____ kHz.

19-43
(*100*) Consider another example. An incoming signal has a frequency of 3,500 kHz, while the local oscillator injects a signal of 3,955 kHz into the mixer at the same time. For this case, $f1$ is 3,500 kHz and $f2$ is _____ _____.

19-44
(*3,955 kHz*) The intermediate frequency is found by subtracting $f1$ from _____.

19-45
(*f2*) Thus, the IF for this case is 3,955 kHz minus 3,500 kHz or _____ kHz.

19-46
(*455*) If the modulating frequency used at the studio to modulate the carrier is 632 Hz, then the carrier will have superimposed on it a frequency of 632 Hz. Also, the modulation will be found superimposed on the IF. Thus, the 455-kHz signal will be carrying a modulation frequency of _____ Hz.

19-47
(*632*) Summarizing: In the basic heterodyne process, an incoming signal is heterodyned by a local oscillator to produce a mixer output of each of the two original signals—a sum frequency, and a _____ frequency.

19-48
(*difference*) We will discard the original frequencies after mixing or frequency conversion. We will also discard the sum frequency, $f2 + f1$, after frequency conversion. The frequency we will use is the difference frequency, $f2 - f1$, also called the _____.

19-49
(*IF*) The IF is then always found by _____ the smaller frequency from the larger one, whether the local oscillator frequency is above or below the incoming frequency.

19-50
(*subtracting*) In most superheterodyne receivers, the local oscillator is adjusted to operate at a frequency *higher* than that of the incoming signal. Thus, for this design, $f1$ will always be the incoming signal and $f2$ will always be the _____ oscillator frequency.

19-50
(*local*)

SUPERHETERODYNE SELECTIVITY — IF

We continue in the next section with some numerical examples that show how the intermediate frequency of a superheterodyne is held constant despite the fact that the receiver may be successively tuned to different incoming signals. The objective here is to carry on amplification at one particular frequency so that the optimum *L/C* tuning ratio may be used throughout the band of frequencies to be covered.

19-51
It will be recalled that we showed that greater amplification is possible if the tuned circuits operate at _____ rather than at high frequencies.

19-52
(*low*) This was pointed out as one of the defects of a TRF receiver. That is, its radio-frequency amplification is poor because its tuned circuits are required to work at _____ frequencies.

19-53
(*high*) In a superheterodyne receiver, the incoming signal is heterodyned by the signal from the local _____.

19-54
(*oscillator*) Of the 2 heterodyne frequencies thus generated, we always select the _____ rather than the sum frequency for further amplification.

19-55
(*difference*) Since the difference frequency is obtained by subtracting the incoming signal frequency from the local oscillator frequency, the difference frequency may be made considerably lower than the original incoming _____.

19-56
(*frequency*) As another example, consider an incoming frequency of 1,500 kHz heterodyned by a local oscillator frequency of 1,650 kHz. The intermediate or difference frequency for this case is _____ _____.

19-57
(*150 kHz*) Thus, the IF is considerably _____ in frequency than the signal carrying the desired modulation as it arrives over the air.

19-58
(*lower*) Since 150 kHz is much lower than 1,500 kHz, the tuned circuits are more effective, with the result that _____ amplification may be expected.

19-59
(*more, better, etc.*) Thus, the superheterodyne receiver overcomes one of the two important defects of the TRF receiver: it changes the incoming high frequency to a much _____ frequency.

19-60
(*lower*) We showed that the second defect of the TRF receiver was that its RF amplifier tuned circuits were _____ rather than fixed-tuned.

19-61
(*variable*) We also showed that the use of variable tuned circuits could not give consistent selectivity because the Q of the tuned circuit did not remain the _____ while the set was tuned from one end of the band to the other.

19-62
(*same*) To see how a superheterodyne receiver overcomes this defect, we first note that the tuning capacitor of the RF amplifier and the tuning capacitor of the local oscillator are connected together so that one changes as the other is changed. This is shown by the dashed line in Fig. 19-62 labeled "_____ tuning capacitors."

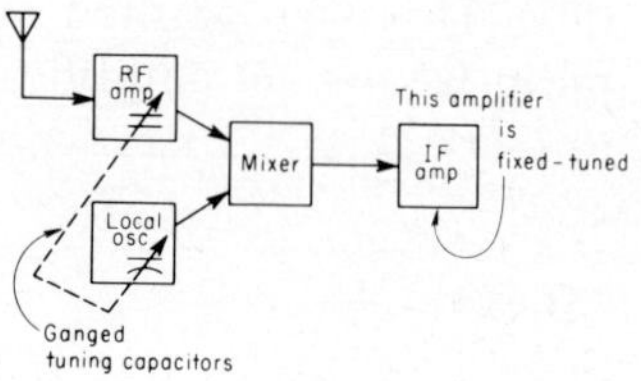

Fig. 19-62

19-63
(*ganged*) Ganged capacitors have a common rotor shaft. Thus, as the capacitance of one of them is decreased, the _____ of the other also decreases.

19-64
(*capacitance*) If the tuning coils of these two stages are correctly matched, then as the resonant frequency of the RF amplifier is changed (by varying capacitance), the oscillation _____ changes by the same amount in the local oscillator.

19-65
(*frequency*) An example will clarify this. Suppose that the RF amplifier is tuned to an incoming signal of 3,500 kHz, and the local oscillator is tuned to an incoming signal of 3,955 kHz. In this case, the IF is _____ _____.

19-66
(*455 kHz*) Now suppose that we change the tuning of the variable capacitors so that the RF amplifier resonates to 3,700 kHz. To select a higher frequency, it must have been necessary to _____ the capacitance of the RF tuning capacitor.

19-67
(*reduce*) While the capacitance of the RF tuning capacitor was being decreased, the ganged oscillator capacitor must also have been _____ in capacitance.

19-68
(*reduced*) With less capacitance, the local oscillator will oscillate at a _____ frequency than before.

19-69
(*higher*) If the tuning systems are properly designed, a 200-kHz increase in resonant frequency in the RF amplifier will result in a 200-kHz rise in oscillator frequency. If the oscillator previously worked at 3,955 kHz, its new frequency must be _____ _____.

19-70
(*4,155 kHz*) At this point, the RF amplifier is tuned to 3,700 kHz and the oscillator is working at 4,155 kHz. The difference frequency or IF is, therefore, _____ kHz.

19-71
(*455*) Thus, the IF is exactly the _____ as it was before the set was retuned.

SUPERHETERODYNE PRINCIPLES

19-72
(*same*) The identical effect occurs over the entire tuning range of the receiver: the oscillator "tracks" with the RF amplifier tuned circuit so that the IF is always the _____.

19-73
(*same*) Unwanted signals are heterodyned by the local oscillator, too. Since these are of slightly different frequency than the desired signal, the difference frequency for an unwanted signal will not be the same _____ as that produced by the heterodyne of the oscillator and the desired signal.

19-74
(*frequency*) Figure 19-74 illustrates this idea. Suppose that the desired signal is 2,100 kHz and that the oscillator is operating at 2,555 kHz to give an IF of _____ kHz.

No.	Incoming signals	Local oscillator	IF produced
2	2,000 kHz (unwanted)	2,555 kHz	555 kHz
1	2,100 kHz (desired)	2,555 kHz	455 kHz
3	2,200 kHz (unwanted)	2,555 kHz	355 kHz

Fig. 19-74

19-75
(*455*) Two interfering signals (No. 2 and No. 3) are being picked up at the same time. Signal No. 2 has a frequency of 2,000 kHz and, when heterodyned by the same local oscillator frequency of 2,555 kHz, produces an IF of _____.

19-76
(*555 kHz*) Also, signal No. 3 at 2,200 kHz is heterodyned by the same local oscillator frequency of 2,555 kHz and produces an IF of _____ kHz.

19-77
(*355*) An IF amplifier that is *fixed-tuned* to exactly 455 kHz will have no trouble in selecting the desired frequency. It can easily reject the 2 _____ frequencies.

19-78
(*undesired*) The reason for this ease of rejection lies in the idea of *percentage difference*. Signal No. 2 differs in the IF from signal No. 1 by 555 kHz − 455 kHz = _____ kHz.

19-79
(*100*) The percentage difference between No. 1 and No. 2 is, therefore, $\frac{100}{455}$ = _____ percent.

19-80
(22) Similarly, in the IF signal No. 3 differs from signal No. 1 by 455 kHz − 355 kHz = _____ kHz.

19-81
(100) Thus, the percentage difference between signal No. 1 and signal No. 3 is again _____ percent.

19-82
(22) Thus, in our example, the percentage difference between the desired signal and either of the undesired signals *in the IF* is _____ percent.

19-83
(22) This is a large percentage difference. Any reasonably good tuned circuit can easily _____ undesired signals that differ from the desired signal by as much as 22 percent.

19-84
(*reject*) This, then, permits the intermediate-frequency amplifier tuned to 455 kHz to have sharp _____ since it can easily separate the desired from the undesired signals.

19-85
(*selectivity*) Now compare the percentage differences between the *original* RF signals at the RF amplifier. The desired signal is 2,100 kHz, and signal No. 2 (undesired) is 2,000 kHz. The difference between these 2 frequencies is 2,100 kHz − 2,000 kHz = _____ kHz.

19-86
(100) The percentage difference is, therefore, 100/2,100 = _____ percent.

19-87
(4.8) Similarly, the percentage difference between the desired signal (No. 1) and the undesired signal (No. 3) at radio frequency is (2,200 − 2,100)/2,100 or 100/2,100 = _____ percent.

19-88
(4.8) This is a very small percentage difference. Hence, if this were a tuned RF set rather than a superheterodyne, the tuned circuit could not separate signals that are so close to each other (4.8 percent difference). Hence, the selectivity would be quite _____.

19-89
(*poor, bad, etc.*) Summarizing: At high frequencies, the percentage difference between adjacent frequencies is small, thus making it difficult for the tuned circuits to separate them. This leads to poor _____.

19-90
(*selectivity*) Summarizing further: When the frequency is converted to a lower fixed value using the heterodyne system, the percentage difference between desired and undesired signals in the IF is made much _____.

19-91
(*larger*) If the percentage difference is increased between desired and undesired signals at the lower IF, the _____ is vastly improved.

19-92
(*selectivity*) In addition, since the IF amplifier is fixed-tuned, its Q can be made the optimum value, thus improving the _____ of the system still more.

19-92
(*selectivity*)

SUPERHETERODYNE RECEIVER—BLOCK DIAGRAM

A brief survey of a block diagram of a complete superheterodyne receiver follows. It should be noted at this time that certain portions of the receiver (such as the power supply and other refinements) will be discussed in much greater detail in the coming chapters. At this point in our studies, we are concerned with comprehension of *general* rather than specific information.

19-93
Before examining the separate circuits of the stages of a superheterodyne receiver, let us study the block diagram of a complete receiver. Starting at the left (Fig. 19-93), we meet the frequency conversion stages which include the RF amplifier, the mixer, and the local _____.

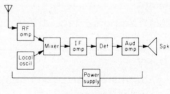

Fig. 19-93

19-94
(*oscillator*) The two stages that are both tuned by a ganged capacitor are the local oscillator and the _____ _____.

19-95
(*RF amplifier*) The heterodyne action between these two stages creates the difference frequency used as the IF in the stage labeled _____.

19-96
(*mixer*) The fixed-tuned stage which then amplifies the new, lower intermediate frequency is the stage labeled _____ amplifier.

19-97
(*IF*) Since the modulation is transferred to the intermediate frequency without change, the output of the mixer contains the IF plus the original _____ superimposed on the carrier at the transmitter.

19-98
(*modulation*) Thus, the input signal to the IF amplifier also carries the same _____ frequencies which originally rode in on the carrier.

19-99
(*modulation, audio, etc.*) The IF amplifier is operated in class A with as little distortion as possible. Its output is therefore an amplified version of the original intermediate frequency and also carries the _____ riding on it.

19-100
(*modulation, audio, etc.*) The only difference, therefore, between the IF signal and the original incoming RF signal is that the _____ of the IF is substantially lower than that of the original RF.

19-101
(*frequency*) Since the IF amplifier works as a fixed-tuned low-frequency amplifier (compared with the RF), it is an effective and selective amplifier. Thus, most of the amplification and selectivity of the superheterodyne is handled by the _____ amplifier.

19-102
(*IF*) All that is left to do is to recover the audio component from the IF signal and then amplify it enough to operate a loudspeaker or other reproducer. Recovery of the audio component is handled by the _____ stage (Fig. 19-93).

19-103
(*detector*) The detector treats the IF as just another radio signal. This means that the detector demodulates the signal, feeding the audio component to the _____ _____ stage and bypassing the carrier component out of the main stream.

19-104

(*audio amplifier*) The audio amplifier then increases the voltage and power of the _____ component to a point where the loudspeaker is satisfactorily activated.

19-105

(*audio*) The dc requirements are provided for by the common _____ _____ used for all stages.

19-105

(*power supply*)

CRITERION CHECK TEST

_____19-1 In a TRF receiver, the RF and detector stages are tuned to the (*a*) radio frequency, (*b*) intermediate frequency, (*c*) audio frequency, (*d*) USB.

_____19-2 As an *LC* tank circuit is tuned, the (*a*) capacitance always goes up, (*b*) inductance always goes down, (*c*) *Q* varies, (*d*) *L/C* ratio remains constant.

_____19-3 The input of the mixer stage is the (*a*) IF and RF signals, (*b*) RF and AF signals, (*c*) RF and local oscillator signals, (*d*) AF signal only.

_____19-4 If the RF signal is 600 kHz and the local oscillator is 800 kHz, then at the output of the mixer will be found (*a*) 600 kHz, 800 kHz only; (*b*) 200 kHz only; (*c*) 200 kHz, 600 kHz, 800 kHz, 1,400 kHz; (*d*) 800 kHz only.

_____19-5 At the output of the mixer, (*a*) only the sum frequency is modulated, (*b*) only the difference frequencies are modulated, (*c*) only the intermediate frequency is modulated, (*d*) all the above frequencies are modulated.

_____19-6 In a superheterodyne receiver, the difference frequency is chosen as the IF rather than the sum frequency because (*a*) lower frequencies are easier to amplify, (*b*) only the difference frequency is modulated, (*c*) the sum frequency signal is 20 dB down from the difference frequency, (*d*) the difference frequency is closer to the local oscillator frequency.

_____19-7 The IF amplifier stage will reject unwanted signals better than the RF stage because of the (*a*) larger percentage difference between signals, (*b*) smaller percentage difference between signals, (*c*) fixed percentage difference between signals, (*d*) higher sensitivity.

_____19-8 In a superheterodyne receiver, *perfect tracking* means that the (*a*) RF and local oscillator always have identical frequencies, (*b*) power of the RF and local oscillator remain identical, (*c*) amplification of the RF amplifier and mixer remain identical, (*d*) RF and local oscillator frequency difference remains constant.

_____19-9 The *Q* of the IF amplifiers (*a*) depends on the local oscillator frequency, (*b*) depends on the RF signal, (*c*) remains constant for all RF signals, (*d*) is always less than 5.

_____19-10 Most of the amplification in a superheterodyne receiver occurs in the (*a*) RF amplifier stage, (*b*) IF stage, (*c*) detector stage, (*d*) audio amplifier stage.

20 superheterodyne stage analysis

OBJECTIVES

(1) Sketch a single-stage transistor RF amplifier circuit and indicate the RF transformers. **(2)** State the function of a preselector, the ferrite rod in the RF tuned circuits, and the ganged capacitors in the circuit. **(3)** Draw several curves depicting the selectivity as a function of the number of preselector stages. **(4)** Sketch a 2-transistor mixer-oscillator stage. **(5)** Indicate the RF mixer stage, the local oscillator stage, the place in the circuit where the difference frequency is obtained, the ganged capacitors, the components which provide dc bias, and the ac coupling capacitors. **(6)** State the circuit elements which select the difference frequency. **(7)** State the method generally used to tune IF circuits. **(8)** Sketch a 1-transistor IF amplifier and indicate the IF transformers. **(9)** State the function of the IF transformers and the ferrite slugs in an IF transformer. **(10)** Indicate the components which provide dc bias and bias stability. **(11)** Define and give an example of the development of an image frequency. **(12)** State the function of the IF stage in removing the image frequency. **(13)** Derive an expression for the ac voltage gain of a single-stage transistor amplifier. **(14)** Draw a graph showing the ac input resistance of a transistor as a function of bias current. **(15)** Demonstrate that the dc bias of a transistor affects the ac gain. **(16)** Sketch a complete diode detector circuit and its rectified output. **(17)** Demonstrate the relation between the dc and ac components of the detector output. **(18)** Relate the strength of the IF signal to the dc output of the detector stage. **(19)** Indicate and discuss those components of the detector which separate the ac and dc components of the detector output. **(20)** Determine the polarity of the dc voltage output of the detector stage. **(21)** Sketch a transistor amplifier using AVC voltage to control gain. **(22)** Sketch a modulated carrier wave and state the reaction of the AVC system on it. **(23)** Indicate the AVC filter capacitor. **(24)** Compute the time constant of the AVC filter and the period of the AF signal coming out of the detector. **(25)** State the function of each component of an AVC system.

RF AMPLIFIER STAGE

Each stage of a superheterodyne has its individual function or functions. The operation of the complete receiver can best be analyzed by considering the function of each stage separately and the interrelationships between separate stages. In addition to analyzing each stage in this chapter, we shall also introduce *automatic volume control* (AVC) when we discuss the superheterodyne diode detector. This feature of the modern superheterodyne is particularly important in receiving areas or under receiving conditions where RF signal level varies. We begin our stage-by-stage analysis with the *preselector*, or radio-frequency amplifier.

20-1
Many superheterodyne receivers are equipped with an RF amplifier ahead of the mixer stage. As shown in Fig. 20-1, such an RF amplifier is also called a _____.

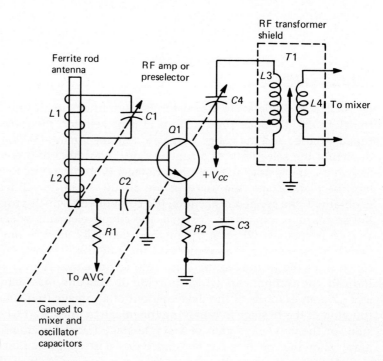

Fig. 20-1

20-2
(*preselector*) The preselector is a tuned RF amplifier (or more than one such amplifier) intended to improve the selectivity of the receiver. A tuned RF amplifier helps to _____ undesired signals, making it easier to have a single IF produced for further amplification and detection.

20-3
(*reject*) In the preselector of Fig. 20-1, a transistor, denoted by _____, is used as the amplifying device.

20-4
(*Q1*) In order that the preselector have sharp selectivity, its tuned circuits should have as high a Q as possible. Recall that the Q of a series tuned circuit is given by $Q = \sqrt{L/C} \times 1/R$, where C is the capacitance, L is the inductance, and R is the series _____.

20-5
(*resistance*) To maximize the Q, we wish to make the inductance as _____ as possible and the resistance as _____ as possible.

20-6
(*large, small*) Refer to Fig. 20-1. The tuned input circuit of the preselector consists of the components $L1$ and _____.

20-7
(*C1*) Basically, the only resistance of this tuned circuit is contained in the internal resistance of the windings of the inductor _____.

20-8
(*L1*) Hence, it is desirable to minimize the numbers of windings. That is the purpose of the _____ _____ antenna depicted in Fig. 20-1.

20-9
(*ferrite rod*) The ferrite rod is made of a ferromagnetic material which *increases* the effective inductance of the coil. The rod permits a coil with only a few turns of wire to produce enough inductance to properly tune the $L1$ _____ circuit.

20-10
(*C1*) Utilizing the ferrite core, the $L1C1$ circuit has a Q of about 50. Without the ferrite core, the Q would be about 5. Thus the ferrite core has increased the Q by a factor of _____.

20-11
(*10*) The RF energy from the $L1C1$ circuit is magnetically coupled to the coil denoted by _____. This coil is also wound on the ferrite rod.

20-12
(*L2*) The RF energy of $L2$ is then fed into the _____ of the transistor, $Q1$. This transistor produces an amplified RF signal at its collector.

20-13
(*base*) The emitter-swamping resistor of $Q1$ is denoted by $R2$, and its associated bypass capacitor is labeled by _____.

20-14
(C3) The dc bias for Q1 is not provided in the usual manner. In the past, the resistor R1 was returned to the dc supply, _____.

20-15
(V_{cc}) Instead, resistor R1 is directed toward a piece of circuitry called _____.

20-16
(AVC) In subsequent sections, we shall study the nature of the AVC circuit. We shall simply note at this point that AVC provides variable gain, that is, _____ gain for weak signals and _____ gain for strong signals.

20-17
(*increased, decreased*) The collector load of Q1 consists of the tuned circuit, C4 _____.

20-18
(L3) Coupling of selected and amplified voltage from the output of Q1 to the mixer stage is handled by the RF transformer labeled _____. The arrow between L3 and L4 means that this transformer uses a ferrite rod (or slug).

20-19
(T1) Capacitor C1 tunes the input of the preselector and C4 tunes the output. These must *track* with each other so that both circuits are always tuned to the same incoming signal. Tracking is accomplished by _____ the 2 capacitors as shown in Fig. 20-1.

20-20
(*ganging*) The improvement in selectivity to be expected from the use of a preselector is shown in Fig. 20-20. These curves are plotted to show how the input _____ to the receiver must be changed to maintain a constant output voltage at various frequencies off resonance.

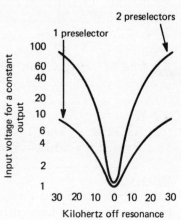

Fig. 20-20

20-21
(*voltage*) The lower of the 2 curves shows the selectivity effect of 1 preselector. Note that the voltage input at resonance (0 *kHz off resonance* means on resonance) for the given output is 1 V. This is seen by referring to the bottom of the vertical _____ which shows the voltages.

20-22
(*axis*) Referring to the lower curve, we see that the voltage needed to maintain the same output when the set is tuned 15 kHz off resonance is about 4 V. Thus, when tuned 15 kHz off resonance, the signal input needed to maintain constant output is _____ times as large as at resonance.

20-23
(*4*) Similarly, the selectivity is such that when the set is tuned 20 kHz off resonance, the required signal voltage is about _____ times as large as it is at resonance to maintain constant output.

20-24
(*6*) At 30 kHz off resonance, the signal has to be _____ times larger than its value at resonance to keep the output constant.

20-25
(*9*) The upper curve illustrates the response of 2 preselectors before the mixer. Here again, the voltage used to establish reference output at resonance is _____ V.

20-26
(*1*) In this case, however, when we go to 10 kHz off resonance, the signal voltage needed to maintain constant output must be approximately _____ times as large as it was at resonance.

20-27
(*12*) If we look at the signal voltage required to maintain constant output 30 kHz off resonance for 2 preselectors, we see that it has to be _____ times as great as it was at resonance.

20-28
(*90*) This shows the effect on selectivity of using 2 rather than 1 preselector. The use of 2 preselectors effects a tremendous improvement in _____.

20-29
(*selectivity*) In high-quality communication receivers, 2 preselector stages are usually used; in good-grade broadcast receivers, only 1 is used. Thus, the selectivity of a communication receiver may be expected to be considerably _____ than that of a good broadcast receiver.

20-29
(*better*)

MIXER-OSCILLATOR STAGE

Some superheterodynes do not have preselectors at all; the incoming signal is fed directly to the mixer stage which we are about to discuss. The omission of the preselector is a matter of economy. The ordinary tabletop radio is an example of a broadcast receiver without a preselector. Since this type of receiver is so common, we shall discuss the mixer action in terms of such a receiver, where the antenna leads directly to the mixer. It should be noted, however, that the performance of the mixer is not affected by the stage that precedes it. Thus, anything said about the mixer in the next section applies equally well to receivers in which there is a preselector stage.

20-30
Reviewing the basic superheterodyne principle (Fig. 20-30), we note that this circuit contains a mixer-RF amplifier stage and a(n) _____ stage.

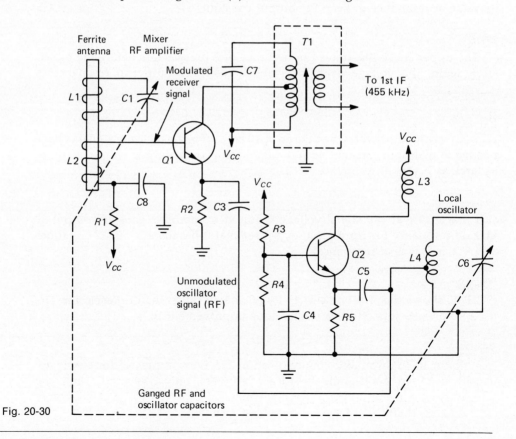

Fig. 20-30

20-31
(*oscillator*) The incoming radio-frequency signal, assumed to be voice-modulated, is picked up by the antenna tuned circuit, $L1C1$, and fed to the base of the mixer-RF amplifier by inductive coupling that occurs through the medium of the coil identified as _____.

20-32
(*L2*) The local oscillator signal is produced in the circuit containing the transistor identified as _____.

20-33
(*Q2*) This signal is capacitance-coupled through the component identified as _____ to the emitter of $Q1$.

20-34
(*C3*) Mixing occurs in $Q1$. The output of this transistor, therefore, contains the sum and difference frequencies of the heterodyne process. The difference frequency will be used. Hence, $C7$ and the primary of $T1$ resonate to the _____ frequency.

20-35
(*difference* or *IF*) The dashed lines connecting $C1$ and $C6$ indicate that these capacitors are _____ to each other so that their shafts must rotate as one.

20-36
(*ganged* or *coupled*) $C1$ tunes the mixer input circuit while $C6$ tunes the _____ _____ circuit.

20-37
(*local oscillator*) Assume that the mixer input circuit and the oscillator circuit are adjusted in such a way that there is a constant frequency difference of 455 kHz between the signals to which they resonate. The IF of this superheterodyne is, therefore, _____ kHz.

20-38
(*455*) The local oscillator uses a tuned emitter which receives feedback energy from $L3$. Should the mixer be tuned to 1,000 kHz at a given time, the oscillator will be tuned 455 kHz *above* the incoming frequency to produce the given IF. Hence the oscillator frequency at this time is _____ _____.

20-39
(*1,455 kHz*) The oscillator output is coupled to the emitter of Q1. The incoming signal is coupled to the base of Q1. Mixing occurs because the changes in the transistor are acted upon simultaneously by the signals on the base and the _____.

20-40
(*emitter*) The IF produced is then fed to the resonant circuit consisting of C7 and the _____ of transformer T1.

20-41
(*primary*) The resonant circuit consisting of C7 and the primary of T1 must be resonant at _____ kHz.

20-42
(*455*) The output voltage from the secondary of T1 is, therefore, the modulated _____ signal.

20-43
(*IF*) Thus *frequency conversion* has occurred. Analyzing the remainder of the circuit, we can first state that the dc bias of Q1 comes from V_{cc} through the resistor _____.

20-44
(*R1*) The dc bias for Q2 is provided by the 2 resistors _____ and _____.

20-45
(*R3, R4*) The capacitor identified as _____ provides an ac ground return for coil L2.

20-46
(*C8*) The capacitor denoted by _____ provides an ac ground return for the base of transistor Q2.

20-47
(*C4*) The capacitor labeled as _____ provides an ac path between the L4C6 tuned circuit and the emitter of Q2.

20-48
(*C5*) The arrow in between the primary and secondary of _____ indicates a ferrite slug is being used.

20-49
(*T1*) In fact, the resonant circuit consisting of C7 and the primary of T1 is tuned by moving the ferrite slug in or out of the transformer. This is almost universally the method used to tune IF transformers.

IF AMPLIFIER STAGE

After frequency conversion has taken place, the IF is then amplified in one or more intermediate-frequency stages. The IF amplifier to be discussed at this time is basic. In typical superheterodynes several refinements, particularly the inclusion of automatic volume control, may be found. We shall confine ourselves at the moment to the fundamental circuit as presented in the following section.

20-50
A typical IF amplifier is shown in Fig. 20-50. The input transformer to this stage is the transformer whose primary is fed by the output of the mixer. The input transformer is identified as _____ in Fig. 20-50.

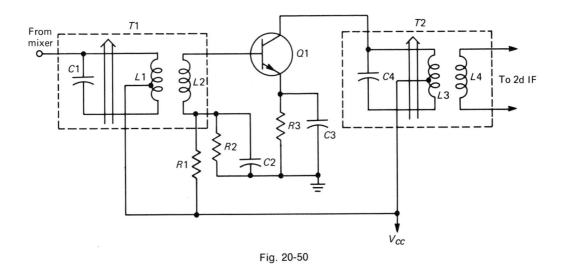

Fig. 20-50

20-51
(*T1*) A ferrite slug (indicated by the vertical arrow) is shown at the primary of T1. The dashed line around *T1* represents a metal shield can. The presence of only one capacitor in *T1* indicates that only the coil _____ is tuned.

20-52
(*L1*) Assume that the IF of this receiver is 455 kHz. The tuned circuit in *T1* is adjusted to peak the signal at the intermediate frequency. Hence, the ferrite slug is adjusted so that *L1* resonates with *C1* at _____ kHz.

20-53
(*455*) Because of the ferrite slug, the *Q* of the *L1C1* tuned circuit will be high. This results in good selectivity and voltage output from *T1*. The output voltage is taken from the _____ winding of *T1*.

20-54
(*secondary*) The output from the secondary winding of *T1* is then applied directly to the base of *Q1*. The other lead from *L2* is returned to ground by capacitor _____.

20-55
(*C2*) If capacitor *C2* were not present, then the signal voltage developed by *L2* would be seriously attenuated by a resistor, _____, before reaching the input of *Q1*.

20-56
(*R2*) The collector load of *Q1* is the tuned circuit consisting of _____ and _____.

20-57
(*C4, L3*) These 2 components are contained within the RF transformer denoted by _____. The *C4L3* resonant circuit is tuned to the IF frequency of 455 kHz.

20-58
(*T2*) The ferrite slug in *T2* will produce a high *Q*, further enhancing the IF amplifier's ability to amplify the IF frequency and _____ other frequencies.

20-59
(*reject*) The ferrite slugs of *T1* and *T2* are adjusted at the factory when the radio is manufactured. If no trouble is encountered with the radio, no readjustment of *T1* and *T2* will normally be required for the life of the set. Thus, the primaries of *T1* and *T2*, denoted by *L1* and _____, will remain constant during the operation of the set.

20-60
(*L3*) This constancy is another aid in giving a high Q. Recall that $Q = \sqrt{L/C} \times 1/R$. Since neither L nor C changes in the IF stage, L and C can be adjusted for optimum Q. This is one of the big advantages of the superheterodyne receiver over the T_____ receiver.

20-61
(*RF*) Analyzing the rest of the circuit, we see that the emitter-swamping resistor is given as _____.

20-62
(*R3*) The emitter-resistor bypass capacitor is denoted by _____.

20-63
(*C3*) The dc bias for $Q1$ is provided by the 2 resistors, _____ and _____.

20-64
(*R1, R2*) The output of $T2$ is fed to the _____ _____ stage.

20-65
(*second IF*) From the second IF, the signal passes to successive IF stages. Some radio sets have as few as 1 IF stage, while some expensive sets have as many as 5 IF stages. The more IF stages, the better the _____ of the receiver.

20-66
(*selectivity*) The final IF output is then fed to a _____ stage for demodulation and audio recovery. From this point on, the superheterodyne is exactly like a TRF receiver.

20-67
(*detector*) Thus, the IF amplifier is followed by a detector and then by an _____ amplifier system that terminates in a reproducer such as a set of headphones or a loudspeaker.

20-67
(*audio*)

IMAGE-FREQUENCY REJECTION

20-68
Figure 20-68 illustrates the input circuits of a _____ type of receiver. This type of receiver is recognizable by the fact that it contains a mixer and local oscillator.

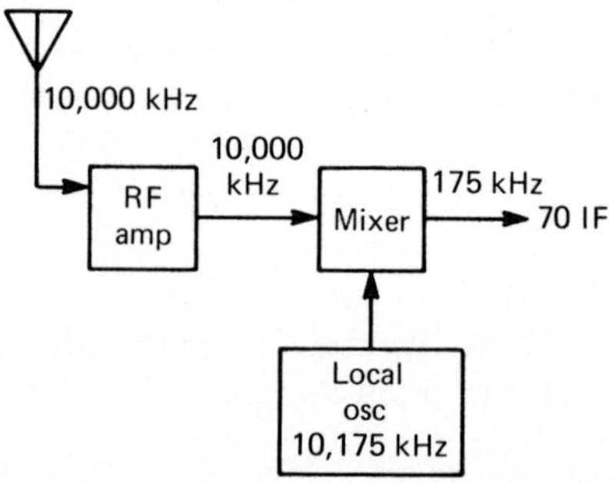

Fig. 20-68

20-69
(*superheterodyne*) An incoming signal to the RF amplifier of the receiver (Fig. 20-68) has a frequency of 10,000 kHz (10 MHz). The local oscillator frequency in this particular superheterodyne is _____ kHz.

20-70
(*10,175*) The 10-MHz input signal and the output signal of the local oscillator are mixed in the mixer stage. The difference frequency is 175 kHz. Hence, the IF amplifier must be resonant to _____ kHz.

20-71
(*175*) If the 10-MHz signal were the only signal allowed to pass through the RF amplifier, then it would be the only signal that could heterodyne the local oscillator signal. Hence, only the 10-MHz signal would produce an _____ of 175 kHz for further amplification.

20-72
(*IF*) Even if a signal of 10,050 kHz were present at the input to the mixer, the IF stage would not amplify it because the difference frequency it would produce when heterodyned by the 10,175-kHz oscillator would be _____ kHz.

20-73
(*125*) Since the IF amplifier is tuned to 175 kHz and since this amplifier has very good selectivity, it will reject any signal that differs from _____ kHz by more than a few kilohertz.

498 A PROGRAMMED COURSE IN BASIC ELECTRONICS

20-74
(*175*) Now suppose that the receiver is tuned to 10,000 kHz so that its oscillator is operating at 10,175 kHz and that a second signal of _____ kHz, as shown in Fig. 20-74, also reaches the antenna.

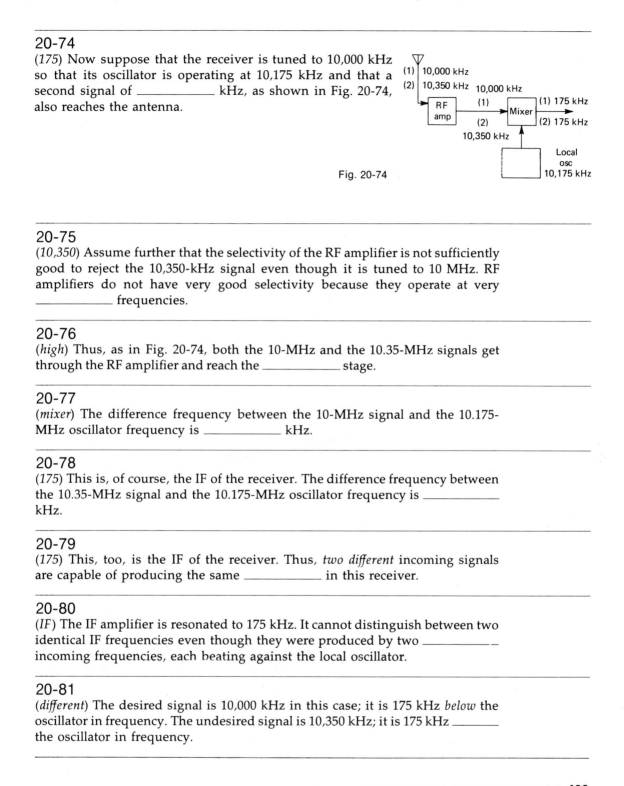

Fig. 20-74

20-75
(*10,350*) Assume further that the selectivity of the RF amplifier is not sufficiently good to reject the 10,350-kHz signal even though it is tuned to 10 MHz. RF amplifiers do not have very good selectivity because they operate at very _____ frequencies.

20-76
(*high*) Thus, as in Fig. 20-74, both the 10-MHz and the 10.35-MHz signals get through the RF amplifier and reach the _____ stage.

20-77
(*mixer*) The difference frequency between the 10-MHz signal and the 10.175-MHz oscillator frequency is _____ kHz.

20-78
(*175*) This is, of course, the IF of the receiver. The difference frequency between the 10.35-MHz signal and the 10.175-MHz oscillator frequency is _____ kHz.

20-79
(*175*) This, too, is the IF of the receiver. Thus, *two different* incoming signals are capable of producing the same _____ in this receiver.

20-80
(*IF*) The IF amplifier is resonated to 175 kHz. It cannot distinguish between two identical IF frequencies even though they were produced by two _____ incoming frequencies, each beating against the local oscillator.

20-81
(*different*) The desired signal is 10,000 kHz in this case; it is 175 kHz *below* the oscillator in frequency. The undesired signal is 10,350 kHz; it is 175 kHz _____ the oscillator in frequency.

20-82
(*above*) In both cases—the desired and undesired frequency—the *difference* is the IF of the receiver. That is, both incoming signals produce a heterodyne frequency of _____ kHz.

20-83
(*175*) The undesired 10,350-kHz signal is called the *image frequency*. An image frequency is defined as a frequency that is as far above the oscillator frequency as the desired signal is below the _____ frequency. It could also be a frequency that is as far below the oscillator frequency as the desired signal is above the _____ frequency.

20-84
(*oscillator, oscillator*) A mixer stage cannot separate the image frequency from the desired frequency; neither can an IF amplifier separate the two because both of them produce the _____ intermediate frequency.

20-85
(*same*) Separation of the image from the desired frequency must occur, therefore, in a stage that precedes the mixer stage. That is, separation must occur in the _____ stage.

20-86
(*RF amplifier, preselector, etc.*) But if the preselector or RF amplifier does not have good enough selectivity to separate an image from a desired frequency, the two will interfere with each other and both will be heard simultaneously because both of them produce the _____ IF when heterodyned by the local oscillator.

20-87
(*same*) There are two possible answers. First, we might add additional preselector stages to improve the overall _____ of the receiver.

20-88
(*selectivity*) Given enough preselector stages, the selectivity could be improved sufficiently to pass the desired signal but reject the _____ frequency.

20-89
(*image*) Since this is expensive, the second method may be used. Suppose we change the IF tuning of the receiver to 455 kHz (a standard frequency now in use). If the incoming signal is 10 MHz, then the oscillator frequency for *this* IF would be _____ kHz (Fig. 20-89).

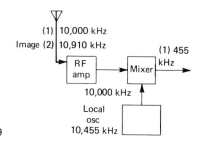

Fig. 20-89

20-90
(*10,455*) Thus, a 10,000-kHz signal beating with a 10,455-kHz oscillator will produce an IF of _____ kHz.

20-91
(*455*) Since the image frequency must be as far above the oscillator as the desired frequency is below it in order to produce a difference frequency equal to the IF, in this case the image frequency must be _____ kHz.

20-92
(*10,910*) In the previous case, the image and real signal were separated by 10,350 kHz − 10,000 kHz, or _____ kHz.

20-93
(*350*) This separation is small; hence the preselector selectivity was not good enough to select the desired signal and reject the _____ frequency.

20-94
(*image*) In the second case, however, the separation of desired and image frequencies is 10,910 kHz − 10,000 kHz, or _____ kHz.

20-95
(*910*) This separation is almost 3 times as great as it formerly was. The preselector selectivity will now be good enough to pass the desired frequency and _____ the image frequency.

20-96
(*reject*) Thus, the image is eliminated. Note that the method of eliminating the image involves raising the IF from its former low value of 175 kHz to its new _____ value of 455 kHz.

20-97
(*higher*) If the IF were made still higher, say 1,000 kHz, then the separation between desired signal and image would be even _____ than it was in the last example.

20-98
(*larger, greater, etc.*) But we have seen that one of the chief advantages of the superheterodyne is that its selectivity occurs in a _____-frequency stage like the IF. (Remember percentage separation between signals?)

20-99
(*low*) Thus, we would like to make the IF as _____ as possible in order to realize the best selectivity for separating adjacent frequencies.

20-100
(*low*) On the other hand, we should make the IF as _____ as possible in order to get best image rejection.

20-101
(*high*) This leads to a compromise in choosing the best IF. The value of 455 kHz has become standard because it is low enough to give good adjacent station selectivity and high enough to give good _____ rejection.

20-102
(*image*) To test your understanding, try the following: A radio receiver has an IF of 85 kHz and is tuned to a 5,000-kHz signal. The local oscillator frequency, therefore, is _____ kHz. (Assume oscillator frequency higher than RF.)

20-103
(*5,085*) The image frequency in this case is 85 kHz higher than the oscillator frequency. In other words, the image frequency is _____ kHz.

20-104
(*5,170*) Thus, the frequency separation between the desired signal and image signal is only _____ kHz.

20-105
(*170*) This is considered a small separation between image and desired signal. The reason that this separation is small is because the IF has been selected at too _____ a frequency.

20-106
(*low*) To increase the separation between image and desired frequency, therefore, it would be necessary to raise the _____ frequency of the receiver.

20-107
(*intermediate*) The IF must not be raised too much, however, because doing this would begin to affect adjacent-frequency selectivity. Thus, the _____ of the receiver must be chosen neither too low nor too high, but at some compromise figure.

20-107
(IF or *intermediate frequency*)

VARIABLE TRANSISTOR GAIN

In this section we shall derive the ac voltage gain of a transistor and show that this gain can be varied.

20-108
In Fig. 20-108, we have drawn a typical single-stage transistor amplifier. The input voltage is an ac signal denoted by _____.

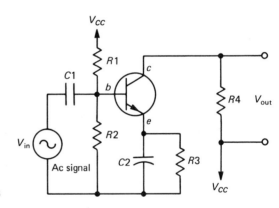

Fig. 20-108

20-109
(V_{in}) A capacitor, _____, couples the ac signal to the base of the transistor.

20-110
(C1) The emitter-swamping resistor is denoted by R3 and its bypass capacitor by _____.

20-111

(C2) If we operate this amplifier at a fairly high frequency, then the reactance of C1 and C2 will be very small. In Fig. 20-111, we have approximated the small reactance of C1 and C2 by _____ circuits.

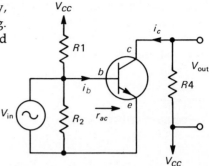

Fig. 20-111

20-112

(*short*) With this simplification, we can compute the base alternating current, i_b. From Ohm's law, $V_{in} = i_b r_{ac}$ or $i_b = V_{in}/r_{ac}$, where r_{ac} is the ac resistance looking into the _____-_____ junction of the transistor.

20-113

(*base-emitter*) Recall that the collector current of a transistor is much larger than the base current of a transistor. In fact, i_c is related to i_b by the _____ parameter h_{fe}.

20-114

(*hybrid*) Specifically, $i_c = h_{fe} i_b$, where h_{fe} is typically between 30 and 200. If we substitute the previous expression for i_b, then the result is

$$i_c = \frac{h_{fe} V_{in}}{?}$$

20-115

(r_{ac}) Again using Ohm's law, this time on the collector circuit, we can compute V_{out}:

$$V_{out} = i_c R4 = ? \times \frac{V_{in}}{r_{ac}} R4$$

20-116

(h_{fe}) Recalling that the voltage gain, A_v, is defined as $A_v = V_{out}/V_{in} = (h_{fe} V_{in} R4/r_{ac})/V_{in}$, we find

$$A_v = h_{fe} \left(\frac{R4}{?} \right)$$

20-117
(r_{ac}) Summarizing: The voltage gain of this transistor amplifier for ac signals is $h_{fe}(R4/r_{ac})$. The only assumption made in the derivation is that the ac reactance of the capacitors is _____.

20-118
(*small*) For our purposes, the important term in the gain expression is r_{ac}. Figure 20-118 shows that r_{ac} _____ (does, does not) change as the operating dc changes.

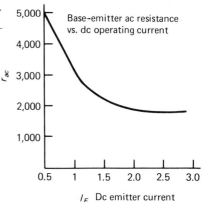

Fig. 20-118

20-119
(*does*) For example, when the emitter dc is 0.5 mA, r_{ac} is 5 kΩ, whereas at $I_E = 1.5$ mA, r_{ac} is about _____ kΩ.

20-120
(*1.3*) But if r_{ac} changes with the dc operating point, doesn't that mean that the ac voltage gain will also change? Yes; if r_{ac} gets smaller, the gain gets larger; if r_{ac} gets larger, the gain becomes _____.

20-121
(*smaller*) Since increasing I_E decreases r_{ac}, and decreasing r_{ac} increases the gain, we may conclude that _____ I_E increases the ac gain of a transistor.

20-122
(*increasing*) Summarizing: The _____ gain of a transistor can be varied by changing its dc bias. The more the direct current, the more the gain; the less the direct current, the less the gain.

20-123
(*voltage*) In subsequent sections, it will be seen that the varying dc bias will be provided by the AVC voltage. This will enable the amplifier to adjust its _____ to very weak or very strong signals.

20-123
(*gain*)

AVC VOLTAGE

Radio receivers are subject to a variety of signal conditions. Some signals are strong; others are weak. These variations of signal input may be due to any one or a combination of circumstances: transmitting stations differ widely in their power usage; some signals arrive at the receiver over longer paths than others, with consequent attenuation; signal fading occurs as transmission conditions change; the receiver may be in motion, as in an automobile, and therefore passing from good to poor reception zones. Whatever the cause may be, a receiver cannot perform satisfactorily unless it is equipped with some form of *automatic volume control* which can compensate for changing signal conditions. An AVC system is expected to increase the *gain* of a receiver when the signal becomes attenuated and decrease the gain as the signal picks up in amplitude. Clearly, such an automatic system must be controlled by the signal itself; in some way, the signal intensity must provide "information" to a special circuit which will then react quickly to alter the receiver gain in a compensatory fashion.

20-124
Refer to Fig. 20-124. The entire stage, including the transformer, diode, resistors, and capacitors, is a diagram of a _____ detector.

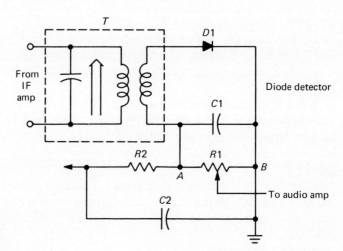

Fig. 20-124

20-125
(*diode*) The signal input to this stage will be taken to be an amplitude-modulated voltage which comes from the _____ stage that precedes the detector.

20-126
(*IF* or *intermediate frequency*) The IF signal is coupled to the diode through a _____ identified as T.

20-127
(*transformer*) The transformer is enclosed in a shield can, indicated by the dashed box around T. Assume that the intermediate frequency of this receiver is 455 kHz. The tuned circuit in the can then must be resonated to a frequency very close to _____ kHz.

20-128
(*455*) As the signal voltage (ac carrier) is applied to the diode, D1, rectification occurs. This process changes the ac to pulsating _____.

20-129
(*dc*) The rectified waveform in the absence of modulation may be represented as in Fig. 20-129. The waveform at the intermediate frequency ranges from the zero axis to a level identified as the _____ current.

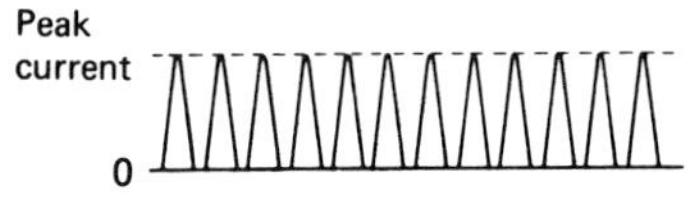

Fig. 20-129

20-130
(*peak*) This waveform pictures the *diode current*. Starting with the secondary of transformer T, the current flows through the parallel group C1R1 (Fig. 20-124), to the cathode of the diode, to the _____ of the diode, and back to the secondary.

20-131
(*anode*) A pulsating dc can be shown to comprise two components as in Fig. 20-131. There is a dc component, which when added to the _____ component results in the pulsating dc waveform.

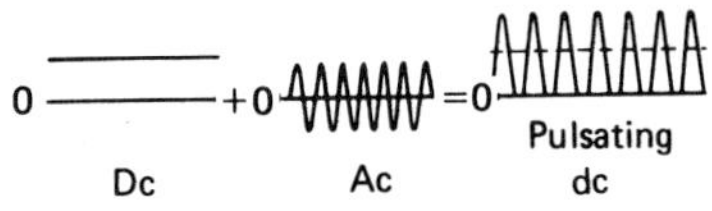

Fig. 20-131

20-132
(*ac*) This means that a pulsating dc waveform such as that in Fig. 20-129 can be *resolved* into two components. One of these is an ac component of the same frequency as the original waveform, while the other is a _____ component whose magnitude depends upon the intensity of the original signal.

20-133
(*dc*) Refer to Fig. 20-133. In *a* we see a pulsating dc waveform of relatively small amplitude. The dc component of this waveform is also relatively _____.

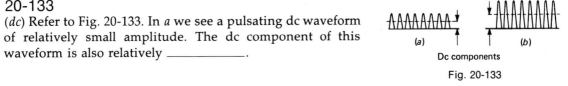

Dc components

Fig. 20-133

20-134
(*small*) In *b*, the pulsating dc waveform is of larger amplitude; hence its dc component is also relatively _____.

20-135
(*large*) Thus, the magnitude of the dc component of a pulsating waveform depends upon the peak _____ of the original waveform.

20-136
(*amplitude* or *magnitude*) In reference to the original IF signal, if this is a strong signal, the peak amplitude of the diode current it produces will be large; if it is a weak signal, the peak amplitude of the diode current will be _____.

20-137
(*small*) Hence, for a strong IF signal, the dc component of the rectified waveform will be _____, and for a weak signal, the dc component will be _____.

20-138
(*large, small*) Now consider capacitor C1. In a standard receiver, this capacitance is of the order of 200 pF, a relatively small capacitance but still sufficiently large to give it a low reactance for signals of the intermediate frequency. Thus, the _____ component of the rectified diode current sees a low reactance in this capacitor (Fig. 20-124).

20-139
(*ac*) Since C1 has a low reactance for the ac component, the voltage drop across it due to this component is negligibly _____.

20-140
(*small*) C1, however, blocks dc. That is, the reactance of C1 for the dc component of the rectified diode current is very _____.

20-141
(*large*) Note that R1 is in parallel with C1. If the reactance of C1 is very large for a given current component, then this current will tend to flow through _____.

20-142
(*R1*) Since the reactance of C1 for the dc component is large, then we would expect that the dc component would flow through _____.

20-143
(*R1*) Thus, a voltage drop develops across *R1* due to the dc component of the diode current flowing through it. As we have seen, the larger the peak amplitude of the received signal, the _____ the magnitude of the dc component of the rectified diode current.

20-144
(*larger*) This means that a strong IF signal will cause a _____ voltage drop across *R1*, and a weak IF signal will cause a _____ voltage drop across *R1*.

20-145
(*large, small*) We must now determine the polarity of the voltage drop across *R1* due to the dc component of the rectified diode voltage. Electron current flows from cathode to anode of the diode, through the secondary of *T*, through *R1* from point _____ to point _____, and then back to the cathode (Fig. 20-124).

20-146
(*A, B*) Since the direction of the dc component is from *A* to *B*, then point *A* must become more _____ (positive, negative) in potential than point *B*.

20-147
(*negative*) Note that point *B* is connected directly to a common ground point. Thus, point *A* is more _____ than ground potential.

20-148
(*negative*) Refer to Fig. 20-148. This depicts a portion of the AVC circuit. Ignore *R2* for the moment. Since the top of *C2* is connected to point *A*, then the top of *C2* must also have a negative potential with respect to _____.

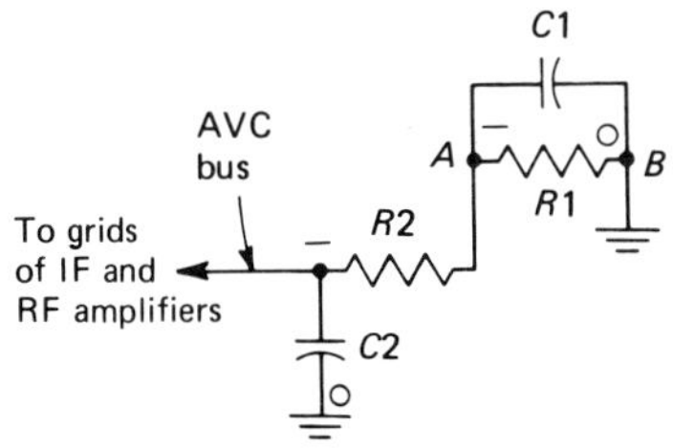

Fig. 20-148

20-149
(*ground*) The voltage that appears across *C2* is the AVC voltage. The negative voltage, as shown in the diagram, is then applied to the _____ of the IF and RF amplifier transistors through the AVC bus.

20-150
(*bases*) If the IF signal is strong, this negative voltage is _____; if the IF signal is weak, this negative voltage is _____.

20-150
(*large, small*)

AVC OPERATION

Refer back to Fig. 20-50. This is a diagram of an IF stage without AVC. The important section is reproduced in Fig. 20-151. We will compare this with the modified section of an IF amplifier as shown in Fig. 20-152.

20-151
Refer to Fig. 20-151. The base of $Q1$ is made positive by the bias resistor network consisting of $R4$ and _____.

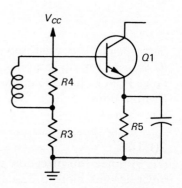

Fig. 20-151

20-152
(*R3*) Now refer to Fig. 20-152. Here, besides the bias resistors $R1$ and $R2$, the _____ of the transistor is connected to the negative AVC potential point.

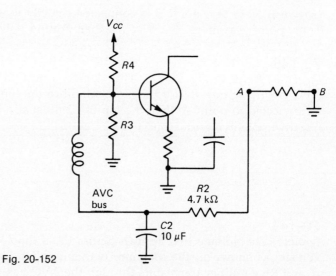

Fig. 20-152

20-153
(*base*) This also establishes bias by tending to make the base _____ with respect to the emitter.

510 A PROGRAMMED COURSE IN BASIC ELECTRONICS

20-154
(*negative*) But remember that this negative voltage is generated by the signal itself. If the signal is strong, this voltage is large; if the signal is weak, this voltage is small. Thus, for strong signals the IF base is made _____ positive than for weak signals.

20-155
(*less*) The same AVC voltage may also be applied to the RF amplifier or amplifiers. The same situation applies; that is, for strong signals, these transistors are supplied a more _____ AVC voltage.

20-156
(*negative*) Recall that in a previous section it was shown that the gain of a transistor increases with increasing _____ (forward, reverse) bias.

20-157
(*forward*) Since these transistors have a portion of their bias supplied by the AVC voltage appearing across C2 (Fig. 20-152), their gain will change as the AVC _____ changes.

20-158
(*voltage*) A strong signal causes a large negative AVC voltage, thereby reducing the positive bias on the bases of the IF and RF amplifiers. A reduced forward bias on a transistor reduces its gain. Hence, a strong signal reacts on the system in such a way as to lower the _____ of the RF-IF section.

20-159
(*gain*) A weak signal produces a small negative AVC voltage, thereby producing little effect on the positive bias of the transistor. This results in _____ gain in the RF-IF sections.

20-160
(*increased, high,* or *large*) In this way, the signal itself provides the information required to control the gain of the receiver. If the received signal is _____, the gain of the set is low; if the received signal is _____, the gain of the set is high.

20-160
(*strong, weak*)

AVC FILTER NETWORK

It is now clear why this type of gain control is called *automatic.* The gain of the RF-IF section of the receiver automatically adjusts to the signal; as a result of this action, the output of the receiver tends to stay at the same level regardless of variations in signal strength (within limits). It remains now to account for the presence of $R2$ and $C2$ in the AVC system since, from what has been said so far, it might appear as though the AVC system would operate if point A, Fig. 20-152, were connected directly to the RF and IF transistors.

20-161
Refer to Fig. 20-161. Until now we have been discussing an RF carrier without modulation. This figure shows a carrier modulated by an _____-frequency signal.

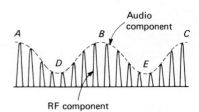

Fig. 20-161

20-162
(*audio*) The modulating signal causes the _____ of the carrier to increase and decrease as the audio rises to its positive peaks and falls to its negative peaks.

20-163
(*amplitude*) If the AF signal has a frequency of, say, 1,000 Hz, then the amplitude of the carrier rises and falls _____ times per second.

20-164
(*1,000*) This is the equivalent of a situation in which the signal reaching the detector *increases and decreases in strength 1,000 times per second.* At the audio positive peaks (A, B, C in Fig. 20-161), the signal is strong and the AVC system would then tend to produce a _____ negative bias on the RF-IF transistors.

20-165
(*large*) This bias would tend to _____ the gain of the set at each of these instants.

20-166
(*reduce* or *lower*) Also, at the negative audio peaks (D and E) the signal would appear weak to the AVC system, which would then tend to produce a smaller _____ bias on the RF-IF transistors.

20-167
(*negative*) The effect of this would be to _____ the gain of the set at points D and E.

20-168
(*increase*) Thus, the gain of the set would be small at points _____ and large at points _____ in Fig. 20-161.

20-169
(*A, B, C; D, E*) This would tend to *wash out* the modulation and defeat the purpose of the receiver. Components R2 and C2 prevent this from occurring. Note that C2, once charged, must discharge through the path consisting of R2 and _____ in Fig. 20-124.

20-170
(*R1*) But C2 is a very large capacitor. When a large capacitor discharges through a resistor, it may take a relatively long _____ to do so.

20-171
(*time*) For an RC circuit, one has the expression $T = RC$ where T is the time in seconds for a capacitor to discharge to a value that is 37 percent of its initial full-charge voltage. In our AVC circuit, $C2 = 10$ μF and $R2 =$ _____ Ω. Their product is 0.047 s.

20-172
(*4,700*) This means that if C were initially charged to 10 V, it would require 0.047 s for the charge to drop to approximately _____ V.

20-173
(*3.7*) Assume that the lowest audio frequency our receiver will have to reproduce is 50 Hz. The *period* of a waveform is the reciprocal of the frequency ($1/f$). For this audio frequency, then, the period is _____ s.

20-174
($\frac{1}{50}$) Or, in decimals, the period is 0.02 s. This is a _____ time than 0.047 s, which was found to be the discharge time of C2 in the AVC circuit.

20-175
(*shorter*) For higher audio frequencies, the period of the waveform is even _____ than it is for 50 Hz.

20-176
(*shorter*) Thus, the discharge time of C2 is always longer than the longest period of the audio waveform. This means that C2 cannot charge or discharge appreciably during the short intervals between the peaks and troughs of the modulated wave (Fig. 20-161). Thus, the _____ voltage developed across C2 cannot follow or wash out the modulation.

20-177
(*AVC* or *negative*) The combination consisting of C2 and R2 is called the *AVC filter*. Its purpose is to prevent the audio modulation from varying the _____ voltage, and hence to prevent a smoothing out of the modulation.

20-178
(*AVC*) Let us summarize the action of an AVC system by stating the function of each of its components. Refer to Fig. 20-124 for this summarization. Rectification of the IF waveform is accomplished by the component identified as _____.

20-179
(*D1*) The dc component of the rectified IF waveform is forced to appear as a voltage drop across the resistor R1 as a result of the blocking action of the component identified as _____.

20-180
(*C1*) Since electron current flows from cathode to anode in D1, point _____ of resistor R1 becomes more negative than point _____.

20-181
(*A, B*) This negative voltage is applied to the bases of the _____ and _____ amplifiers through resistor R2.

20-182
(*RF, IF*) The RF and IF amplifiers are transistors whose gain increases for more _____ bias and decreases for less _____ bias.

20-183
(*forward, forward*) A strong signal produces a _____ negative AVC voltage and hence a reduced positive bias on the transistors; a weak signal produces a _____ negative AVC voltage.

20-184
(*large, small*) Thus, for a _____ signal, the gain of these transistors is low, and for a _____ signal the gain is high.

20-185
(*strong, weak*) Thus, the overall gain of the RF-IF section of the receiver is controlled by the strength of the incoming _____.

20-186
(*signal*) AVC action is prevented from responding to the _____ on the signal by the action of R2 and C2.

20-187
(*modulation*) For this reason the combination comprising R2 and C2 is called the _____ filter.

20-187
(*AVC*)

CRITERION CHECK TEST

____20-1 Another name for the front-end RF amplifier of a receiver is the (*a*) IF stage, (*b*) detector, (*c*) preselector, (*d*) driver.

The following 4 questions refer to Fig. 20-1.

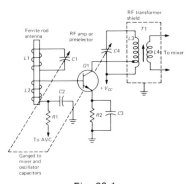 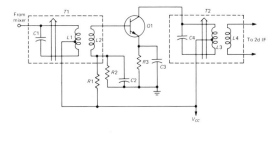

Fig. 20-1   Fig. 20-50

____20-2 The capacitors which track together are (*a*) C1C2, (*b*) C2C3, (*c*) C1C4, (*d*) C2C4.

____20-3 The emitter-swamping resistor is denoted by (*a*) R1, (*b*) R2, (*c*) R3, (*d*) R4.

____20-4 The collector load of Q1 consists of (*a*) R2C3, (*b*) C4L4, (*c*) C4L3, (*d*) L3L4.

____20-5 The function of the ferrite antenna rod is to (*a*) increase the Q of the tuned circuit, (*b*) provide better tracking, (*c*) stabilize the dc bias, (*d*) reduce stray capacitance.

The following 4 questions refer to Fig. 20-50.

____20-6 Forward bias for Q1 is provided by (*a*) R1R2, (*b*) R1L1, (*c*) L1L2, (*d*) R2R3.

____20-7 The circuit C4L3 is tuned to the (*a*) local oscillator frequency, (*b*) radio frequency, (*c*) intermediate frequency, (*d*) modulating frequency.

____20-8 The ferrite slug in T2 is denoted by (*a*) the dashed lines around T2, (*b*) a vertical arrow, (*c*) V_{cc}, (*d*) L4.

SUPERHETERODYNE STAGE ANALYSIS

_____20-9 A function of T1 is to match the (a) high impedance of the tuned circuit to the high impedance of the base-emitter circuit, (b) high impedance of the tuned circuit to the low impedance of the base-emitter circuit, (c) low impedance of the tuned circuit to the low impedance of the base-emitter circuit, (d) low impedance of the tuned circuit to the high impedance of the base-emitter circuit.

_____20-10 If a kitchen-type superheterodyne receiver is tuned to 800 kHz, then the image frequency is (a) 455 kHz, (b) 1,255 kHz, (c) 1,710 kHz, (d) 2,155 kHz.

_____20-11 In question 20-10, the difference between the RF signal and the image frequency is (a) 455 kHz, (b) 910 kHz, (c) 1,710 kHz, (d) 2,155 kHz.

_____20-12 Figure 20-118 shows that as the emitter dc changes, the (a) forward current gain changes, (b) ac input resistance changes, (c) ac output resistance changes, (d) collector load-resistance changes.

_____20-13 The ac voltage gain of a transistor (a) is proportional to the emitter dc, (b) is inversely proportional to the emitter dc, (c) does not depend on the emitter dc, (d) is always greater than 1.

_____20-14 A pulsating dc signal is the sum of (a) a positive and negative dc signal, (b) 2 ac signals 90° out of phase, (c) a dc and an ac component, (d) 2 ac signals 270° out of phase.

_____20-15 The letters AVC stand for (a) audio voltage control, (b) automatic volume control, (c) approximate voltage change, (d) abrupt volume change.

_____20-16 The function of the AVC circuit in a receiver is to (a) stabilize the local oscillator, (b) make the output level independent of signal strength, (c) sharpen the preselector Q, (d) prevent burnout of the driver stage.

_____20-17 It is the function of the AVC filter to (a) eliminate hum, (b) prevent voltage variations, (c) eliminate intermediate frequencies, (d) eliminate audio frequencies.

_____20-18 If the lowest modulating frequency is 100 Hz, then the time constant of the AVC filter should be about (a) 20 μs, (b) 0.2 ms, (c) 2 ms, (d) 20 ms.

21 basic power supplies

OBJECTIVES

(1) Sketch a half-wave rectifier circuit and trace the output signal for a complete cycle of ac. **(2)** Sketch the output of a half-wave rectifier. **(3)** State the condition for obtaining maximum voltage across the rectifier. **(4)** Relate rms voltage to peak voltage. **(5)** Define and give examples of PIV rating of a diode. **(6)** Describe the function of a capacitor and of an inductor in filtering power supply ripple. **(7)** Trace the flow of an unfiltered dc signal as it passes through a *CLC* filter. **(8)** Sketch a full-wave rectifier circuit using two diodes. **(9)** Describe the primary and secondary windings of a center-tapped transformer. **(10)** Describe the phase and magnitude relation of the voltages from a center-tapped transformer. **(11)** Trace the current flow through a full-wave rectifier for a complete input cycle of ac. **(12)** Contrast the outputs of the half-wave and full-wave rectifiers. **(13)** Sketch a two-capacitor voltage-doubler circuit. **(14)** Indicate the charging path of each capacitor. **(15)** State the relative phase of the voltages in the capacitors. **(16)** Compute the voltage output of the doubler circuit. **(17)** State the factors which affect the output of the voltage doubler.

INTRODUCTION

A power supply is an assembly of electrical and electronic components which converts one form of electrical energy into another. For civilian and industrial use, most power supplies are designed to provide one or more dc outputs for application to transistors or vacuum tubes, all from a primary source of 120-V 60-Hz ac in the United States. In some situations, the primary source of power is a 220-V 60-Hz ac line.

Medium- and high-voltage dc power is normally obtained from a stepup transformer and then filtered to remove the ac ripple. Low voltage for transistors can be obtained by using a stepdown transformer plus filtering. However, some transistors have been developed to operate at high voltage, so that devices using such transistors do not need a stepdown transformer.

Filament power for vacuum tubes is obtained in one of two ways: (1) for tubes whose filaments are designed for parallel operation, the heater power is obtained from the secondary of a stepdown transformer; and (2) for tubes designed to operate with their heaters in series, heater power is obtained by connecting the series string across the 120-V ac line.

We will start our study of power supplies with the simplest of all systems: *half-wave rectification* directly from the power lines.

HALF-WAVE RECTIFIER
21-1
Figure 21-1 shows the symbol for a diode. The material used in this type of rectifier is either germanium, silicon, or selenium. The symbol has been standardized so that it shows the direction of the *electron* current to be from the terminal identified as the _____ to the terminal identified as the _____.

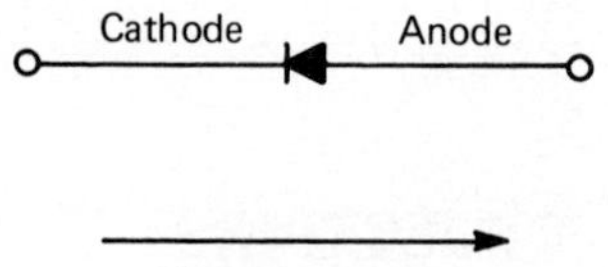

Fig. 21-1

21-2
(*cathode, anode*) Refer to Fig. 21-2. This is a simple rectifier system using a _____ as a rectifier.

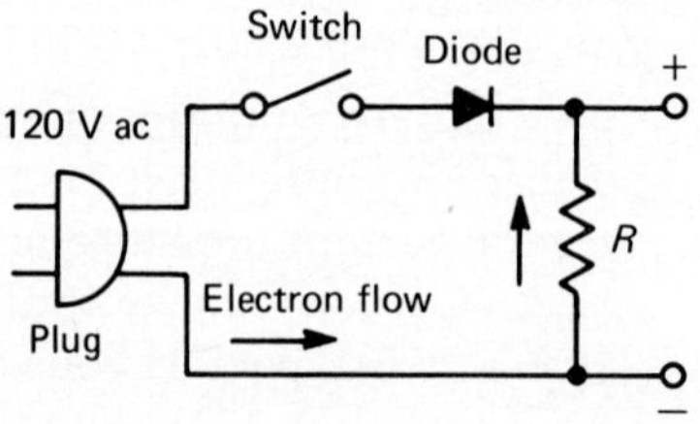

Fig. 21-2

21-3
(*diode*) Power is taken through the *plug* directly from the _____-V ac line.

21-4
(*120*) During half of any cycle, the anode of the diode is given a potential that is positive with respect to the cathode. During the succeeding half-cycle, the anode becomes _____ with respect to the cathode.

21-5
(*negative*) Current can flow in the diode only when the anode is _____ with respect to the cathode.

21-6
(*positive*) Thus, an electron current flows in the diode for _____ of each cycle when the anode is positive with respect to the cathode.

21-7
(*half*) The direction of the electron current flow in the diode is from the _____, where the electrons are emitted, to the _____, where the electrons are collected.

21-8
(*cathode, anode*) The electron current flowing during the conduction half-cycle has an upward direction in the load R. For this reason, the bottom of R becomes _____ in potential with respect to the top.

21-9
(*negative*) The voltage drop across R can now serve as a source of pulsating dc. Refer to Fig. 21-9. Waveform *a* illustrates an electric current (or voltage) with positive and negative excursions. This is a representation of an _____ waveform.

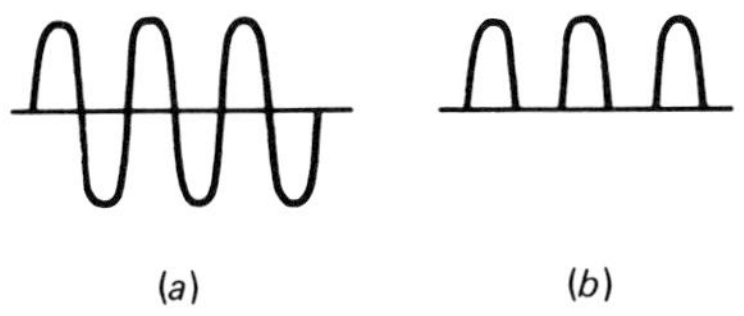

Fig. 21-9

21-10
(*ac*) This is the waveform of the 120-V 60-Hz line. Waveform *b* shows a pulsating _____ waveform that results from the rectification process carried on by the diode in the circuit of Fig. 21-2.

21-11
(*dc*) This is called *half-wave* rectification because only _____ of each full ac cycle appears as rectified output.

21-12
(*half*) Refer to Fig. 21-12. During half of the input ac cycle the anode is made positive with respect to the cathode. During this half of the cycle, the diode is in a _____ condition as indicated by the arrow above the diode symbol.

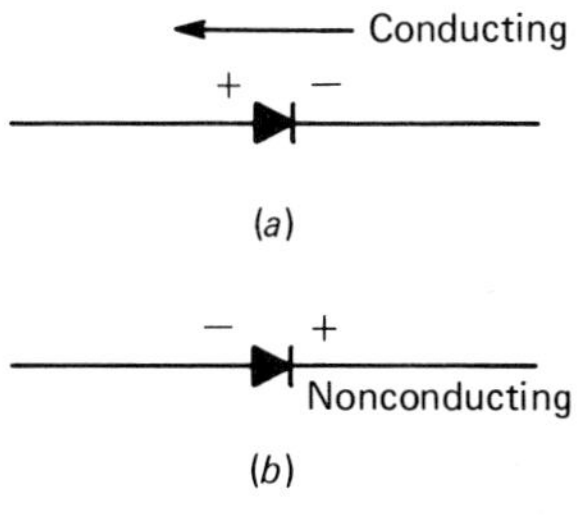

Fig. 21-12

21-13
(*conducting*) While the diode is conducting, it behaves like a *low* resistance. If the current through the diode is properly limited by the load, then the voltage drop across the low resistance between cathode and anode will be _____ (large, small).

21-14
(*small*) That is, the voltage that appears between cathode and anode is small while the diode is conducting. In Fig. 21-12b, however, we have shown the anode _____ (positive, negative) with respect to the cathode.

21-15
(*negative*) For this polarity, the diode cannot conduct. That is, the diode is in a _____ state.

21-16
(*nonconducting*) Refer to Fig. 21-16. This presents the voltage picture when the diode is in the nonconducting state. As indicated, the voltage output of the generator is _____ V rms.

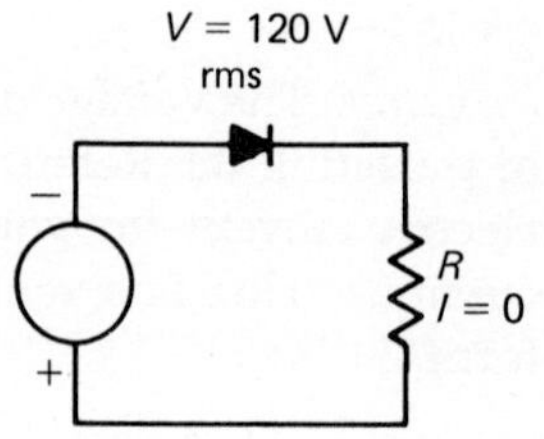

Fig. 21-16

21-17
(*120*) Since the diode is nonconducting, the current in the circuit must be _____.

21-18
(*zero*) If the current through the load is zero, the voltage drop across the load must be _____.

21-19
(*zero*) Hence, the full line voltage must appear across the diode. That is, the voltage that appears across the diode when it is in the nonconducting state in this example is _____ V rms.

21-20
(*120*) Peak voltage is 1.41 times the rms value. In this case, the rms value of the voltage across the diode is 120 V; hence, the peak voltage under these conditions is _____ V.

21-21
(*169.2*) The peak inverse voltage (PIV) that appears across the elements of the diode in Fig. 21-16 is 169.2 V. It must therefore have a PIV rating somewhat higher than _____ V.

520 A PROGRAMMED COURSE IN BASIC ELECTRONICS

21-22
(*169.2*) Suppose a diode were operated in a circuit where the source voltage is 220 V rms. The peak voltage that would be anticipated across the diode elements, then, would be _____ V.

21-23
(*310.2*) This diode would have to have a _____ rating higher than 310.2 V if satisfactory operation is to be obtained.

21-24
(*PIV* or *peak inverse voltage*) A 1N566 diode has a peak inverse rating (maximum) of 400 V. If it were used in a circuit such as that of Fig. 21-16, the maximum rms voltage the generator would be permitted to have would be _____ V. (Remember that rms = peak × 0.707.)

21-24
(*382*)

FILTER NETWORKS

The pulsating output of a half-wave rectifier is not suitable for most electronic applications. For transistor and vacuum tube power, the pulsations must be minimized. The removal of the *residual ripple*, or the smoothing out of the pulsations to produce as pure dc as possible, is accomplished by 1 or more capacitors in combination with 1 or more resistors or inductors. The *RC* or *LC* network which removes the ripple is called the *filter system* or *filter network*.

21-25
The pulsating direct current that flows through the diode in Fig. 21-25 has the waveform shown in Fig. 21-9b. In our study of AVC, we saw that such a waveform could be considered to comprise 2 components: an ac component and a _____ component.

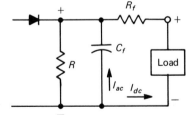

Fig. 21-25

21-26
(*dc*) These 2 components are presented to the filter network in Fig. 21-25 consisting of C_f and R_f. Capacitor C_f is a large unit; assume its capacitance to be 40 μF. A capacitance of this large value has a relatively _____ capacitive reactance even at 60 Hz.

BASIC POWER SUPPLIES 521

21-27

(*small*) Hence, the ac component of the rectified waveform tends to take this path. The path of the ac component through the capacitor is identified by the arrow labeled _____.

21-28

(I_{ac}) The dc component of the rectified waveform, however, is blocked by C_f. Thus, the dc component of the current must flow through the external load, which might be the plate and screen circuits of a radio, a TV set, or any other electronic device. The dc component is identified by the arrow labeled _____.

21-29

(I_{dc}) Filter resistor R_f assists in this action by presenting a relatively high impedance to the ac component, thus encouraging this component to flow back to the source through capacitor _____.

21-30

(C_f) Filter resistor R_f does not distinguish between ac and dc; hence it causes a voltage drop for the dc component. This _____ (increases, decreases) the voltage available for the load.

21-31

(*decreases*) This is an undesirable action. It can be reduced by substituting an inductor called a choke for R_f as in Fig. 21-31. The inductance of the choke shown in this circuit is _____ H.

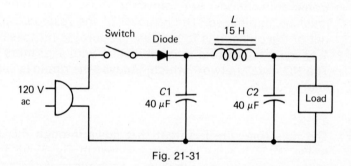

Fig. 21-31

21-32

(15) This is a *large* inductance. Since inductive reactance is $X_L = 2\pi f L$, if L is large then X_L will be _____ even for 60 Hz.

21-33

(*large*) A second capacitor C2 has been added on the remote side of the choke. Let us examine the filtering action of the entire unit, since this is a representative, practical power supply. The diode presents a _____ dc voltage to the filter network.

21-34
(*pulsating*) This pulsating dc voltage sets up a current consisting of a dc component and an _____ component.

21-35
(*ac*) A large part of the ac component tends to flow up through C1 and return to the diode circuit without ever reaching the load. This occurs because C1 has a very low capacitive _____ to 60 Hz.

21-36
(*reactance*) The choke, by presenting a large series inductive reactance, helps confine the ac component to the capacitor path. However, a small amount of ripple voltage still manages to pass the choke. On the other side it comes to another low-reactance _____ identified as C2.

21-37
(*capacitor*) The load represents another relatively large impedance to 60 Hz. Hence, the residual ripple current tends to return to the power supply through _____ rather than through the load.

21-38
(*C2*) The voltage finally applied to the load, therefore, has most of the residual _____ removed and is virtually pure dc.

21-38
(*ripple*)

FULL-WAVE RECTIFIER

Filter networks may take many forms. One often finds 2 chokes and 3 capacitors used; or one may find the choke placed before rather than after the capacitors. The various practical filter networks have specific advantages and disadvantages that dictate which of them is selected for a specific job. A discussion of the multitude of possible filter networks is outside the scope of a basic principles program such as this.

The power supply shown in Fig. 21-31 may be described as a half-wave rectifier followed by a single-section π filter. The physical arrangement of the 2 capacitors and choke in the diagram suggest the Greek letter π, hence the name. A half-wave rectifier, although perfectly practical, produces an output waveform that is somewhat difficult to filter because of the wide "valleys" between the successive peaks. The rectified waveform can be improved from this point of view by utilizing a *full-wave* rectifier.

21-39
Refer to Fig. 21-39. In this full-wave rectifier circuit, 2 diodes are used. The 2 diodes are identified as _____ and _____.

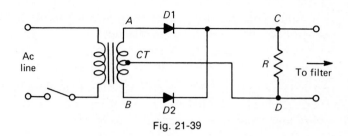

Fig. 21-39

21-40
(*D1, D2*) The ac power is handled by a 2-winding _____ identified as *T* in the figure.

21-41
(*transformer*) The primary winding of the transformer is connected to the ac line. The transformer has _____ secondary winding.

21-42
(*1*) The secondary winding may have any predetermined output voltage desired. It is a 3-terminal winding. The terminals are identified as *A*, *B*, and _____.

21-43
(*CT*) The terminal identified as *CT* is a *center tap*, or a connection to the electric center of this winding. Roughly, if there are 1,000 turns included between *A* and *CT*, there will be _____ turns included between *B* and *CT*.

21-44
(*1,000*) Even more important, this secondary is wound so that equal *voltages* appear across both halves. If the total secondary voltage, say, is 400 V, then the voltage between *A* and *CT* and that between *B* and *CT* will be _____ V each.

21-44
(*200*)

CENTER-TAPPED TRANSFORMER

To appreciate the reason for the center tap, let us first reexamine a convention which has been implicit in our thinking about voltages between points. Refer to Fig. 21-45. Here is a transformer secondary of the usual kind. The voltage that appears across the winding is ac. This means that the potential difference between terminals A and B is constantly varying, with A sometimes "+" with respect to B and sometimes "−" with respect to B. Of course, we can say with equal validity that B is sometimes "+" with respect to A and sometimes "−" with respect to A.

The usual practice, however, is to *assume* 1 of the 2 terminals to be a *reference zero*. That is, we assume 1 terminal to remain at a fixed voltage while the voltage of the other varies above "+" and below "−" the reference zero. It helps somewhat to picture a stick pivoted around the reference zero; as the potential of the other terminal changes, we then picture the end of the stick as being in vibratory motion as shown in the figure.

21-45
In Fig. 21-45, the reference zero is assumed to be terminal B. The potential of terminal _____ is then considered to vary between some "+" potential and an equal "−" potential around the reference zero.

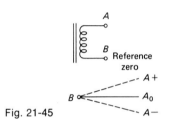

Fig. 21-45

21-46
(A) When terminal A is "+" with respect to B, the stick is visualized in position BA+; when A is "−" with respect to B, the stick is seen in position BA−; when the stick is at position A_0, terminal A must be at the _____ potential as B.

21-47
(same) The highest point reached by the moving end of the stick is thus considered the peak positive voltage reached by end A with respect to B; the lowest point reached by the stick is considered the peak _____ voltage reached by end A with respect to B.

21-48
(negative) Refer to Fig. 21-48. When a secondary is center-tapped, the center-tapped terminal is normally taken as the reference _____.

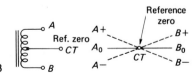

Fig. 21-48

21-49
(*zero*) The stick then is pictured as a child's seesaw, the pivot being considered to be at the _____ as shown in the figure.

21-50
(*CT* or *reference zero*) While terminal A is "moving" up into the "+" voltage region, terminal B is "moving" down into the _____ voltage region.

21-51
("−") On the next half-cycle, while terminal A "moves" into the _____ voltage region, terminal B "moves" into the _____ voltage region.

21-52
("−," "+") Thus, the stick oscillates like a seesaw. The reference zero (or *CT*), however, like the pivot of a seesaw, does not _____ either up or down.

21-53
(*move*) Now let us transfer this idea to the center-tapped transformer. Regardless of the potentials at A and B, the potential of the *CT* is considered to have _____ magnitude.

21-54
(*zero*) If terminal A is positive with respect to zero, then terminal B must be _____ with respect to zero, and vice versa.

21-55
(*negative*) *Twice* during each cycle, both terminal A and B are at zero potential with respect to the center tap. This condition is pictured as line _____ in the seesaw drawing.

21-56
(A_0B_0) At the instant that A reaches *peak positive* voltage with respect to *CT*, B reaches _____ *negative* voltage with respect to *CT*.

21-57
(*peak*) Now let us refer back to Fig. 21-39. Since A can never be positive when B is positive, and since B can never be positive when A is positive, then only 1 diode can be conducting at a given instant. Say A is positive and B is negative at time *t*. Then at time *t*, diode _____ must be conducting.

21-58
(*D1*) During the next half-cycle, *B* will be positive and *A* will be negative. Thus, during this half-cycle the anode of *D*1 will be negative so that *D*1 will be non-conducting; also, the anode of *D*2 will be positive so that _____ will be conducting.

21-59
(*D2*) Thus, each diode conducts alternately on successive half-cycles of ac input. When *D*1 is conducting, *D*2 is _____; when *D*1 is _____, *D*2 is conducting.

21-60
(*nonconducting, nonconducting*) Refer to Fig. 21-60. This drawing depicts a condition at an instant when *A* has gone positive with respect to reference zero, *CT*. A complete circuit exists through diode *D*1. Since the anode of *D*1 is positive with respect to the cathode, _____ will flow from the cathode to the anode.

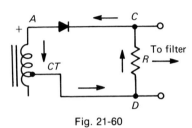

Fig. 21-60

21-61
(*current* or *electrons*) The circuit current is shown by the arrows as going from the anode of *D*1 to *A*, down to *CT*, across to *D*, up through _____, and back to the cathode.

21-62
(*R*) With reference to the load *R*, the electron current flows from end *D* to end *C*. This makes end *D* _____ in polarity with respect to *C*.

21-63
(*negative*) Thus, the output voltage that could be taken across *R* for use in other circuits for this half-cycle is _____ at point *C* and _____ at point *D* (positive, negative).

21-64
(*positive, negative*) Refer to Fig. 21-64. This shows the next half-cycle. That is, during this half-cycle, terminal _____ of the transformer is positive with respect to the center tap while terminal *A* is negative.

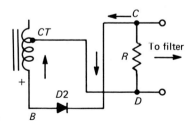

Fig. 21-64

21-65
(*B*) Diode *D*1 is nonconducting during this half-cycle and therefore has been omitted from the picture. Diode *D*2, however, will be in the _____ state since its anode is made more positive than its cathode by the voltage at terminal *B*.

21-66
(*conducting*) The current in the circuit is again illustrated by the arrows. The flow occurs from the anode of *D*2 to *B*, up to *CT*, across to *D*, then up through _____ to *C*, and back to the cathode.

21-67
(*R*) Note that the current direction in this case is exactly the same relative to the load *R* as it was in the previous case. Thus, for this half-cycle, the potential of *C* becomes _____ with respect to *D* just as it did in the previous half-cycle.

21-68
(*positive*) Hence, for *each* ac input half-cycle the output voltage across *R* has the same polarity. Not only has this circuit rectified the ac, but it has made use of both _____-_____ of the input.

21-69
(*half-cycles*) Refer to Fig. 21-69. Part *a* shows the typical waveform obtained from a rectifier which rectifies only _____ of the input ac cycle.

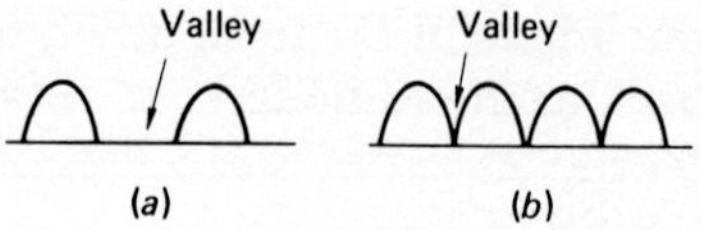

Fig. 21-69

21-70
(*half*) This is a half-wave rectifier. Part *b* shows the waveform of a rectifier which rectifies both halves of the input ac cycle. This is a _____-wave rectifier.

21-71
(*full*) Note the difference in valley width in both diagrams. The valley width of the full-wave system is much _____ than that of the half-wave system.

21-72
(*smaller*) This difference makes it much easier to *filter* or smooth out the output of the _____-wave rectifier system.

21-73
(*full*) The full-wave power supply, to be complete and ready for use, would have to have added to it a _____ system consisting of one or more capacitors, a choke, or a suitable resistor as discussed previously.

21-73
(*filter*)

VOLTAGE MULTIPLIERS

The voltage output at the dc terminals of the power supply just discussed is determined by the selection of the ac characteristics of the transformer. If the transformer is chosen properly, virtually any dc voltage between wide limits can be obtained from this system. Power transformers, aside from being comparatively expensive, are bulky and heavy; there are some applications where the use of a transformer would be an inconvenience. For situations like this, increased dc voltage—that is, voltage higher than that of the ac lines—may be obtained by a *voltage-multiplier* rectifier. We shall discuss a voltage-doubler circuit next. In this arrangement, no transformer is used, yet it is possible to obtain a dc output voltage in the order of 300 V from the 120-V ac lines.

21-74
The complete voltage-doubler circuit is illustrated in Fig. 21-74. Part of the voltage doubler is a _____ (series, parallel) circuit containing 2 capacitors connected across the dc output terminals.

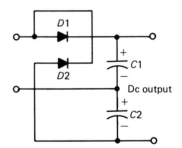

Fig. 21-74

21-75
(*series*) Refer to Fig. 21-75. We will first study the action when the top lead of the ac is positive and the bottom lead is negative. For this condition, the anode of $D1$ is connected directly to the _____ terminal of the ac line.

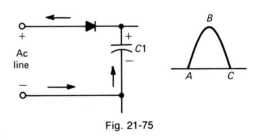

Fig. 21-75

21-76
(*positive* or *top*) Since the diode D1 is *forward*-biased, electrons can flow into the lower plate of C1. This drives electrons out of the upper plate of C1 through D1 to the _____ terminal of the ac line.

21-77
(*positive*) Capacitor C1 thus charges to the *peak* line voltage in an extremely short time. This charge accumulates during the interval when the input voltage is rising from A to _____.

21-78
(B) As this half of the ac input cycle falls off (B to C), the capacitor retains *all* its charge since the dc output circuit is open and the capacitor has no discharge path. If the line voltage is 120 V rms, then the charge on the capacitor is now _____ V.

21-79
(169.2) Refer to Fig. 21-79. This shows the action during the second half of the input cycle. The top terminal has gone negative, and the bottom terminal has gone _____.

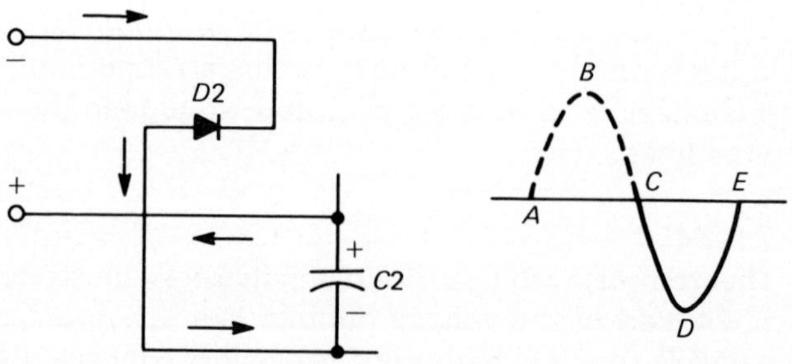

Fig. 21-79

21-80
(*positive*) Diode D2 is now forward-biased. Electrons now flow from the negative input terminal to diode _____ whence they go into the lower plate of C2.

21-81
(D2) Electrons are then driven from the upper plate of C2 back into the _____ terminal of the power supply, leaving C2 charged with the polarity shown.

21-82
(*lower* or *positive*) C2 also charges to the _____ voltage of the ac line. That is, with the dc output circuit open as shown, the voltage on C2 is 169.2 volts when the input voltage passes point D (see sine curve).

21-83
(*peak*) From D to E, as the ac input voltage falls toward zero, the capacitor retains all its charge, since there is no _____ path.

21-84
(*discharge*) Thus, at the end of the full input cycle (from A to E) both capacitors are charged to 169.2 V. As mentioned previously, these capacitors are in _____ connection.

21-85
(*series*) Refer to Fig. 21-85. The capacitor voltages are additive as shown by the "+" and "−" signs of their charges. Thus, the total output voltage V is the sum of the 2 capacitor voltages, or _____ V.

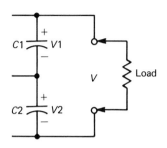

Fig. 21-85

21-86
(*338.4*) Now let us connect a load across the output terminals as in Fig. 21-85. This load offers a discharge path for both C1 and C2. That is, when the ac input cycle nears point C or E, both capacitors can _____ a part of their stored charge through the load.

21-87
(*discharge*) Thus, the capacitors do not retain their peak charge in the doubling process. This means that the output voltage across the load will be somewhat _____ than 338.4 V.

21-88
(*smaller* or *lower*) The actual dc output voltage will depend upon two things: (1) the resistance of the load and (2) the capacitances of _____ and _____.

21-89
(*C1, C2*) We are again dealing with a time constant situation for which $T = RC$. If the load R is a large resistance, then the time required for partial discharge of C1 and C2 will tend to be correspondingly _____.

21-90
(*long*) If both C1 and C2 are very large capacitances, then the _____ for discharge will also tend to be correspondingly long.

BASIC POWER SUPPLIES 531

21-91

(*time*) In both cases, the capacitors will retain a large portion of their peak charge since the time available for discharge is a short one, being only about one-fourth of a cycle. If the ac frequency is 60 Hz, then 1 cycle requires $\frac{1}{60}$ s to be completed; hence one-fourth of 1 cycle represents about _____ s in discharge time available.

21-92

($\frac{1}{240}$) Thus, in applications where voltage doublers are used, it is advisable to make C1 and C2, as well as the load resistance, as _____ as possible if an output voltage approaching twice the line voltage is to be realized.

21-92
(*large*)

CRITERION CHECK TEST

___21-1 For the diode below, electron current flow (*a*) is right to left, (*b*) is left to right, (*c*) is both left to right and right to left, (*d*) depends on whether the diode is silicon or selenium.

___21-2 In Fig. 21-16, when the diode is conducting, (*a*) no current flows through the load, (*b*) all the supply voltage appears across the diode, (*c*) all the supply voltage appears across the load, (*d*) the diode is in a high-resistance condition.

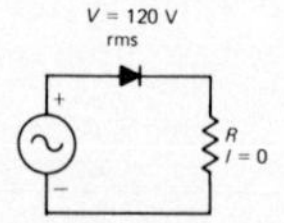

Fig. 21-16

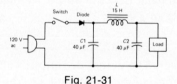

Fig. 21-31

___21-3 If the line voltage is 240 rms, then the peak voltage is (*a*) 260 V, (*b*) 310 V, (*c*) 340 V, (*d*) 410 V.

___21-4 The function of a filter network in a power supply is to (*a*) limit the current in the rectifiers, (*b*) limit the peak voltage to the rectifiers, (*c*) separate the ac and dc components of voltage, (*d*) alternate the voltage.

___21-5 The function of C2 in Fig. 21-31 is to (*a*) limit the alternating-current flow of the diode, (*b*) limit the maximum voltage to the diode, (*c*) shunt the ac signal to ground, (*d*) shunt the dc signal to ground.

____21-6 The disadvantage of a half-wave rectifier circuit is that the (a) components are expensive, (b) diodes must have a large PIV rating, (c) diodes must have a high power rating, (d) output is difficult to filter.

____21-7 The primary function of a center-tapped transformer in a power supply is to (a) step up the voltage, (b) step down the voltage, (c) cause the diodes to conduct alternately, (d) isolate the load from ground.

____21-8 In the circuit below, (a) D1 and D2 are conducting, (b) only D1 is conducting, (c) only D2 is conducting, (d) neither D1 nor D2 is conducting.

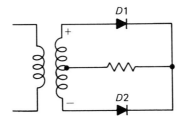

____21-9 A semiconductor diode may be used as a rectifier because it offers a (a) low resistance in both directions, (b) high resistance in both directions, (c) high resistance in one direction and a low resistance in the opposite direction, (d) solid current path.

____21-10 To get the maximum voltage from a voltage doubler, the (a) discharge time constant should be longer than the period of the ac source, (b) load resistor should be as small as possible, (c) capacitors should be as small as possible, (d) PIV rating of the diodes should be at least 4 times the rms rating of the ac source.

22 analysis of a superheterodyne receiver

OBJECTIVES

(1) Analyze the power supply of a transistor radio. (2) Identify the stepdown transformer and the filter circuit. (3) State the function of the surge resistor and the type and characteristics of the rectifier. (4) Relate the full-load current to the no-load current for class A and B outputs. (5) State the purpose of the decoupling network in the power supply. (6) State the type and characteristics of the transistor used in the output stage. (7) Identify the emitter-swamping resistor, bypass capacitor, and bias network. (8) Estimate the size of the emitter bypass capacitor. (9) State the relative current drain of the output stage. (10) Identify the collector load of the output transistor. (11) Identify the collector load of the driver transistor. (12) State the type and specifications of the driver transistor. (13) Discuss the biasing arrangement of the driver stage. (14) Identify the volume control. (15) Identify the detector diode and demodulating filter. (16) Trace the signal level as the wiper arm of the volume control is moved. (17) Indicate the AVC filter circuit. (18) Locate the stage which receives AVC voltage. (19) State the type and specifications of the transistor in the second IF stage. (20) Indicate the IF transformer of this stage and the method of tuning it. (21) Indicate the emitter-swamping resistor, bypass capacitor, and bias resistors. (22) State the type of transistor used in the first IF stage. (23) Indicate the IF tuned circuit, emitter-swamping resistor, bypass capacitor, and biasing network in this stage. (24) Sketch and discuss the functions of a ferrite loopstick. (25) Locate and state the range of the RF tuning capacitor. (26) Describe the functions of a trimmer capacitor in adjusting the dial setting of a receiver. (27) Locate and state the range of the local oscillator tuning capacitor. (28) Relate the resonant frequencies of the RF and local oscillator circuits. (29) State the function of the oscillator trimmer capacitor. (30) Indicate the component which provides feedback for the local oscillator. (31) Indicate the function of the oscillator lead.

POWER SUPPLY

At first glance, a complete diagram of a superheterodyne receiver appears hopelessly complicated to the uninitiated. Even a student who understands the functioning of each individual stage may tend to get lost among the interconnections the first time he or she begins serious study of the whole receiver. The fact that every electronic draftsman tends to represent circuits a little differently does not help this situation. After a while, however, the student learns to isolate individual sections of the receiver for study and ignore the rest until the time for tracing the interconnections arrives.

In selecting the circuit for analysis in this chapter, we intentionally chose one that is similar to the vast majority of transistor radios on the market today. Small pocket radios would have essentially the same circuitry, but a battery would replace the rectifier power supply.

22-1
The first step in our analysis will be to locate the power supply. This is most easily done by noting, in Fig. 22-1, the 117 V ac power plug. The power plug feeds the primary of transformer $T1$ through an _____-_____ _____.

22-2
(*on-off switch*) The purpose of transformer $T1$ is to step down the 117 V ac to a value of _____ V ac.

22-3
(*8*) The secondary of transformer $T1$ is connected to a resistor denoted by $R18$. From the parts list, it is seen that this resistor has a value of _____ Ω.

22-4
(*27*) The function of this resistor is to protect the diode rectifier. When the switch is first turned on, a large surge of current could flow through the diode to charge capacitor $C20$. This surge could easily exceed the _____ rating of the diode.

22-5
(*current*) Hence, $R18$ ensures that the surge current stays within the current rating of the diode. The rectifier diode, denoted by $D2$, is given as type _____ in the parts list.

22-6
(*1N2858*) The type 1N2858 is a medium-power silicon rectifier. Since there is only 1 rectifying diode, we are dealing with a _____-_____ rectifier circuit.

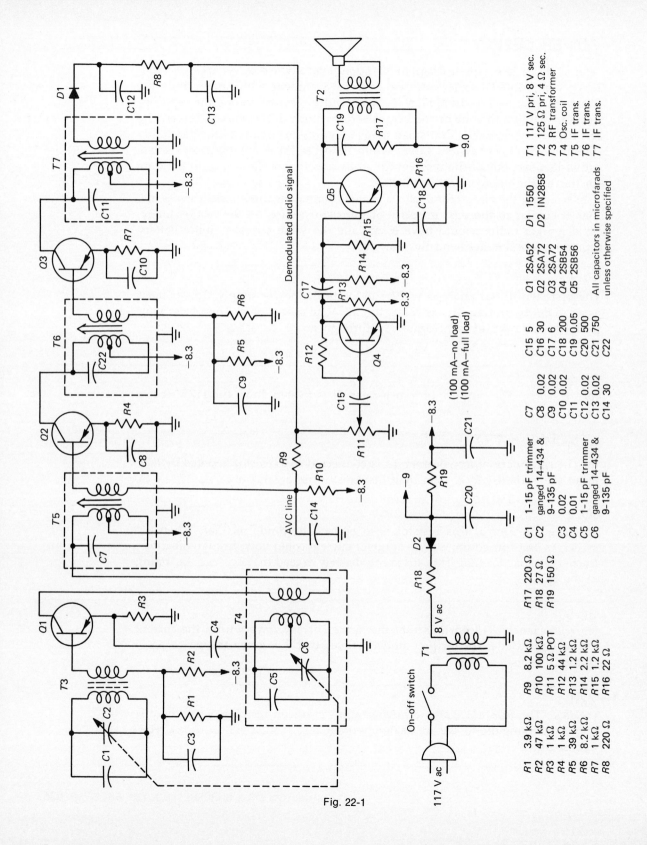

Fig. 22-1

22-7
(*half-wave*) In order to smooth out the pulsating dc coming from the rectifier, a filter circuit is necessary. The filter in this diagram consists of capacitors C20 and _____, and resistor R19.

22-8
(*C21*) It is seen from the parts list that C20 and C21 are very large capacitors. C20 is given as 500 µF and C21 as _____ µF.

22-9
(*750*) These large capacitors guarantee a small ac ripple. The current output from the supply is _____ mA no load and _____ mA full load. The reason that the no-load and full-load currents are the same is that the output stage is class A.

22-10
(*100, 100*) If the radio receiver were to be battery powered, then the output stage would be class B push-pull. This would cause the no-load current to be much _____ (less, greater) than the full-load current.

22-10
(*less*)

DECOUPLING NETWORK

This completes the analysis of the power supply. We see that it is a half-wave rectifier system with an *RC* π-network filter containing provision for surge protection. One point we must explore further is why the power supply produces 2 dc voltages rather than just 1.

22-11
Two output leads are seen to come from the power supply. The one coming from capacitor C20 produces −9 V dc, while the lead from C21 produces _____ V.

22-12
(*−8.3*) Inspecting the diagram, Fig. 22-1, reveals that the −8.3-V lead feeds all the stages except the _____ stage.

22-13
(*output*) The output stage is fed by the _____-V supply. Thus, the resistor-capacitor combination R19-C21 serves to separate, or *decouple*, the supply for the output stage and the remainder of the circuit.

22-14
(−9) A perfect power supply, by definition, would produce a *constant* voltage independent of the current demand. The reason for the R19-_____ decoupling network is that the receiver's power supply is not perfect.

22-15
(C21) The output stage draws about 95 percent of the total dc used by the receiver. The heavy signal current drawn by the output stage will cause the dc supply voltage to _____.

22-16
(*vary*) If the other stages of the receiver used the same dc voltage as the output, then those stages would see a _____ dc supply.

22-17
(*varying*) The variation in dc supply voltage caused by the output stage could be as great as the signals developed by the _____ stages. This would seriously hamper the proper functioning of the receiver circuit.

22-18
(*input*) Thus, nearly all receiver circuits, whether they use 117 V ac or battery supplies, employ 2 or more dc voltages. These dc voltages are separated by 1 or more _____ networks.

22-19
(*decoupling*) Summarizing the functions of the components in the power supply, we have (1) the stepdown transformer, T1; (2) the rectifier, a type _____; (3) the filter components, C20, R19, and _____; (4) surge protection provided by resistor _____; (5) the decoupling network consisting of R19 and C21.

22-19
(*1N2858, C21, R18*)

OUTPUT STAGE

Next, we shall analyze the power output stage. For the reader who might wonder why we are working through the set from the output toward the input rather than the inverse, we point out that this enables us to discuss special circuits, such as the AVC circuit, in the proper order.

22-20
The transistor used in the power output amplifier is denoted by Q5, whose type number is _____.

538 A PROGRAMMED COURSE IN BASIC ELECTRONICS

22-21
(*2SB56*) The 2SB56 is a medium-power, PNP, germanium, audio-frequency transistor. The designation *audio-frequency* means that the transistor provides useful gain at _____ (low, high) frequencies.

22-22
(*low*) The emitter-swamping resistor is denoted by R16. According to the parts list, this resistor has a value of _____ Ω.

22-23
(*22*) The bypass capacitor is given on the schematic as C18. This capacitor has the large value of _____ μF.

22-24
(*200*) This large value of capacitance is necessary because the output stage is amplifying audio frequencies. Recall that the impedance of a capacitor _____ (increases, decreases) as the frequency goes down.

22-25
(*increases*) Hence, if the capacitor is to effectively bypass the emitter resistor at low frequencies, the capacitor must have a _____ value.

22-26
(*large*) The bias resistors for this stage are R14 and R15, whose values are _____ kΩ and _____ kΩ, respectively.

22-27
(*2.2, 1.2*) This bias network produces a collector dc of about 95 mA in Q5. Recall that the total no-load current from the supply is 100 mA; this supports the contention that the output stage takes about _____ percent of the total current.

22-28
(*95*) The collector load of Q5 consists of the primary of transformer _____.

22-29
(*T2*) From the parts list, it is seen that T2 produces an impedance of 125 Ω at the primary if a _____-Ω speaker is used.

22-30
(*4*) The impedance of speakers ranges from 2 Ω to about 32 Ω, so that 4 Ω is a typical value. Across the primary of T2 are 2 components labeled C19 and _____.

22-31
(*R17*) This capacitor-resistor network is intended to bypass some of the higher audio frequencies away from the output transformer to avoid the "tinny" sound that characterizes many small speakers. This action requires a capacitor that will have a small impedance for the higher frequencies; hence it has a value of _____ μF.

22-32
(*0.05*) Coupling of the audio signal from the previous stage is accomplished through a 6-μF capacitor. This capacitor is identified as _____ in Fig. 22-1.

22-32
(*C17*)

DRIVER STAGE

This completes the analysis of the power output stage. The stage preceding this is an audio driver amplifier consisting of *Q*4 and its associated components.

22-33
The collector load resistor for the audio driver amplifier is identified as _____.

22-34
(*R13*) The value of this resistor from the parts list is _____ kΩ.

22-35
(*1.2*) The transistor *Q*4 is identified as type _____.

22-36
(*2SB54*) The type 2SB54 is a medium-power, audio-frequency, germanium, PNP transistor. *Audio frequency* indicates that the transistor produces useful gain at relatively _____ frequencies.

22-37
(*low*) Note that the emitter of *Q*4 is connected directly to _____.

22-38
(*ground*) This means that there is no emitter-_____ resistor to provide feedback stability.

540 A PROGRAMMED COURSE IN BASIC ELECTRONICS

22-39
(*swamping*) However, bias stability is present because of the bias resistor arrangement. The bias resistor is identified as _____.

22-40
(*R12*) By tying R12 to the collector of Q4 rather than to the supply voltage, a variable bias is provided to the base of Q4. This variable bias acts to stabilize dc variations in Q4. In the parts list, R12 is given as _____ kΩ.

22-41
(*44*) Q4 is biased in class A operation by R12. The no-load current for this stage is about 1.6 mA. Hence, because operation is class A, the full load current is _____ mA.

22-42
(*1.6*) R11 is the volume control, which is coupled to the driver stage by capacitor C15. Capacitor C15 is given as _____ μF.

22-43
(*5*) The volume control is a potentiometer having a resistance value of _____ kΩ.

22-44
(*5*) Note the connections of R11. The top of R11 is connected to the line labeled *demodulated audio signal*; the bottom of R11 is connected to _____.

22-44
(*ground*)

DETECTOR AND AVC

We are finished with the analysis of the audio driver amplifier and will now examine the diode detector and AVC.

22-45
The secondary of IF transformer _____ is connected to diode D1.

22-46
(*T7*) The function of D1 is to rectify or detect the IF signal. The output of the rectifier is not pure dc, but rather _____ dc.

22-47
(*pulsating*) Thus, D1 is followed by a filter consisting of C12, R8, and _____.

22-48
(*C13*) Note that C12 and C13 are rather small capacitors, each having the value of _____ μF.

22-49
(*0.02*) The capacitors have a small impedance for RF, but a _____ impedance for audio frequencies.

22-50
(*large*) Thus the CRC filter removes the _____ frequencies and lets the modulating AF pass on.

22-51
(*radio*) The component across which the modulation voltage appears for further amplification is the resistor we have already mentioned as the volume control. This resistor is identified as _____.

22-52
(*R11*) When the wiper of R11 is moved to the position closest to ground, the voltage being impressed on the base of the driver amplifier is zero. Hence, in this setting, the volume of the signal on the loudspeaker is _____.

22-53
(*zero* or *smallest*) The sound from the speaker is loudest when the wiper is moved to the _____ of volume control R11.

22-54
(*top*) Locate the AVC *line*. The AVC filter consists of a resistor and a capacitor. The resistor portion of the filter is identified as _____.

22-55
(*R9*) The capacitor portion of the filter is the capacitor connected from R9 to ground. This capacitor is identified as _____.

22-56
(*C14*) If we trace the AVC voltage from capacitor C14, we are led to transistor _____.

22-57
(*Q2*) Transistor Q2 is in the first IF stage. Thus, in this receiver, the automatic gain control feature is applied to only 1 stage, an _____ stage.

22-57
(*IF*)

IF STAGES

We are ready to proceed now to the IF amplifier.

22-58
The transistor of the second IF stage is denoted by _____.

22-59
(Q3) Q3 is a type _____, a PNP, germanium transistor which gives useful gain at radio frequencies.

22-60
(2SA72) The collector load of Q3 is the tuned circuit consisting of the capacitor C11 and the primary of _____. This tuned circuit resonates at the intermediate frequency of 455 kHz.

22-61
(T7) The arrow between the coils of T7 indicates a ferrite slug which can be moved in and out to adjust the tuning. The _____ lines around T7 indicate a metal shield can.

22-62
(dotted) Note that the metal shield can is electrically connected to _____.

22-63
(ground) Figure 22-63 depicts a typical IF transformer. Note that the transformer is surrounded by a metal can and that a _____ is used to move the ferrite slug.

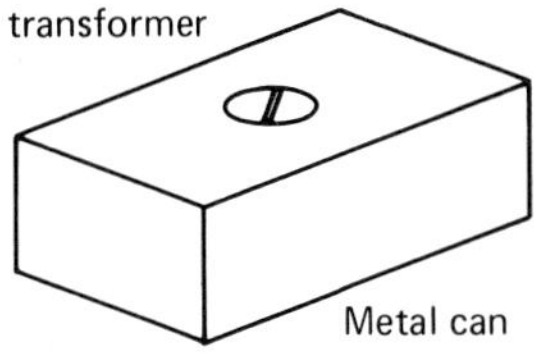

Fig. 22-63 — Screw to tune transformer; Metal can

22-64
(screw) The emitter-swamping resistor of Q3 is labeled _____.

22-65
(R7) The emitter bypass capacitor is C10. Note that C10, which is given as 0.02 µF, is much _____ (smaller, larger) than the bypass capacitor in the audio stages.

22-66
(smaller) This is because the IF is a relatively _____ (low, high) frequency, while the AF is a relatively _____ (low, high) frequency.

ANALYSIS OF A SUPERHETERODYNE RECEIVER

22-67
(*high, low*) The bias network for Q3 consists of resistors R5 and _____.

22-68
(*R6*) Capacitor _____ provides a signal ground return for the secondary of transformer T6.

22-69
(*C9*) The first IF amplifier stage uses a transistor denoted by Q2. From the parts list, it is seen that Q2 is the same type of transistor as used in the _____ _____ stage.

22-70
(*second IF*) The collector load for Q2 is the combination of the primary of T6 and capacitor _____.

22-71
(*C22*) This resonant circuit is tuned to the IF of 455 kHz. According to the schematic, transformer T6 is surrounded by a metal _____ _____.

22-72
(*shield can*) The emitter-swamping resistor and bypass capacitor are identical to those of the _____ _____ stage.

22-73
(*second IF*) Forward bias from the supply is brought to the base of Q2 by resistor R10. As we have previously noted, the base of Q2 is also fed by the _____ line.

22-74
(*AVC*) This AVC voltage produces automatic volume control by producing _____ (more, less) gain for weak signals and _____ (more, less) gain for strong signals.

22-74
(*more, less*)

OSCILLATOR-CONVERTER STAGE

We next look into the construction and functioning of the oscillator-converter stage.

544 A PROGRAMMED COURSE IN BASIC ELECTRONICS

22-75
Locate the input coils, indicated by T3. Note that between the primary and secondary coils are two _____ lines.

22-76
(*dotted*) Contrast these dotted lines with the solid lines found in transformer T2. The solid lines indicate that a transformer is wound on an iron core. From Fig. 22-76, we see that T3 is wound not on an iron core, but around a ferrite _____.

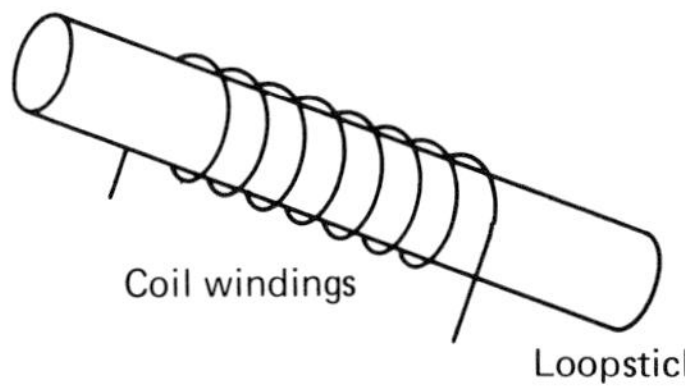

Fig. 22-76

Ferrite rod
Coil windings
Loopstick antenna

22-77
(*loopstick*) In most transistor receivers, the loopstick acts as both the tuned RF circuit and the _____. The long ferrite rod provides a large area to intercept the incoming signal and produces a high-Q tuned circuit.

22-78
(*antenna*) The main input tuning capacitor for the RF section of the converter is C2. The broken line indicates that C2 is ganged to capacitor _____ so that a common shaft will vary both capacitors simultaneously.

22-79
(C6) Capacitor C2 has a range of variation from _____ to _____ pF.

22-80
(*14, 434*) When C2 is set near the 14-pF end of its rotation, the RF section of the converter will be resonant to the _____-frequency end of the broadcast band.

22-81
(*high*) The low-frequency end of the broadcast band is tuned in when C2 is set near _____ pF.

22-82
(*434*) Referring to the list of parts, we see that C1, a capacitor connected in parallel with C2, is called a _____.

22-83
(*trimmer*) This capacitor has a very small maximum capacitance. It is usually semivariable and is provided with a screwdriver adjustment. It is used to make up for small differences in receiver assembly and adds or subtracts capacitance in small amounts from capacitor _____.

22-84
($C2$) In use, $C2$ is adjusted to the point where the receiver dial reads, say, 1,400 kHz. Then _____ is adjusted until a signal of 1,400 kHz is fully peaked. Once capacitor _____ is adjusted to peak the signal at the proper dial setting, it is left that way and is not readjusted for long periods of time.

22-85
($C1, C1$) Dc bias for $Q2$ is provided by the resistors $R1$ and _____.

22-86
($R2$) The main tuning capacitor for the oscillator section is ganged to the main tuning capacitor of the RF section. Thus, the capacitor identified as _____ must be the main tuning capacitor of the oscillator section.

22-87
($C6$) Note that the range of possible capacitor values for the oscillator tuning capacitor goes from _____ to _____ pF.

22-88
(*9, 135*) Both the minimum and maximum capacitance values for $C6$ are smaller than those for $C2$. This indicates that the oscillator will operate at a _____ frequency than the incoming RF signal.

22-89
(*higher*) Since the IF of this receiver is 455 kHz, the frequency of the oscillator should be _____ kHz higher than any incoming signal tuned in at any point in the band.

22-90
(*455*) For example, if a station at 890 kHz is tuned in, the oscillator frequency should be _____ kHz if the receiver is correctly aligned.

22-91
(*1,345*) To correct for small errors in oscillator tracking (production of a constant difference of 455 kHz between oscillator and the station tuned in), a small, semivariable capacitor is connected in parallel with $C6$. This is the oscillator trimmer and is identified as _____ in the schematic.

22-92
(C5) The oscillator transformer is identified as _____ in the figure.

22-93
(T4) Sustained oscillation is produced by inductive feedback from the secondary of T4 to the _____ of T4.

22-94
(*primary*) The local oscillator signal is coupled to the emitter of Q1 by capacitor _____.

22-95
(C4) The RF signal and the local oscillator signal are mixed together by Q1. The tuned circuit consisting of _____ and T5 picks out the difference frequency, or IF.

22-95
(C7)

CRITERION CHECK TEST

All questions refer to first diagram in Chapter 22.

____ 22-1 The function of transformer T1 is to (a) isolate the chassis from ground, (b) match impedances, (c) step down the line voltage, (d) give high efficiency.

____ 22-2 The decoupling network of this receiver consists of (a) T1, (b) R18-C20, (c) C20-R19, (d) R19-C21.

____ 22-3 The function of C18 is to (a) bypass radio frequencies, (b) bypass intermediate frequencies, (c) bypass audio frequencies, (d) prevent spurious oscillations.

____ 22-4 The function of Q4 is to (a) amplify IF signals, (b) amplify RF signals, (c) amplify AF signals, (d) detect the modulated signal.

____ 22-5 Rectification of the modulated IF signal occurs in (a) D1, (b) D2, (c) Q2, (d) Q3.

____ 22-6 The AVC voltage is applied to (a) the RF stage, (b) only 1 IF stage, (c) both IF stages, (d) the driver stage.

____ 22-7 Capacitor C10 is bypassing (a) IF signals, (b) RF signals, (c) AF signals, (d) local oscillator frequencies.

____ 22-8 The tuning of T6 is changed by (a) a variable capacitor, (b) a trimmer capacitor, (c) changing the numbers of secondary windings, (d) a threaded ferrite slug.

____ 22-9 In the receiver, there (a) are two RF stages, (b) is a combined RF and local oscillator stage, (c) are two driver stages, (d) is a zener diode power supply.

____ 22-10 Most small AM receivers use as an antenna (a) a 20-ft folded dipole, (b) a ferrite loopstick, (c) a 300-Ω coaxial cable, (d) an air-core transformer.

____ 22-11 If the receiver is tuned to 855 kHz on the AM dial, the local oscillator frequency is (a) 400 kHz, (b) 455 kHz, (c) 1,310 kHz, (d) 1,520 kHz.

answers to criterion check tests

CHAPTER 1

1. c	11. d	21. b	31. d
2. b	12. c	22. b	32. b
3. a	13. c	23. c	33. c
4. c	14. d	24. d	34. d
5. c	15. a	25. c	35. d
6. c	16. a	26. c	36. c
7. c	17. b	27. d	37. b
8. a	18. d	28. b	38. a
9. d	19. d	29. b	39. c
10. c	20. c	30. d	

CHAPTER 2

1. a	8. c	15. d	22. c
2. c	9. b	16. a	23. d
3. c	10. b	17. b	24. d
4. d	11. d	18. a	25. b
5. a	12. a	19. d	26. d
6. d	13. d	20. b	27. d
7. b	14. b	21. b	

CHAPTER 3

1. a	6. c	11. c	16. d
2. c	7. c	12. b	17. d
3. a	8. b	13. c	18. d
4. c	9. c	14. c	19. b
5. d	10. c	15. c	20. c

CHAPTER 4

1. b
2. a
3. c
4. d
5. d
6. c
7. c
8. d
9. d
10. c
11. c
12. d
13. c
14. d
15. a
16. c
17. c
18. c
19. b
20. b
21. c
22. c

CHAPTER 5

1. c
2. d
3. a
4. d
5. d
6. b
7. a
8. c
9. d
10. d
11. c
12. c
13. c
14. c
15. c
16. a
17. b
18. c
19. c
20. c
21. b
22. b
23. c
24. a

CHAPTER 6

1. c
2. c
3. c
4. b
5. b
6. c
7. c
8. a
9. c
10. b
11. a
12. c
13. c
14. b
15. b
16. b
17. a
18. b
19. b
20. c
21. d
22. b
23. b
24. a
25. c

CHAPTER 7

1. b
2. c
3. a
4. d
5. b
6. c
7. a
8. c

CHAPTER 8

1. b
2. d
3. a
4. a
5. a
6. a
7. b
8. c
9. c
10. c
11. d
12. d
13. a
14. c
15. a
16. c
17. d
18. b

CHAPTER 9

1. d
2. d
3. a
4. a
5. b
6. c
7. b
8. d
9. b
10. b
11. b
12. b
13. a

CHAPTER 10

1. d
2. a
3. b
4. a
5. a
6. b
7. c
8. a
9. a
10. d
11. c
12. c
13. a
14. a
15. c
16. c
17. a
18. d
19. a
20. c
21. a
22. a
23. a
24. c
25. a
26. c
27. a
28. a
29. c
30. a
31. a
32. c

CHAPTER 11

1. c
2. b
3. b
4. c
5. d
6. b
7. a
8. b
9. a
10. b
11. d
12. b
13. a
14. a
15. d
16. c
17. a
18. b
19. c
20. a
21. b

CHAPTER 12

1. b
2. d
3. a
4. c
5. d
6. a
7. c
8. b
9. d
10. c
11. d
12. c
13. d
14. b
15. c
16. d
17. b
18. a
19. a
20. a

CHAPTER 13

1. d
2. b
3. a
4. b
5. a
6. d
7. a
8. b
9. c

CHAPTER 14

1. c
2. a
3. d
4. a
5. d
6. c
7. c
8. a
9. b
10. c
11. c
12. c

CHAPTER 15

1. c
2. d
3. d
4. d
5. b
6. b
7. c
8. b
9. d
10. a
11. c
12. c

CHAPTER 16

1. d
2. c
3. a
4. a
5. b
6. c
7. b
8. d
9. c
10. b
11. b
12. a
13. a

CHAPTER 17

1. d
2. c
3. b
4. b
5. c
6. c
7. c
8. b
9. a
10. b
11. a
12. a
13. b
14. c
15. c

CHAPTER 18

1. b
2. c
3. d
4. c
5. d
6. d
7. b
8. b
9. c
10. d
11. c
12. b

CHAPTER 19

1. a
2. c
3. c
4. c
5. d
6. a
7. a
8. d
9. c
10. b

CHAPTER 20

1. *c*	6. *a*	11. *b*	15. *b*
2. *c*	7. *c*	12. *b*	16. *b*
3. *b*	8. *b*	13. *b*	17. *d*
4. *c*	9. *b*	14. *c*	18. *d*
5. *a*	10. *c*		

CHAPTER 21

1. *a*	4. *c*	7. *c*	9. *c*
2. *c*	5. *c*	8. *b*	10. *a*
3. *c*	6. *d*		

CHAPTER 22

1. *c*	4. *c*	7. *a*	10. *b*
2. *d*	5. *a*	8. *d*	11. *c*
3. *c*	6. *b*	9. *b*	

INDEX

Acceptor, 21
Admittance, 191
Aiding, series, 436
Alpha, 52, 114
Aluminum, 21
AM (amplitude modulation), 435
  standard broadcast, 357
Ampere, 13
Amplification, voltage, 54
Amplification factor, 229
Amplifier(s):
  audio, 248
  behavior of wideband, 335
  classes of, 160
    A, 160, 248
    AB, 161, 249
    B, 160, 249
    C, 162
  common-base transistor, 64
  current, 6
  direct-coupled, 268
  driver, 282
  frequency compensation in a wideband transistor, 344
  impedance-coupled, 267
  intermediate frequency, 485
  nonlinear, 476
  phase relationships: in a CB, 66
    in a CC, 71
    in a CE, 69
  power, 248, 292, 295
  push-pull, 163, 293
  radio-frequency, 358
  RC-coupled, 264
  single-ended, 249, 293
  transformer-coupled, 266
  wideband, 336
Analysis, signal, 157
Antenna, 434, 456
Antimony, 19
Arsenic, 19

Atom:
  charged, 22
  germanium, 16
Audio-frequencies, 355
AVC (automatic volume control), 488, 490
  circuit, 509
  filter, 514
  filter network, 512
  operation, 510
  system, 506
  voltage, 506

Bandpass, 364
Bands:
  broadcast radio, 357
  standard, frequency-modulation, 357
  super-high-frequency, 357
  television broadcast, 357
  ultra-high-frequency, 357
  very high-frequency, 357
Bandwidth, 367
Barrier, 29
  height of the, 30, 31
  PN junction, 27
  potential, 41
Base, transistor, 40
Battery:
  bias, 235
  solar, 73
Beta, 113
Bias:
  cutoff, 163
  grid, 225
  PN junction, 30
Biasing, triode, 234
Bonds, electron-pair, 17
Breakdown, crystal, 35
Bridge, balanced, 305
Bridge circuit:
  advantages of, 316
  two-transistor, 314

553

Capacitance:
  collector-base, 169
  collector-emitter, 173
  emitter-base, 169
Capacitor:
  cathode bypass, 237
  ganged tuning, 480
  swamping, 400
  thin-film, 89
Carrier:
  majority, 34
  minority, 34
  RF, 461
Cathode, 207
Cathode follower, 256
Celsius, 15
Ceramic, 89
Channel, 75
  pseudo-N, 79
Characteristics:
  bandpass, 365
  cutoff, 195
  dynamic transfer, 154
  frequency discrimination, 273
  high-frequency, 193
    and dc, 193
  linear, 51
  nonlinear, 51
  push-pull dynamic, 296
  small-signal, 190
  static, 50
    transfer, 154
  $v-i$, 94
Charge carriers, 18
Chip, IC, 82
Choke, 522
Circuit:
  basic oscillator, analysis of, 400
  common-base, 53
  common-collector, 53
  common-emitter, 53
  complementary-symmetry, 306
  complete hybrid, 108
  crystal oscillator, 427
  diode stabilizing, 132
  equivalent, 93, 96

Circuit (*Cont.*):
  flip-flop logic, 87
  general bias, 128
  high-frequency attenuation, 261
  hybrid equivalent, 96
  improved volume control, 279
  input and output, 51
  integrated, 79
  oscillation in an LC electric, 383
  plate, 201, 210
  RF tuned, 358
  single-battery transistor, 63
  tank, 390
  triode equivalent, 228
  volume-control, 271
Cloud, electron, 209
Code, International Radio, 433
Coil:
  feedback, 393
  tickler, 393, 400
Collector, 41
Collector voltage, dc quiescent, 117
Compensation, 133
  direct-coupled preamplifiers with
    frequency, 259
  high-frequency, 344
  series, 345
  shunt, 345
Complementary symmetry, 301, 302, 306
Compounds, 7
Conditions:
  general test, 190
  zero bias, 293
Conductance, 111, 191
Conductors, insulators, and semiconductors, 14
Configurations, transistor:
  common-base, 53, 58
  common-collector, 53, 60
  common-emitter, 53, 59
  grounded-base, 59
  grounded-collector, 60
  grounded-emitter, 59
  printed circuit board, 80
  with respect to gain, comparison of, 176

Configurations, transistor (*Cont.*):
  with respect to resistances, comparison of, 173
Control:
  automatic volume, 488
  tone, 281
  volume and tone, 268
Conversion, frequency, 478, 494
Core, 17
Coulomb, 13
Coupling:
  coefficient of, 373
  critical, 372
  degree of, 372
  inductive, 400, 444
    oscillator-amplifier, 443
  link, 446, 447
  loose, 372
  over, 374
  RC oscillator-amplifier, 441
  RF, 360
  tight, 372
Coverage, frequency, 455
Crossover points, 295
Crystal, 16
  AT cut, 416
  axes, 415
  BT cut, 421
  CT-cut, 419
  GT-cut, 422
  mother, 413
  negative temperature coefficient, 416
  parallel resonant, 423
  positive temperature coefficient of, 416
  Q of a, 426
  quartz, 396, 412
  semiconductor, 2
  series resonant, 423
  sets (radio receivers), 2
  temperature-frequency characteristics of, 416
  temperature stability of, 414
  X-cut, 415
  Y-cut, 416
Current:
  amplification factor, 190

Current (*Cont.*):
  avalanche, 195
  base, 45
  collector, 45
  diode, 203
  drain, 76
  electric, 14
  emitter, 45
  forbidden point, 166
  grid, 224
  ideal, 124
  internal emitter-to-collector, 45
  leakage, 196
  maximum permissible, 188
  in an NPN transistor, 48
  plate, 201
  in a PNP transistor, 44
  pulsating dc, 204
  saturation, 196
    plate, 213
  signal, 54
  space, 224
  temperature saturation, 213
  transfer ratio, 190
Curves:
  breakdown voltage, 136
  combined, 295
  family of, 99, 119
  frequency response, 256
  ideal, 125
  plate characteristic, 227
  selectivity, 363
  static, 97
  triode characteristic, 226

Decibels, 153
DeForest, Lee, 221
Degeneration, signal, 104
Delay:
  phase, 339
  time, 339
Demodulation, 2, 6, 459
Depletion region, 29
Derating factor, 189
Detection, 2, 461

Detector, 6
  and AVC, 451
  basic diode, 463
  field-effect transistor, 468
Diffusion, 28
Dimensions:
  maximum, 187
  nominal, 187
Diode, 200
  anode and cathode of, 518
  breakdown, 136
  compensating, 134
  conducting state, 519
  directly heated, 206
  indirectly heated, 207
  junction, 4
  N-type, 5
  nonconducting state, 520, 533
  P-type, 4
  point-contact, 3
  shunt limiting action of a, 140
  vacuum tube, 198
  zener, 134
Direct current:
  pulsating, 204
  quiescent base, 117
Dissipation, plate, 232
  maximum allowable, 233
  triode gain and, 231
Dissipation line, constant-power, 164
Distortion, 456
  crossover, 295
  diode detector, 466
  meaning and cause of phase, 339
  phase, 339
Divider:
  current, 270
  voltage, 270
Donor, 19
Drain, 75
Drift, electron, 14
Driver, 248
  one-stage phase-inverter, 282
Driver stage, 540

Dropoff:
  high-frequency, 261
  low-frequency, 258
Dynamic system, 155

Efficiency, power, 6
EHF (extremely high-frequency), 357
Electric field, 26, 29
Electron, 10
  free, 12
  migration of an, 21
  thermionic, 199
  unpaired, 21
Electrostatic field, 222
Elements, 7
Emission:
  field, 215
  thermionic, 198, 199
Emitter, 40
Emitter-base potential, 50
Emitter follower, 179, 181, 254
Energy:
  heat, 28
  kinetic, 378
  law of conservation of, 378
  potential, 378
  thermal, 28
Envelope, wave, 366
Equalizer, 261
Equation:
  conversion, 113
  power-gain, 106
Equipotential surface, 207

Feedback, 128
  capacitances and, 169
  in-phase, 383
  negative, 128, 383
  out-of-phase, 383
  positive, 171, 382
  regenerative, 395
FETS (field-effect transistors), 73
Fidelity, 456
  diode detector, 466
  FET detector, 470

Filament, 199
Filter system, 521
  π type, 523
Fleming, John Ambrose, 200
FM (frequency modulation), 357
Forward-bias, 34
Forward-current amplification factor, 100, 110
Frequencies:
  aeronautical, 357
  government, 357
  radio, 355
Frequency:
  alpha cutoff, 193, 398
  crystal resonant, 422
  cutoff, 397
  difference, 476
  extremely high, 357
  heterodyne, 477
  image, 500
  intermediate, 478
  lower side, 366
  sum, 476
  upper side, 366

Gain:
  current, 101, 176
  current and voltage, 101
  numerical evaluation of, 104
  power, 106, 177, 192
  resistance, 55
  variable transistor, 503
  voltage, 102, 176
Galena, 2
Gallium, 21
Gate, 75
Generator:
  audio, 257
  constant current, 94
  constant voltage, 94
Germanium, 3
  atom, 12
  P-type, 22
Graphs (*see* Curves)

Grid, 222
  screen, 238
  suppressor, 238
  tuned, 402

Harmonic, 331
Hearing, range of human, 256
Heater:
  thermionic, 4
  vacuum tube, 205
Hertz, 248
Heterodyning, 477
  frequency, 475
Hexagons, 16
Hexode, 238
HF (high frequency), 357
High-Q tuned stages, 369
Holder, crystal, 422
Holes, 21
  as current carriers, 25
  properties of, 23
Hz, 248

IC (see Integrated circuits)
IF (intermediate frequency), 478
IF amplifier stages, 495
IF stages, 543
Image rejection, 501
Impedance:
  battery, 52
  input, 52, 191
  matching, 174, 179
    device, 180
  output, 52
  reflected, 278
  source, 95
Impurities, 19
Indium, 21
Inductance, mutual, 373
Inertia, thermal, 207
Input, power, 106
Instability, causes of bias, 119
Insulators, 14

Integrated circuits (IC), 81
   monolithic, 81
   thin-film, 87
Interface, PN, 76
Intermediate frequency (*see* IF)
International Radio Code, 433
Inverter, phase, 282
    two-stage, 285
Ion, 20, 22
Isolation, dc, 277

JFET (junction field-effect transistor), 75
Joule, 382
Junction:
   high-resistance, 41
   low-resistance, 40
   PN, 6

kHz, 253
Kirchhoff's law, 146

Law of electric charges, 23
Layer, insulating oxide, 78
Lead sulphide, 2
LF (low frequency), 356
Line:
   high-impedance, 443
   load, 146, 148, 151, 231
   transmission, 447
Loss, current, 52
LSB (lower sideband), 367

Magnetic field, 23, 24
   collapsing, 386
   growing, 384
Marconi, Guglielmo, 221
Mechanical systems, oscillation in, 376
Medium, 432
MF (medium frequency), 357
Mho, 241
Microamperes, 35
Micromhos, 241

Microwatts, 248
Milliwatts, 248
Mixer, 476
Mixer-oscillator stage, 492
Mode:
   parallel resonant, 426
   series resonant, 426
Modulation, 6, 366, 432
   amplitude, 435
   pulsed, 432
Molecule, 8
MOSFET (metal-oxide-semiconductor field-effect transistor), 78
Motion, periodic, 376
Multiplier, voltage, 529

Nanofarad, 174
Networks:
   coupling, 262
   decoupling, 537, 538
   direct coupling, 267
   feedback, 397
   filter, 521
   impedance coupling, 266
   RC coupling, 263
   transformer coupling, 264
Neutralization, 439
Neutron, 10
NF (noise factor), 193, 250
Nichrome, 88
Noise factor, 193, 250
Noise power, 249
NPN transistor, 6
Nucleus, 10

Ohms, reciprocal, 109
Ohm's law, 55, 93, 122
Operating point, 117
   dc, 99
   stability of, 117
Opposition, space, 201
Orbits, electron, 11
Oscillating system, feedback in an, 382

Oscillation:
   damped, 380, 381
   electric, 376, 383
   mechanical, 376
   sustained, 382
      electric, 390
   undamped, 379
Oscillator, 6
   Clapp, 407
   Colpitts, 405
   converter stage, 544
   Hartley, 409, 428
   local, 477
   radio-frequency, 434
   series-fed, 403
   shunt-fed, 403
   standard transistor, 405
   tuned-base, 402
   tuned-collector, 404
   tuned-grid, tuned-plate, 440
   tuned-plate, 405
Oscillatory system, 380
Output, power, 106
Output curves and calculation of gain, 150
Output stage, 538
Overdriving, 158
Oxides:
   barium, 205
   strontium, 205
   thorium, 205

Parameters, 93
   running, 97
Passband, 364
Patterns, crystal, 16
Peaking:
   combination, 346
   series, 346
   shunt, 346
Pendulum, 380
Pentode, 238
   plate characteristics of, 239
Phosphorus, 19
Picofarad, 194
Picowatts, 248

Piezoelectricity, 412, 413
Pinchoff, 76
PIV (peak inverse voltage), 218, 520
Plate, 199, 200
PNP transistor, 6
Potential, emitter-base, 50
Power, 164
   capability, 189
   maximum collector, 164
Power gain unity, 399
Power rating, 164
Power supply, 535
Preamplifier, 248
Preselector, 488
Protection, surge, 538
Proton, 10
Push-pull, 249

Quiescent conditions, 118

Radio receiver criteria, 454
Range, frequency, 455
Ratings:
   absolute maximum, 187
   dissipation, 188
   temperature range, 190
   thermal, 188
Ratio:
   L/C, 445
   reverse-voltage-transfer, 191
   signal-to-noise, 249
Reactance, leakage, 266
Receiver, 432
   single-transistor, 467
   tuned radio-frequency, 472, 473
Recombinations, 29
Rectification, 2
Rectifier, 203
   full-wave, 523, 524
   half-wave, 518
   simple diode, 203
Region, temperature saturation, 213
Regulation, bandwidth, 371
Regulator, voltage, 137

Rejection, image-frequency, 498
Resistance:
   collector saturation, 196
   equivalent, 202
   grid, 235
   input, 55
   load, 56
   plate, 216, 228
Resistor:
   loading, 371
   "swamping," 124
Response:
   flat, 258
   square-wave analysis of frequency and phase, 251
Reverse-biased, 31
Reverse-voltage-amplification factor, 109, 111, 191
RF (radio frequencies), 355
   amplifier stage, 488
Rhomboids, 16
Ripple current, residual, 523
Ripple residual, removal of, 521
Rod, ferrite, 489

Saturation:
   plate, 214
   temperature, 210, 212
Schottky effect, 215
Selectivity, 363, 455
   diode detector, 466
   FET detector, 470
   TRF receiver, 474
Semiconductors, 2, 14
Sensitivity, 455
   diode detector, 465
   FET detector, 470
   TRF receiver, 474
Series opposing, 436
SHF (super-high frequency), 357
Sidebands, 367
Signal rejection, percentage difference, 482
Silicon, 15
Sinusoid, 328
Slope, 94

Sources, high-impedance, 254
Space charge, 208, 209
Spectrum, frequency, 355
Stability, frequency, 399
Stabilization:
   resistor bias, 121
   temperature, 104
Structure:
   atomic, 10
   matter and its, 7
Substrate, 83
Superheterodyne, 492
   receiver: advantages of, 475
      block diagram, 484
   selectivity, 479
Symbols, designations and, 56

Tetrode, 238
Time constant, 338, 513, 531
Tourmaline, 412
Tracer, curve, 101
Tracking, 482
Transconductance, 240, 241
Transformer:
   air-core, 444
   center-tapped, 525
   double-tuned coupling, 371
   double-tuned RF, 362
   impedance matching, 175
   single-tuned RF, 360
Transistors, 38
   compensating, 134
   compound-connected, 310, 311
   junction, 5
      field-effect, 75
   metal-oxide-semiconductor field-effect, 78
   NPN, 43, 56
   PNP, 39, 56
   point-contact, 4
Transit time, 27
Transmission, single-frequency, 365
Transmitter, 432
   transistorized AM, 449
TRF (tuned radio frequency), 473
Trimmer, 546
   oscillator, 546

Triode:
    directly heated, 222
    fundamentals, of 238
    indirectly heated, 222
Tube(s):
    electron, comparison of transistors and, 242
    gas, 209
    hard, 208
    soft, 209
    triode vacuum, 222
    2-element vacuum, 200

UHF (ultra-high frequency), 357
USB (upper sideband), 368

Valence, 17
VHF (very high frequency), 357
VLF (very low frequency), 356
Voltage:
    breakdown, 136
    collector breakdown, 195
    collector saturation, 196
    complex, 333
    doubler, 529
    effective supply, 451
    grid, 225
    maximum dc collector, 188
    nonsinusoidal, 330
    peak inverse, 218, 520
    plate, 210

Voltage (*Cont.*):
    point, forbidden, 166
    rectifier peak, 217
    signal input, 54
    sinusoidal, 329
    surge, 141
    zener, 136

Wave:
    carrier, 434
    complex, 335
    components of, 460
    composite, 459
    continuous, 365
    rectangular, 328
    recurrent, 328
    sawtooth, 328
    sine, 328
    square, 328, 351
    triangular, 328
Waveform:
    asymmetrical, 159
    input voltage, 66, 67
    output voltage, 66, 67
    rectangular, 337
    recurrent, 328
    single-frequency, 332
    standard, 332
Work function, 205

Zero, reference, 525